STP NATIONAL CURRICULUM MATHEMATICS

9A

L. BOSTOCK, B.Sc.

S. CHANDLER, B.Sc.

A. SHEPHERD, B.Sc.

E. SMITH, M.Sc.

Nelson Thornes

First published in 1997 by:
Stanley Thornes (Publishers) Ltd

Revised edition printed in 2001 by:
Nelson Thornes Ltd
Delta Place
27 Bath Road
CHELTENHAM
GL53 7TH
United Kingdom

11 12 / 15 14 13

A catalogue record of this book is available from the British Library.

ISBN 978 0 7487 6495 2

Illustrations by Peters & Zabransky, Linda Jefferey.
Page make-up by Tech-Set

Printed by Multivista Global Ltd

Front cover image produced using material kindly supplied by I LOVE LOVE CO,
makers of The Happy Cube © Laureyssens/Creative City Ltd 1986/91.
Distributed in UK by: RIGHTRAC, 119 Sandycombe Road, Richmond,
Surrey TW9 2ER Tel. 020 8940 3322.

CONTENTS

INTRODUCTION

To the pupil

This book continues to help you to learn, enjoy and progress through Mathematics in the National Curriculum. As well as a clear and concise text the book offers a wide range of practical and investigational work that is relevant to the mathematics you are learning.

Everyone needs success and satisfaction in getting things right. With this in mind we have divided many of the exercises into three types of questions.

The first type, identified by plain numbers, e.g. **15**, helps you to see if you understand the work. These questions are considered necessary in every chapter.

The second type, identified by an underline, e.g. **15**, are extra, but not harder, questions for quicker workers, for extra practice or for later revision.

The third type, identified by a coloured square, e.g. **15** , are for those of you who like a greater challenge.

Most chapters have a 'mixed exercise' after the main work of the chapter has been completed. This will help you to revise what you have done, either when you have finished the chapter or at a later date. All chapters end with some mathematical puzzles, practical and/or investigational work. For this work you are encouraged to share your ideas with others, to use any mathematics you are familiar with, and to try different approaches, appreciating the advantages and disadvantages of each method.

The book starts with a summary of the main results from Books 7A and 8A. At intervals throughout the book you will find further summaries. These list the most important points that have been studied in the previous chapters and conclude with revision exercises that test the work you have studied up to that point.

At this stage you will find that you use a calculator frequently but it is unwise to rely on a calculator for work that you should do in your head. Remember, whether you use a calculator or do the working yourself, always estimate your answer and always ask yourself the question, 'Is my answer a sensible one?'

Mathematics is an exciting and enjoyable subject when you understand what is going on. Remember, if you don't understand something, ask someone who can explain it to you. If you still don't understand, ask again. Good luck with your studies.

To the teacher

This is the third book of the STP National Curriculum Mathematics series. It is based on the ST(P) Mathematics series but has been extensively rewritten and is now firmly based on the Programme of Study for Key Stages 3 and 4.

The majority of scientific calculators now on sale use direct keying sequences for entering functions such as tan 33, $\sqrt{2}$. This is the order used in this book.

The A series of books aims to prepare pupils for about Level 8 at Key Stage 3 and for the higher tier at GCSE.

SUMMARY 1

**MAIN RESULTS
FROM BOOKS 7A
AND 8A**

**TYPES OF
NUMBER**

A *factor* of a number will divide into that number exactly.
A *multiple* of a number has that number as a factor,

e.g.   3 is a factor of 12   and   12 is a multiple of 3.

A *prime number* has only 1 and itself as factors, e.g. 7.
Remember that 1 is not a prime number.

Square numbers can be drawn
as a square grid of dots, e.g. 9:

Rectangular numbers can be
drawn as a rectangular grid of dots, e.g. 6:

Triangular numbers can be drawn
as a triangular grid of dots, e.g. 6:

**OPERATIONS OF
×, ÷, +, −**

The sign in front of a number refers to that number only.

When a calculation involves a *mixture of operations*, start by calculating anything inside brackets, then follow the rule 'do the multiplication and division first.'

FRACTIONS

Equivalent fractions are formed by multiplying or dividing the top and the bottom of a fraction by the same number,

e.g. $\frac{1}{2} = \frac{3}{6}$ (multiplying top and bottom of $\frac{1}{2}$ by 3)

Fractions can be *added* or *subtracted* when they have the *same* denominator,

e.g.   to add $\frac{1}{2}$ to $\frac{1}{3}$ we first change them into equivalent fractions with the same denominators,

i.e. $\frac{1}{2} + \frac{1}{3} = \frac{1}{2} \times \frac{3}{3} + \frac{1}{3} \times \frac{2}{2} = \frac{3}{6} + \frac{2}{6} = \frac{5}{6}$

To *multiply one fraction by another fraction*, we multiply their numerators and multiply their denominators,

e.g. $\frac{1}{2} \times \frac{5}{3} = \frac{1 \times 5}{2 \times 3} = \frac{5}{6}$

To *divide by a fraction*, we multiply by the fraction turned upside down,

e.g. $\quad \frac{1}{2} \div \frac{5}{3} = \frac{1}{2} \times \frac{3}{5} = \frac{3}{10}$

To multiply or divide with *mixed numbers*, e.g. $1\frac{3}{4}$, first change the mixed numbers to improper fractions.

A *fraction can be changed to a decimal* by dividing the bottom number into the top number,

e.g. $\quad \frac{3}{8} = 3 \div 8 = 0.375$

A *fraction can be expressed as a percentage* by multiplying the fraction by 100,

e.g. $\quad \frac{2}{5} = \frac{2}{5} \times 100\% = \frac{2}{5} \times \frac{100}{1}\% = 40\%$

To find *a fraction of a quantity*, we multiply that fraction by the quantity,

e.g. $\quad \frac{1}{2}$ of $\frac{3}{4}$ means $\frac{1}{2} \times \frac{3}{4}$, and $\frac{3}{8}$ of £24 $= £\left(\frac{3}{8} \times 24 \right)$

To express *one quantity as a fraction of another*, first make sure that both quantities are in the same unit, then place the first quantity over the second,

e.g. $\quad$ 24 p as a fraction of £2 is $\frac{24}{200} \left(= \frac{3}{25} \right)$

DECIMALS

Decimals can be added or subtracted using the same methods as for whole numbers, provided that the decimal points are placed in line.

To multiply a decimal by 10, 100, 1000,..., we move the point 1, 2, 3,... places to the right,

e.g. $\quad 2.56 \times 10 = 25.6,$ and $\quad 2.56 \times 1000 = 2560(.0)$

To divide a decimal by 10, 100, 1000,..., we move the point 1, 2, 3,... places to the left, (equivalent to moving the figures to the right),

e.g. $\quad 2.56 \div 10 = 0.256,$ and $\quad 2.56 \div 1000 = 0.002\,56$

To multiply decimals without using a calculator, first ignore the decimal point and multiply the numbers. Then add the number of decimal places in each of the decimals being multiplied together; this gives the number of decimal places in the answer,

e.g. $\quad 7.5 \times 0.5 = 3.75 \hspace{4cm} (75 \times 5 = 375)$

$\quad\quad [(1) + (1) = (2)]$

To divide by a decimal, move the point in *both* numbers to the right until the number we are dividing by is a whole number,

e.g. $\quad 2.56 \div 0.4 = 25.6 \div 4 = 6.4$

A decimal can be expressed as a percentage by multiplying the decimal by 100,

e.g. $\quad 0.325 = 32.5\%$

SIGNIFICANT FIGURES

The first significant figure in a number is the first non-zero figure when reading from left to right.

The second significant figure is the next figure to the right, whether or not it is zero, and so on.

For example, in 0.0205, the first significant figure is 2,

and the second significant figure is 0.

ROUNDING NUMBERS

To round (that is, to correct) a number to a specified place value or number of significant figures, look at the figure in the next place: if it is 5 or more, add 1 to the specified figure, otherwise leave the specified figure as it is,

e.g. $13|7 = 140$ to the nearest 10, or 2 s.f.

$2.|564 = 3$ to the nearest whole number, or 1 s.f.

$2.56|4 = 2.56$ correct to 2 decimal places, or 3 s.f.

STANDARD FORM

A number written in standard form is a number between 1 and 10 multiplied by a power of ten,

e.g. 1.2×10^5

PERCENTAGES

'Per cent' means 'out of one hundred'.

Hence *a percentage can be expressed as a fraction* by placing the percentage over 100,

e.g. $33\% = \frac{33}{100}$,

and *a percentage can be expressed as a decimal* by dividing the percentage by 100, that is, by moving the decimal point two places to the left,

e.g. $33\% = 0.33$

To find *one quantity as a percentage of another quantity*, we place the first quantity over the second quantity and multiply this fraction by 100,

e.g.   24 p as a percentage of £2 is $\frac{24}{200} \times \frac{100}{1}\% = 12\%$

To find *a percentage of a quantity*, change the percentage to a decimal and multiply it by the quantity,

e.g. 32% of £18 $= 32\% \times £18 = £0.32 \times 18 = £5.76$

To *increase* a quantity by 15%,

we find *the increase* by finding 15% of the quantity;

we find *the new quantity* directly by finding $100\% + 15\%$, i.e. 115%, of the original quantity; so we multiply it by 1.15.

To *decrease* a quantity by 15%,
we find *the decrease* by finding 15% of the quantity,
we find *the new quantity* directly by finding 100% − 15%, i.e. 85%, of the
original quantity; so we multiply it by 0.85.

**DIRECTED
NUMBERS**

Positive and negative numbers are collectively known as directed
numbers. They can be represented on a number line.

The *rules for multiplying and dividing directed numbers* are

when the signs are the same, the result is positive.
when the signs are different, the result is negative.

For example, $(+2) \times (+3) = +6$ and $(-2) \times (-3) = +6$
$(+2) \times (-3) = -6$ and $(-8) \div (+2) = -4$

The same rules apply to adding and subtracting directed numbers,

e.g. $3 + (+2) = 3 + 2 = 5$ and $3 - (-2) = 3 + 2 = 5$
$3 + (-2) = 3 - 2 = 1$ and $3 - (+2) = 3 - 2 = 1$

SQUARE ROOTS

A square root of 20 is a number which, when multiplied by itself, gives
20. A number has two square roots, one positive and one negative, for
example if $x^2 = 4$, then $x = \pm 2$
2 is the positive square root of 4 and −2 is the negative square root of 4.
$\sqrt{20}$ means the positive square root of 20.

RATIO

Ratios are used to compare the relative sizes of quantities.
For example, if a model car is 20 cm long and the real car is 200 cm long,
we say that their lengths are in the ratio 20 : 200.

Ratios can be simplified by dividing the parts of the ratio by the same
number,

e.g. $20 : 200 = 1 : 10$ (dividing 20 and 200 by 20).

A *map ratio* is the ratio of a length on the map to the length it represents
on the ground. When expressed as a fraction (or sometimes as a ratio),
it is called the *Representative Fraction*.

**DIRECT
PROPORTION**

Two quantities are directly proportional when they are always in the
same ratio.

For example, if 1 kilogram of apples costs 96 p then n kg of apples cost 96n pence, so the ratio of weight to cost is $n : 96n = 1 : 96$, that is, the weight of these apples is directly proportional to their cost.

INDICES

When a number is written in the form 3^4, 3 is called the *base* and 4 is called the *index* or *power* and 3^4 means $3 \times 3 \times 3 \times 3$.
a^2 is called a squared, a^3 is called a cubed.

Negative index

3^{-4} means $\dfrac{1}{3^4}$

Zero index

$3^0 = 1$, in fact $a^0 = 1$ whatever number a stands for.

Rules of indices

We can multiply different powers of the same base by adding the indices,

e.g. $3^4 \times 3^2 = 3^{4+2} = 3^6$

We can divide different powers of the same base by subtracting the indices,

e.g. $3^4 \div 3^2 = 3^{4-2} = 3^2$

UNITS

Metric units of length in common use are the kilometre, metre, centimetre and millimetre, where

$$1\,\text{km} = 1000\,\text{m}, \qquad 1\,\text{m} = 100\,\text{cm}, \qquad 1\,\text{cm} = 10\,\text{mm}$$

Metric units of mass are the tonne (t), kilogram (kg), gram (g) and milligram (mg), where

$$1\,\text{tonne} = 1000\,\text{kg}, \qquad 1\,\text{kg} = 1000\,\text{g}, \qquad 1\,\text{g} = 1000\,\text{mg}$$

Imperial units of length in common use are the mile, yard (yd), foot (ft) and inch (in), where

$$1\,\text{mile} = 1760\,\text{yards}, \qquad 1\,\text{yard} = 3\,\text{feet}, \qquad 1\,\text{foot} = 12\,\text{inches}$$

Imperial units of mass still in common use are the ton, hundredweight (cwt), stone, pound (lb) and ounce (oz), where

$$1\,\text{ton} = 2240\,\text{lb}, \qquad 1\,\text{stone} = 14\,\text{lb}, \qquad 1\,\text{lb} = 16\,\text{ounces}$$

For a *rough conversion* between metric and Imperial units, use

$$1\,\text{km} \approx \tfrac{1}{2}\,\text{mile}, \quad 1\,\text{yard} \approx 1\,\text{m}, \quad 1\,\text{kg} \approx 2\,\text{lb}, \quad 1\,\text{tonne} \approx 1\,\text{ton}$$

For a more accurate conversion use

$$5\,\text{miles} \approx 8\,\text{km}, \qquad 1\,\text{inch} \approx 2.5\,\text{cm} \qquad 1\,\text{kg} \approx 2.2\,\text{lb}$$

Area is measured by standard-sized squares.

$$1\,cm^2 = 10 \times 10\,mm^2 = 100\,mm^2$$

$$1\,m^2 = 100 \times 100\,cm^2 = 10\,000\,cm^2$$

$$1\,km^2 = 1000 \times 1000\,m^2 = 1\,000\,000\,m^2$$

Volume is measured by standard-sized cubes.

$$1\,cm^3 = 10 \times 10 \times 10\,mm^3 = 1000\,mm^3$$

$$1\,m^3 = 100 \times 100 \times 100\,cm^3 = 1\,000\,000\,cm^3$$

The *capacity* of a container is the volume of liquid it could hold.
The main *metric units of capacity* are the litre and the millilitre (ml),
where

$$1\,litre = 1000\,ml \quad and \quad 1\,litre = 1000\,cm^3 \quad so \quad 1\,ml = 1\,cm^3$$

The main *Imperial units of capacity* are the gallon and the pint, where

$$1\,gallon = 8\,pints$$

Rough conversions between metric and Imperial units of capacity are
given by

$$1\,litre \approx 1.75\,pints \qquad and \qquad 1\,gallon \approx 4.5\,litres$$

CIRCLES

The *diameter* of a circle is twice the *radius*.

The *circumference* is given by $C = 2\pi r$,
where r units is the radius of the circle
and $\pi = 3.1415\ldots$

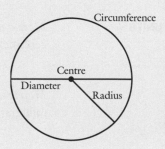

AREA

The *area of a square* = (length of a side)^2.
The *area of a rectangle* = length × breadth.
The *area of a circle* is given by $A = \pi r^2$

The *area of a parallelogram* is given by
A = length × height.

The *area of a triangle* is given by
$A = \frac{1}{2}$ base × height

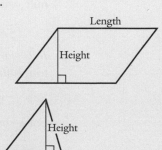

When we talk about the height of a triangle or of a parallelogram, we
mean the perpendicular height.

VOLUME AND CAPACITY

The *volume of a cuboid* = length × breadth × height

A solid with a constant cross-section is called a *prism*.
The *volume of a prism* is given by

$$\text{area of cross-section} \times \text{length}$$

The *volume of a cylinder* is given by $V = \pi r^2 h$

DENSITY

The *density* of a material is the mass of one unit of volume of the material, for example, the density of silver is $10.5 \, \text{g/cm}^3$, that is, $1 \, \text{cm}^3$ of silver weighs $10.5 \, \text{g}$.

DISTANCE, SPEED AND TIME

The relationship between distance, speed and time is given by

$$\text{Distance} = \text{Speed} \times \text{Time}$$

The relationship can be remembered from the triangle: (cover up the one you want to find)

Average speed for a journey $= \dfrac{\text{Total distance covered}}{\text{Total time taken}}$

ANGLES

One complete revolution = 4 right angles = 360°.

An *acute angle* is less than 90°.

An *obtuse angle* is larger than 90° but less than 180°.

A *reflex angle* is larger than 180°.

Vertically opposite angles are equal.

Angles on a straight line add up to 180°.
Two angles that add up to 180° are called *supplementary angles*.

Angles at a point add up to 360°.

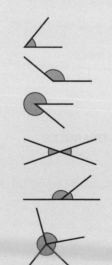

PARALLEL LINES

When two parallel lines are cut by a transversal

the *corresponding angles* are equal,

the *alternate angles* are equal,

the *interior angles* add up to 180°.

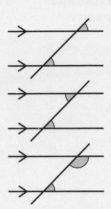

ANGLES OF ELEVATION AND DEPRESSION

If you start by looking straight ahead, the angle that you turn your eyes through to look *up* at an object is called the angle of elevation, the angle you turn your eyes through to look *down* at an object is called the angle of depression.

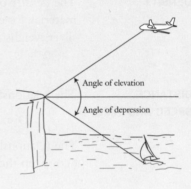

Angle of elevation

Angle of depression

THREE-FIGURE BEARINGS

The three-figure bearing of a point A from a point B gives the direction of A from B as a clockwise angle measured from the north.

For example, in this diagram, the bearing of A from B is 140°.

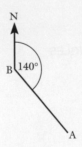

N

B 140°

A

TRIANGLES

The three angles in a triangle add up to 180°.

An *equilateral triangle* has all three sides equal and each angle is 60°.

An *isosceles triangle* has two sides equal and the angles at the base of these sides are equal.

QUADRILATERALS   A quadrilateral has four sides.
The four angles in a quadrilateral add up to 360°.

Special quadrilaterals
In a square
- all four sides are the same length
- both pairs of opposite sides are parallel
- all four angles are right angles.

In a rectangle
- both pairs of opposite sides are the same length
- both pairs of opposite sides are parallel
- all four angles are right angles.

In a rhombus
- all four sides are the same length
- both pairs of opposite sides are parallel
- the opposite angles are equal.

In a parallelogram
- the opposite sides are the same length
- the opposite sides are parallel
- the opposite angles are equal.

In a trapezium
- just one pair of opposite sides are parallel.

POLYGONS   A polygon is a plane figure bounded by straight
lines, e.g.

A *regular polygon* has all angles equal and
all sides the same length.
This is a regular hexagon.

The *sum of the exterior angles* of any polygon
is 360°.

The *sum of the interior angles* of any polygon
depends on the number of sides.
For a polygon with *n* sides, this sum is
$(180n - 360)°$ or $(2n - 4)$ right angles.

**PYTHAGORAS'
THEOREM**

Pythagoras' theorem states that in any right-angled triangle ABC with $\hat{C} = 90°$, $AB^2 = AC^2 + BC^2$

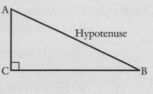

Conversely, if in a triangle PQR, $PR^2 = PQ^2 + QR^2$ then $\hat{Q} = 90°$

CONGRUENCE

Two figures are congruent when they are exactly the same shape and size.

TRANSFORMATIONS

Reflection in a mirror line
When an object is reflected in a mirror line, the object and its image form a symmetrical shape with the mirror line as the axis of symmetry.

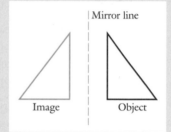

Translation
An object is translated when it moves without being turned or reflected to form an image.

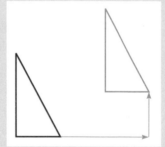

Rotation
When an object is rotated about a point to form an image, the point about which it is rotated is called the *centre of rotation* and the angle it is turned through is called the *angle of rotation*.

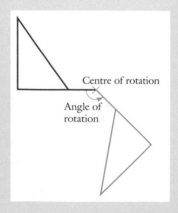

Enlargement

When an object is enlarged by a scale factor 2, each line on the image is twice the length of the corresponding line on the object. The diagram shows an enlargement of a triangle, with centre of enlargement X and scale factor 2. The dashed lines are guide-lines.

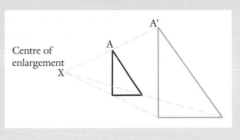

Centre of enlargement X

$$XA' = 2XA$$

When the scale factor is less than one, the image is smaller than the object.

This diagram shows an enlargement with scale factor $\frac{1}{4}$ and centre of enlargement O.

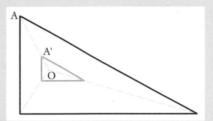

$$OA' = \tfrac{1}{4}OA$$

STATISTICS

A *hypothesis* is a statement that is not known to be true or untrue.

Discrete values are exact and distinct, for example, the number of people in a queue.

Continuous values can only be given in a range of a continuous scale, for example, the length of a piece of wood.

For a list of values,

- the *range* is the difference between the largest value and the smallest value
- the *mean* is the sum of all the values divided by the number of values
- the *median* is the middle value when they have been arranged in order of size, (when the middle of the list is half-way between two values, the median is the average of these two values)
- the *mode* is the value that occurs most frequently.

For a grouped frequency distribution,

- the *range* is estimated as
  the higher end of the last group – the lower end of the first group
- the *modal group* is the group with the largest number of items in it.

We get a *scatter graph* when we plot values of one quantity against corresponding values of another quantity.

When the points are scattered about a straight line, we can draw that line by eye; it is called the *line of best fit*.

We use the word *correlation* to describe the amount of scatter about this line.

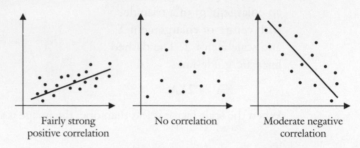

| Fairly strong positive correlation | No correlation | Moderate negative correlation |

PROBABILITY

The probability that an event A happens is $P(A)$ where

$$P(A) = \frac{\text{the number of ways in which } A \text{ can occur}}{\text{the total number of equally likely outcomes}}$$

If A cannot happen, $P(A) = 0$. If A is certain to happen, $P(A) = 1$. In general $P(A)$ lies between 0 and 1.

The probability that an event A does not happen, $P(\bar{A})$, is equal to one minus the probability that it does happen, i.e. $P(\bar{A}) = 1 - P(A)$.

If p is the probability that an event happens on one occasion, then we expect it to happen np times on n occasions, for example, if we toss an unbiased coin 50 times, we expect $\frac{1}{2} \times 50 = 25$ heads.

When we perform experiments to find out how often an event occurs, the *relative frequency* of the event is given by

$$\frac{\text{the number of times the event occurs}}{\text{the number of times the experiment is performed}}$$

Relative frequency is used to give an approximate value for probability.

FORMULAS

A formula is a general rule for finding one quantity in terms of other quantities, for example, the formula for finding the area of a rectangle is given by Area = length × breadth

When letters are used for unknown numbers, the formula can be written more concisely, that is the area, A cm^2, of a rectangle measuring l cm by b cm, is given by the formula $A = l \times b$

ALGEBRAIC
EXPRESSIONS

Terms such as $5n$ mean $5 \times n = n + n + n + n + n$.
Similarly ab means $a \times b$.
$2x + 5x$ can be simplified to $7x$.

Multiplication and division of algebraic fractions
The same rules apply to fractions with letter terms as to fractions with numbers only,

i.e.   to multiply fractions, multiply the numerators and multiply the denominators,

e.g. $\dfrac{2x}{3} \times \dfrac{5x}{7} = \dfrac{2x \times 5x}{3 \times 7} = \dfrac{10x^2}{21}$

and to divide by a fraction, turn it upside down and multiply,

e.g. $\dfrac{x}{6} \div \dfrac{2x}{3} = \dfrac{{}^1x}{{}_2\cancel{6}} \times \dfrac{\cancel{3}^1}{2x_1} = \dfrac{1}{4}$

Simplification of brackets
$x(2x - 3)$ means $x \times 2x + x \times (-3)$.

Therefore $x(2x - 3) = 2x^2 - 3x$

SOLVING
EQUATIONS

An equation is a relationship between an unknown number, represented by a letter, and other numbers, for example, $2x - 3 = 5$
Solving the equation means finding the unknown number.
Provided that we do the same thing to both sides of an equation, we keep the equality; this can be used to solve the equation.
When an equation contains brackets, first multiply out the brackets,

e.g. $3x - 2(3 - x) = 6$

gives $3x - 6 + 2x = 6$ which can be solved easily.

When an equation contains fractions, multiply each term in the equation by the lowest number that each denominator divides into exactly. This will eliminate all fractions from the equation,

e.g.   if $\dfrac{x}{2} + 1 = \dfrac{2}{3}$

multiplying each term by 6 gives

$$\dfrac{6}{1} \times \dfrac{x}{2} + 6 \times 1 = \dfrac{6}{1} \times \dfrac{2}{3}$$ which simplifies to $3x + 6 = 4$

which can be solved easily.

Two equations with two unknown quantities are called *simultaneous equations*. A pair of simultaneous equations can be solved algebraically by eliminating one of the letters; if two letter terms are the same size then, when the signs are different, we add the equations, when the signs are the same we subtract the equations.

For example, to eliminate y from $\qquad 2x + y = 5 \qquad$ [1]

$$\text{and} \qquad 3x - y = 7 \qquad [2]$$

we add [1] and [2] to give $\qquad\qquad 5x = 12 \qquad$ [3]

The value of x can be found from [3]. This value is then substituted for x in [1] or [2] to find y.

Polynomial equations in one unknown contain terms involving powers of x,

e.g. $\quad x^3 - 2x = 4 \quad$ and $\quad 2x^2 = 5 \quad$ are polynomial equations.

Equations containing an x^2 term and a number only may be solved by finding square roots.

More complex equations can be solved by *trial and improvement*, that is, by trying possible values for x until we find a value that fits the equation.

Equations can also be solved by *drawing a graph*, for example to solve $x^3 - x = 10$, we draw the graph of $y = x^3 - x - 10$. The solutions are the values of x where this graph crosses the x-axis (i.e. where $y = 0$).

Inequalities

An inequality remains true when the same number is added to, or subtracted from, both sides,

e.g. $\quad$ if $\quad x > 5 \quad$ then $\quad x + 2 > 5 + 2$

$$\text{and} \quad x - 2 > 5 - 2$$

An inequality also remains true when both sides are multiplied, or divided, by the same *positive* number,

e.g. $\quad$ if $\quad x > 5 \quad$ then $\quad 2x > 10$

$$\text{and} \quad \frac{x}{2} > \frac{5}{2}$$

However multiplication or division by a negative number must be avoided because this destroys the inequality.

GRAPHS

The equation of a line or curve gives the y-coordinate of a point in terms of its x-coordinate. This relationship between the coordinates is true only for points on the line or curve.

Straight lines

The *gradient* of a straight line can be found from any two points, P and Q, on the line, by calculating

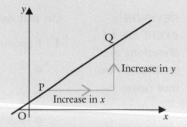

$$\frac{\text{increase in } y \text{ in moving from P to Q}}{\text{increase in } x \text{ in moving from P to Q}}$$

$$= \frac{y\text{-coordinate of Q} - y\text{-coordinate of P}}{x\text{-coordinate of Q} - x\text{-coordinate of P}}$$

When the gradient is positive, the line slopes uphill when moving from left to right.
When the gradient is negative, the line slopes downhill when moving from left to right.

The equation of a straight line is of the form $y = mx + c$

where m is the gradient of the line

and c is the y-intercept,

e.g. the line whose equation is $y = 2x - 3$

has gradient 2 and y-intercept -3.

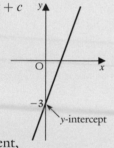

Two lines are parallel when they have the same gradient,

e.g. $y = 2x + 1$ and $y = 2x - 5$ are parallel.

Two lines are perpendicular when the product of their gradients is -1.

e.g. $y = 2x + 4$ and $y = -\frac{1}{2}x + 6$ are perpendicular.

An equation of the form $y = c$ gives a line parallel to the x-axis.

An equation of the form $x = b$ gives a line parallel to the y-axis.

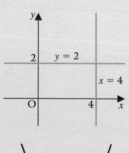

A *parabola* is a curve whose equation is in the form $y = ax^2 + bx + c$.
The shape of this curve looks like this:

When the x^2 term is positive it is the way up shown.
When the x^2 term is negative the curve is upside down.

The exercises that follow are *not* intended to be worked through before starting the main part of this book. They are here for you to use when you need practice on the basic techniques.

Do not use a calculator for any questions in this exercise.

1 Express as a single fraction in its simplest form

a $\frac{2}{3} + \frac{7}{8}$ **d** $\frac{3}{16} + \frac{3}{4} + \frac{5}{12}$ **g** $\frac{3}{4} + \frac{13}{20} + \frac{4}{5}$

b $\frac{3}{5} + \frac{3}{10}$ **e** $\frac{1}{4} + \frac{2}{3}$ **h** $\frac{2}{7} + \frac{1}{9} + \frac{1}{6}$

c $\frac{2}{7} + \frac{1}{2} + \frac{3}{14}$ **f** $\frac{5}{6} + \frac{2}{3} + \frac{1}{4}$ **i** $\frac{8}{21} + \frac{1}{2} + \frac{2}{3}$

2 Express as a single fraction in its simplest form

a $\frac{7}{9} - \frac{5}{12}$ **d** $\frac{1}{10} - \frac{1}{20}$ **g** $2\frac{3}{4} + 1\frac{1}{2} - 1\frac{1}{3}$

b $\frac{1}{4} - \frac{2}{9}$ **e** $2\frac{2}{5} - 1\frac{3}{8}$ **h** $1\frac{3}{8} + 1\frac{1}{4} - 2\frac{1}{2}$

c $\frac{3}{10} - \frac{1}{15}$ **f** $5\frac{5}{6} - 2\frac{5}{9}$ **i** $3\frac{1}{5} - 4\frac{1}{8} + 1\frac{7}{10}$

3 Express as a single fraction in its simplest form

a $\frac{2}{3} \times \frac{5}{6}$ **d** $\frac{2}{5}$ of $1\frac{3}{7}$ **g** $\frac{2}{9} \div 1\frac{2}{7}$

b $\frac{2}{3} \times \frac{1}{4} \times \frac{3}{5}$ **e** $\frac{2}{5} \times \frac{7}{8} \times \frac{3}{7} \times \frac{10}{11}$ **h** $\frac{3}{7} \div 1\frac{3}{4}$

c $\frac{1}{2}$ of $1\frac{1}{3} \times \frac{5}{7}$ **f** $1\frac{2}{3} \div \frac{5}{6}$ **i** $3 \div \frac{2}{3}$

4 Express as a single fraction in its simplest form

a $1\frac{2}{3} \times \frac{1}{2} - \frac{3}{5}$ **g** $2\frac{2}{5} \times 1\frac{7}{8} - 1\frac{2}{3}$

b $\frac{2}{7} + \frac{1}{4} \times 1\frac{1}{3}$ **h** $2\frac{1}{7} - \frac{1}{3}$ of $1\frac{2}{7}$

c $4\frac{1}{2} \div 3 + \frac{3}{4}$ **i** $\left(\frac{1}{2} - \frac{1}{3}\right) \div \left(\frac{3}{4} - \frac{1}{3}\right)$

d $\left(2\frac{1}{5} + 1\frac{2}{3}\right) \div 5\frac{4}{5}$ **j** $\left(4\frac{1}{2} - 3\frac{3}{8}\right) \times 1\frac{1}{3}$

e $\left(2\frac{1}{2} - 1\frac{1}{3}\right) \div 4\frac{2}{3}$ **k** $\left(\frac{11}{12} - \frac{1}{2}\right) \times \frac{2}{5} + \frac{1}{2}$

f $\frac{4}{5} \div \frac{1}{4} + \frac{1}{3} \times 4\frac{1}{2}$ **l** $2\frac{1}{2} \div 1\frac{3}{7} + 1\frac{1}{3}$

5 Find

a $1.26 + 3.75$ **g** $5.3 - 2.1$

b $12.4 + 6.7$ **h** $0.16 - 0.08$

c $5.82 + 0.35$ **i** $1.07 - 0.58$

d $0.04 + 8.86$ **j** $0.37 - 0.009$

e $4.002 + 0.83$ **k** $2 - 0.17$

f $0.000\,32 + 0.0017$ **l** $0.0127 - 0.0059$

6 Find

a 1.2×0.8 e 3.0501×1.1 i $0.18 \div 1.2$

b 0.7×0.06 f 1.002×0.36 j $42.8 \div 200$

c 0.4×0.02 g $1.08 \div 0.4$ k $0.01 \div 0.5$

d 0.5×0.5 h $0.2 \div 2.5$ l $0.0013 \div 1.3$

7 Fill in the blanks (marked with $\&$) in the following calculations.

a $3.7 \times \& = 15.54$ f $\& \times 0.85 = 1.105$

b $22.96 \div \& = 8.2$ g $5.9 \times \& = 0.2537$

c $0.374 \times 0.06 = \&$ h $\& - 4.08 \div 1.7 = 0$

d $\& \div 0.45 = 1.44$ i $0.001\,62 \div 0.045 = \&$

e $0.37 \times 1.92 - \& = 0$ j $\& \div 0.026 = 1.4$

REVISION EXERCISE 1.2 (Using fractions, decimals, ratios and percentages)

1 a Express as a decimal

i $\frac{9}{25}$ ii $\frac{19}{20}$ iii 54% iv $82\frac{1}{2}\%$

b Express as a fraction in its lowest terms

i 85% ii 0.42 iii 65% iv 0.125

c Express as a percentage

i 0.44 ii $\frac{7}{25}$ iii 1.38 iv $\frac{37}{40}$

2 Copy and complete the following table.

	Fraction	Percentage	Decimal
a	$\frac{17}{20}$		
b		$37\frac{1}{2}\%$	
c			0.625
d		$5\frac{3}{4}\%$	
e			1.15
f	$4\frac{3}{4}$		

3 a Put either $<$ or $>$ between each of the following pairs of fractions.

i $\frac{5}{8}$ $\frac{7}{10}$ ii $\frac{2}{5}$ $\frac{1}{3}$ iii $\frac{1}{5}$ $\frac{4}{15}$ iv $\frac{4}{7}$ $\frac{5}{9}$

b Change into an improper fraction

i $2\frac{3}{7}$ ii $5\frac{4}{9}$ iii $3\frac{3}{5}$ iv $9\frac{3}{4}$

c Give as a mixed number

i $\frac{42}{5}$ ii $\frac{17}{4}$ iii $\frac{46}{7}$ iv $\frac{26}{17}$

4 a Find

  i 36% of 50 kg   **ii** 4.5% of 440 g   **iii** 84% of 15 m

b i Increase £480 by 45%   **iii** Increase 150 cm^2 by 56%
  ii Decrease £320 by 55%   **iv** Decrease £44 by 35%

c Find, giving your answer correct to 3 significant figures

  i 37% of 46 km   **ii** $4\frac{3}{4}$% of 12.6 m   **iii** $13\frac{1}{2}$% of 245 mm

d Express

  i 12 mm as a percentage of 6 cm
  ii 650 m as a fraction of 2 km
  iii 56 cm^2 as a percentage of 1 m^2
  iv 6 pints as a fraction of 4 gallons

5 a Give the following ratios in their simplest form.

  i 12 : 18   **iii** 320 : 480   **v** $\frac{1}{2} : \frac{5}{6} : \frac{2}{3}$
  ii 3 : 6 : 9   **iv** 3.5 : 2.5   **vi** 288 : 128 : 144

b Simplify the following ratios.

  i 45 cm : 0.1 m   **iii** 340 m : 1.2 km   **v** 450 mg : 1 g
  ii 42 p : £1.05   **iv** 32 g : 2 kg   **vi** 2.2 t : 132 kg

c Find x if

  i $x : 5 = 2 : 9$ **iii** $x : 6 = 5 : 4$ **v** $3 : 8 = 9 : x$
  ii $x : 3 = 1 : 7$ **iv** $5 : x = 7 : 2$ **vi** $15 : 2 = x : 3$

d i Divide £45 into two parts in the ratio 4 : 5.
  ii Divide 96 m into two parts in the ratio 9 : 7.
  iii Divide 5 kg into three parts in the ratio 1 : 2 : 5.
  iv Divide seven hours into three parts in the ratio 1 : 5 : 8.

6 a Find the map ratio of a map on which 10 cm represents 1 km.

b The map ratio of a map is 1 : 200 000. The distance between two factories is 8 km. What distance is this on the map?

7 a In a sale a pair of trainers priced £35 is reduced by 30%. What is the sale price?

b Sally and Tim bought a portable CD player between them for £44.94. Sally paid $\frac{4}{7}$ of the cost and Tim paid the remainder.

  i What fraction did Tim pay?   **ii** How much did Sally pay?

8 At a concert 64% of the audience were females.

a What fraction of the audience were females?

b Express the part of the audience that was male as

  i a percentage   **ii** a decimal   **iii** a fraction in its lowest terms.

9 Estimate the value of $236.4 \div 48.7$, and then use a calculator to find its value correct to 2 decimal places.

10 A popular leisure club has 2750 members. Of these, 42% are girls, 0.3 are boys, $\frac{4}{25}$ are men and the remainder are women.

a What fraction of the members are girls?

b What percentage of the members are women?

c What decimal fraction of the members are male?

d How many of the members are females?

11 a Find **i** $\frac{2}{3}$ of £36   **ii** $\frac{3}{4}$ of 34 cm   **iii** $\frac{4}{9}$ of 54 kg

b Which is the smaller, $\frac{5}{12}$ of 10 or $\frac{3}{4}$ of 5?

c Which is the larger, $\frac{3}{5}$ of $\frac{9}{10}$ or $\frac{5}{7}$ of $\frac{3}{4}$?

REVISION EXERCISE 1.3 (Number work)

1 Without using a calculator, find

a $349 + 276$ **g** $7 \times 63 - 249$

b $723 - 584$ **h** $6421 - 236 \times 7$

c 66×80 **i** $(19 + 6) \times 5 - 96$

d 48×500 **j** $429 \div 21$ giving the remainder

e $336 \div 6$ **k** $(36 - 14) \times 3 - 49$

f $560 \div 80$ **l** $339 \div 23$ giving the remainder

2 a Find the value of **i** 2^6 **ii** 3^5 **iii** $2^3 \times 3^2 \times 7$

b Express in index form

i 128   **ii** 343   **iii** 625   **iv** 729

c Express as the product of prime factors in index form

i 1080   **ii** 3276   **iii** 1800

3 a Find the lowest number that is a multiple of all the numbers in each set (called the lowest common multiple, LCM).

i 3, 7   **iii** 2, 8, 10   **v** 26, 3

ii 2, 9   **iv** 3, 4, 6   **vi** 5, 4

b Find the highest whole number that divides exactly into all the given numbers (i.e. the highest common factor, HCF).

i 4, 6   **iii** 22, 44, 55   **v** 24, 8, 16

ii 3, 6, 12   **iv** 18, 6, 9   **vi** 14, 28, 56

4 Find the value of

a 5^2	**e** $2^4 \times 3^2$	**i** 3.25×10^2
b 3^4	**f** $8^2 \times 5^2$	**j** 8.01×10^3
c 2^5	**g** $6^3 \times 2^2$	**k** 0.072×10^4
d 5^3	**h** $7^3 \times 2^3$	**l** 1.1×10^6

5 Write, where possible, as a single expression in index form

a $2^3 \times 2^4$	**d** 2×2^4	**g** $3^4 \div 3$
b $5^2 \times 3^5$	**e** $7^3 \div 7^2$	**h** $(2^3)^2$
c $5^1 \times 5^3$	**f** $3^6 \div 3^2$	**i** $(5^3)^2$

6 Find, as a fraction, the value of

a 2^{-1}	**d** $\left(\frac{2}{3}\right)^{-1}$	**g** $\left(\frac{1}{5}\right)^{-3}$	**j** $\left(\frac{2}{7}\right)^{-2}$
b 10^{-1}	**e** 2^{-3}	**h** $\left(\frac{1}{6}\right)^{-2}$	**k** $\left(\frac{3}{10}\right)^{-4}$
c $\left(\frac{1}{3}\right)^{-1}$	**f** 6^{-2}	**i** 4^{-2}	**l** 5^0

7 a Write the following numbers in standard form.

i 265	**iii** 76 700	**v** 450 000
ii 0.18	**iv** 0.000 007	**vi** 0.092

b Write the following numbers as ordinary numbers.

i 3.45×10^{-2}	**iii** 7.3×10^{-1}	**v** 1.4×10^5
ii 5.01×10^{-2}	**iv** 6.37×10^{-4}	**vi** 2.83×10^5

8 Give each of the following numbers correct to

i 3 decimal places   **ii** 3 significant figures.

a 2.7846	**d** 0.150 76	**g** 3.2994
b 0.1572	**e** 254.1627	**h** 0.000 925 8
c 0.073 25	**f** 7.8196	**i** 0.009 638

9 For each calculation, first make a rough estimate of the answer, then use your calculator to give the answer correct to 3 significant figures.

a 78.4×0.527	**d** $(5.09)^3$	**g** $\dfrac{7.21 \times 5.93}{13.74}$
b 842×284	**e** $(0.185)^{-2}$	**h** $\dfrac{849 \times 0.773}{16.34}$
c $9.827 \div 4.731$	**f** $3000 \div 48.66$	**i** $\dfrac{0.515}{6.37 \times 0.717}$

10 Show on a sketch the range in which each of the following measurements lie.

   a 8 m to the nearest metre

   b 15 minutes to the next complete minute

   c 123 cm rounded down to the nearest cm

**REVISION
EXERCISE 1.4
(Shape and space)**

1 Find the size of each marked angle.

a

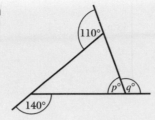

c

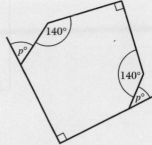

b

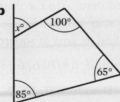

d

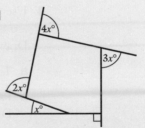

2 a Find the size of each exterior angle of a regular polygon with

   i 15 sides   **ii** 20 sides.

   b Find the size of each interior angle of a regular polygon with

   i 8 sides   **ii** 18 sides.

   c How many sides has a regular polygon

   i if each exterior angle is 15°   **ii** if each interior angle is 162°?

   d Is it possible for each exterior angle of a regular polygon to be

   i 40°   **ii** 70°?

   If it is, give the number of sides.

   e Is it possible for each interior angle of a regular polygon to be

   i 120°   **ii** 160°?

   If it is, give the number of sides.

3 Find the area of each shape.

a

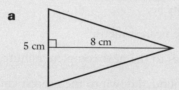

c

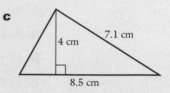

b

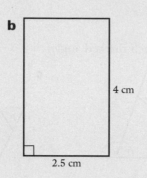

d

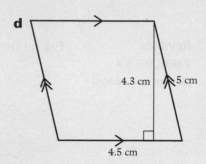

4 Use squared paper and draw axes for *x* and *y* in the ranges
−6 ⩽ *x* ⩽ 6, −6 ⩽ *y* ⩽ 6 using 1 square to 1 unit.
Draw the figure and find its area in square units.

a Triangle ABC with A(0, 6), B(6, 6) and C(5, 2)

b Parallelogram ABCD with A(0, 1), B(0, 6), C(6, 4) and
D(6, −1)

c Rectangle ABCD with A(−4, 2), B(0, 2) and C(0, −1)

d Square ABCD with A(0, 0), B(0, 4) and C(4, 4)

e Triangle ABC with A(−5, −4), B(2, −4), C(−2, 3)

5 For each of the following figures, find the missing measurement.
Draw a diagram in each case.

	Figure	Base	Height	Area
a	Triangle	8 cm		16 cm^2
b	Rectangle	3 cm	15 mm	
c	Parallelogram	4 cm		20 cm^2
d	Square	5 m		
e	Triangle	70 mm		14 cm^2

6 Find the area of the following shapes. Draw a diagram for each question and mark in all the measurements.

a

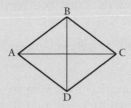

ABCD is a rhombus
AC = 15 cm and
BD = 8 cm

b

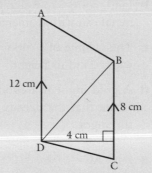

c

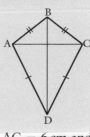

AC = 6 cm and
BD = 10 cm

7 a   i Find, giving your answer in metres, 137 cm + 234 mm + 3.2 m

ii Find, giving your answer in grams, 645 g + 0.37 kg + 960 mg

iii Find, giving your answer in inches, 3 feet + 2 yards + 8 inches

b Express

i	45 mm in cm	**vii**	$0.6\,m^2$ in cm^2
ii	0.56 km in m	**viii**	$3\,ft^2$ in in^2
iii	48 inches in feet	**ix**	$5000\,mm^3$ in cm^3
iv	13 yards in feet	**x**	$0.002\,m^3$ in cm^3
v	$5\,cm^2$ in mm^2	**xi**	$4\,000\,000\,cm^3$ in m^3
vi	$4000\,cm^2$ in m^2	**xii**	$\frac{5}{12}\,ft^3$ in in^3

8 For the shaded part of each diagram, find   **i** the perimeter   **ii** the area.

a

b

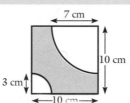

c

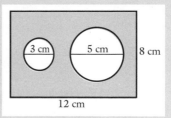

9 In this question, the cross-sections of the prisms and their lengths are given. Find their volumes.

a

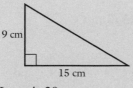

Length 30 cm

b

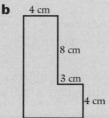

Length 15 cm

c

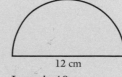

Length 40 cm

10 a Find the volume of a rectangular metal block measuring 4.2 cm by 3.8 cm by 1.5 cm.

b Find, correct to 3 significant figures, the capacity of a cylindrical metal can with diameter 12 cm and height 10 cm.

c The volume of a cuboid is 136 cm^3. It is 8 cm long and 3.4 cm wide. Find its height.

d A gold ingot is a cuboid measuring 5 cm by 5 cm by 3 cm. Given that the density of gold is 19.3 g/cm^3, find the mass of the ingot.

11 a Find AC. **b** Find PR.

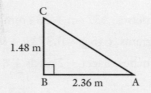

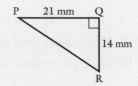

c

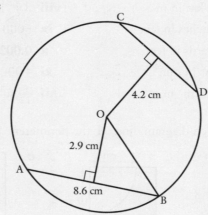

i AB is a chord of length 8.6 cm in a circle, centre O. The distance of the chord from the centre of the circle is 2.9 cm. Find, correct to 3 significant figures, the radius of the circle.

ii A second chord, CD, is 4.2 cm from O. Find the length of CD.

12 Draw x- and y-axes on squared paper for values of x and y from 0 to 10. Mark the points A(2, 1) and B(8, 9) on your graph.

Find **a** the coordinates of the midpoint of AB

   b the length of the line AB.

REVISION EXERCISE 1.5 (Algebra)

1 Simplify the following expressions.

a $3(x+7)+2x$

g $-3(x-2)$

b $5+(3a-7)$

h $-2(4-3x)$

c $3(2x+3)+4(x-2)$

i $-5(2x+7)$

d $5x-3(x+2)$

j $4(a+2)-3(a-4)$

e $4x-2(3+x)$

k $15x-3x-2(4x-3)$

f $8a-(2a+3)$

l $6a-3(2a-5)+3$

2 Simplify

a $2a \times 3b \times 4c$

d $\dfrac{4a}{3} \times \dfrac{7}{20}$

g $(-3a) \times (-2a)$

b $5x \times 2x \times 4y$

e $\dfrac{3x}{5} \div \dfrac{9x}{10}$

h $(-x) \div (-y)$

c $3a \times 4b \times 5b$

f $\dfrac{5x}{3} \div \dfrac{3}{10x}$

i $12x \div (-3x)$

3 a If $P = 2(a-b)$ find P when

i $a=3$ and $b=1$

ii $a=-5$ and $b=2$

iii $a=-4$ and $b=-2$

b If $A = xy$ find A when

i $x=3$ and $y=\frac{1}{2}$

ii $x=4$ and $y=-2$

iii $x=-3$ and $y=-6$

4 a Apples cost x pence each. Write down a formula for C if C pence is the cost of 6 apples.

b Find a formula for u_n in terms of n.

n	1	2	3	4	5
u_n	4	7	10	13	16

5 Solve the following equations.

a $3x = 15$

f $4(x-2) = 5(2x+5)$

b $4x+1 = 17$

g $\dfrac{x}{4} - 3 = 5$

c $7x-3 = 2x+7$

h $\dfrac{x}{2} - \dfrac{1}{4} = \dfrac{7}{12}$

d $\dfrac{x}{5} = 2$

i $0.8x = 5.6$

e $3x-4 = 11-2x$

j $0.03x = 0.42$

6 a Solve the following inequalities and illustrate your solutions on a number line.

  i $x - 6 < 4$ **iv** $4x - 1 \leqslant 15$ **vii** $4x + 1 > 6$

  ii $9 - x \geqslant 4$ **v** $2 > 8 + x$ **viii** $3x + 4 \geqslant 5 - 2x$

  iii $8 > 3 - x$ **vi** $7 > 3 - x$ **ix** $2x + 3 \leqslant 9 - 4x$

b Solve each pair of inequalities and hence find the range of values of x which satisfy both of them.

  i $x - 3 < 5$ and $x + x > 2$ **ii** $2x + 1 > 5$ and $3x - 10 < 2$

c Find the range of values of x for which the following inequalities are true.

  i $x + 3 > 2x - 1 > 3$

  ii $x - 1 < 2x + 3 \leqslant 7$

  iii $4x + 1 < x - 2 < 4$

7 Solve the simultaneous equations

  a $x + y = 4$ **c** $5x + 3y = 25$ **e** $x + 5y = 9$
     $3x + y = 10$ $8x - 3y = 1$ $x - y = 21$

  b $7x - 2y = 22$ **d** $2x - 3y = 15$ **f** $2a + 3b = 9$
     $3x + 2y = 18$ $2x - y = 9$ $2a + 7b = 13$

8 Solve the following equations, giving your answers correct to 3 significant figures.

  a $x^2 = 23$ **b** $x^2 = 0.47$

REVISION EXERCISE 1.6 (Graphical work)

1

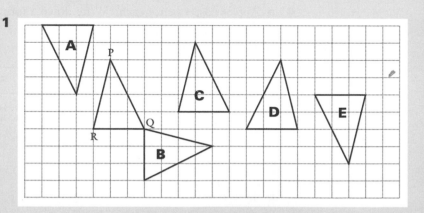

In the diagram, which images of $\triangle$PQR are given by

a a translation   **c** a rotation

b a reflection   **d** none of these.

2 Draw axes for x and y for values from 0 to 10 using 1 cm as 1 unit. Draw $\triangle$XYZ with X(4, 2), Y(4, 4) and Z(5, 1) and $\triangle$X$'$ Y$'$ Z$'$ with X$'$(6, 4), Y$'$(6, 10) and Z$'$(9, 1). Find the centre of enlargement and the scale factor that maps $\triangle$XYZ to $\triangle$X$'$ Y$'$ Z$'$.

3 a Write down the gradient and y-intercept of the line whose equation is

i $y = 3x$ **ii** $y = 2x + 6$ **iii** $y = 3 - \frac{1}{2}x$

b Write down the equation of the line that is parallel to the line $y = 4x$ that goes through the point   **i** (0, 2)   **ii** (0, −3)

c A(2, a) and B(b, 10) are points on the line $y = -3x + 7$. Find a and b.

4 a The table gives the coordinates of three points on a straight line. What is the equation of the line?

x	−3	0	2
y	−4	2	6

b Determine whether each of the following straight lines makes an acute angle or an obtuse angle with the positive x-axis.

i $y = -x + 2$ **ii** $y = 3 - 7x$ **iii** $y = 0.6x$ **iv** $y = 2 + 3x$

c Determine whether each pair of lines is parallel, perpendicular or neither.

i $y = 3x + 1$, $y = 4 - 3x$ **iii** $y = -\frac{1}{2}x + 1$, $y = 2x + \frac{1}{2}$

ii $y = 3x - 1$, $y = 3x + 5$ **iv** $y = -x + 3$, $y = x - 3$

5 The equation of a curve is $y = 3x^2$. Which of these sketches could be the curve?

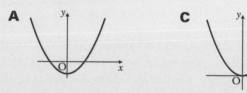

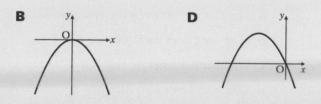

6 Sketch the curve whose equation is   **a** $y = -2x^2$ **b** $y = 5x^2$

7 **a** A car travels at 60 mph. How far will it travel in
   i $1\frac{3}{4}$ hours   **ii** 5 minutes?

b How long will a train travelling at 100 mph take to travel
   i 125 miles   **ii** 240 miles?

c Jim cycles 42 km in $3\frac{1}{2}$ hours. Find his average speed.

8

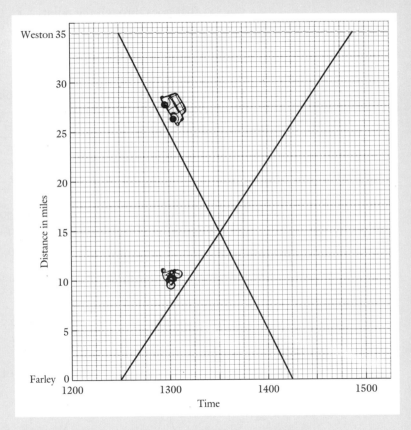

The graph shows two journeys between the villages of Farley and Weston. Nina leaves home on her bicycle to visit her friend who lives at Weston. On her way there she passes her father who is on his way from Weston to Farley in his car.

a How far is it between the two places?

b How long does each journey take?

c Who has the faster average speed and by how much?

d Where and when do they pass?

**REVISION
EXERCISE 1.7
(Probability and
Statistics)**

1 The pie chart shows the breakdown
of John Peters' bill for £120
when his car was serviced.

    a What fraction of the bill was
labour?

    b How much was the charge for VAT?

    c What percentage of the bill was for parts?

2 **a** Find the mode, median and range of the numbers

$$9, \ 8, \ 11, \ 8, \ 9, \ 13, \ 12, \ 11, \ 8, \ 7, \ 9, \ 8, \ 12$$

    b Find the mean of the numbers given in part **a**. Give your answer
correct to 3 significant figures.

3 In a sale 80 dresses are reduced. Of these 16 are Size 8, 28 are
Size 10, 24 are Size 12 and the remainder are Size 14. Ann takes a
dress from the rail at random. What is the probability that the dress
she takes is

    a Size 10   **b** not Size 12   **c** Size 12 or larger?

4 Two ordinary six-sided dice are rolled together 360 times.

    a Draw a table to show the equally likely outcomes when the dice
are rolled once.

    b About how many double threes are there likely to be?

    c About how many times should the score be 10?

5 A number is chosen at random from the first twelve non-zero whole
numbers. What is the probability that the number is

    a a prime number   **c** not exactly divisible by 3

    b exactly divisible by 4   **d** a rectangular number?

6 For each scatter diagram describe the relationship between the
quantities.

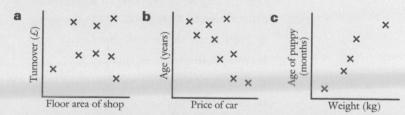

7 Given below is a list of the heights, in centimetres rounded up to the nearest centimetre, of 60 tomato plants. The list is in numerical order.

```
20  20  21  22  22  22  23  24  24  24
24  25  25  25  26  26  26  27  27  27
27  27  27  28  28  28  28  28  28  28
28  29  30  30  31  32  32  33  33  34
34  34  34  34  35  35  35  35  35  35
36  36  36  37  37  38  40  40  41  42
```

a What is the height of   **i** the tallest plant   **ii** the shortest plant?

b Copy and complete this frequency table.

Height, h cm	Frequency
$18 < h \leqslant 23$	
$23 < h \leqslant 28$	
$28 < h \leqslant 33$	
$33 < h \leqslant 38$	
$38 < h \leqslant 43$	

c How many plants have a height that is

   i greater than 28 cm   **ii** 33 cm or less?

d What is the modal group?

e Illustrate this information with

   i a bar chart   **ii** a frequency polygon.

8

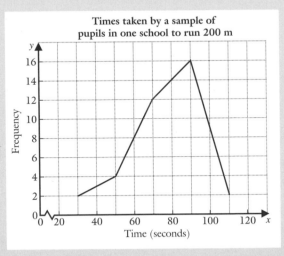

Use the frequency polygon to answer the following questions. If you cannot give an answer say why.

a How many times were measured?

b What is the range of times?

c Is it true to say that half the times were less than 80 seconds?

d How many pupils ran 200 metres?

WORKING
WITH
NUMBERS

David and his father sometimes have communication problems, especially when numbers are involved.

For example, on one occasion, David's father asked him to measure the mat-well inside the front door.
David wrote down '54 cm by 32 cm'.
David's father then had a piece of matting cut to these measurements but, when he got it home, he found that it was too large for the hole; the result was a row in which each unfairly blamed the other.

- When David and his father had calmed down they realised that there were two possible reasons for the error. David had given the measurements correct to the nearest centimetre but the length and width of the well were both slightly under 54 cm and 32 cm. The mat had been cut correct to the nearest centimetre, but its length and width were both slightly over 54 cm and 32 cm. These two roundings had combined to produce a mat that was nearly one centimetre too long and too wide.

EXERCISE 1A   Discuss the possible consequences of using rounded numbers in the following situations.

1   Hasib wanted enough topsoil to cover an area of 12 m² to a depth of 50 cm. He worked out the area correct to the nearest square metre. He ordered 6 cubic metres of top soil.

2 Hannah measured her garden path as **38.5** metres correct to 1 decimal place. She ordered some square paving tiles of side 50 cm, which she then placed edge-to-edge in a single line to pave the path.

3 The operating equipment on a long-range shell launcher needs to be set with the distance and bearing of the target. The gunner worked out these measurements from maps as **2558** m on a bearing of **026.79**°. He assumed that they were accurate to the number of figures given but, because of various reading and calculation errors, they were accurate only to 2 significant figures.

The examples above show that working with numbers that have been rounded gives results that may be less accurate than we realise and that, unless we are aware of this, the consequences can be anything from mildly annoying to disastrous.

The following points may have arisen from your discussions.

- Some numbers are exact and others are rounded.

- It is impossible to give exact numerical values for some quantities, lengths for example, so we need to use and work with corrected numbers.

- We need to be aware that some numbers have been rounded and, where possible, to know how they have been rounded.

- When we use rounded numbers in calculations, we need to appreciate that the results contain errors.

RANGE OF VALUES FOR A CORRECTED NUMBER

Suppose we are told that, correct to the nearest 10, 250 people boarded a particular train. People are counted in whole numbers only. Hence in this case, 245 is the lowest number that gives 250 when corrected to the nearest 10 and 254 is the highest number that can be corrected to 250. We can therefore say that the actual number of people who boarded the train is any whole number from 245 to 254.

Now suppose that we are given a nail and are told that its length is 25 mm correct to the nearest millimetre.

Look at this magnified section of a measuring gauge:

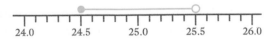

The lowest number that can be rounded up to 25 is 24.5. The highest number that can be rounded down to 25 is not so easy to determine. All we can say is that any number up to, but not including, 25.5 can be rounded down to 25.

The length of the nail is therefore in the range from 24.5 mm up to, but not including, 25.5 mm.

If *l* mm is the length of the nail, we can write

$$24.5 \leqslant l < 25.5$$

To illustrate this on the diagram we use a line segment with a solid circle at the lower end to show that 24.5 is included in the range and an open circle at the upper end to show that 25.5 is not included in the range.

24.5 is called the *lower bound of l* and 25.5 is called the *upper bound of l*.

EXERCISE 1B

Illustrate on a number line the range of values of *x* given by
$0.1 < x \leqslant 0.8$

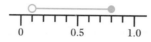

Use a number line like this for questions **1** to **6**.

In each case illustrate the range on your number line.

1 $5 \leqslant x \leqslant 10$ **4** $5 \leqslant x < 15$

2 $0 < x \leqslant 15$ **5** $0 \leqslant x < 10$

3 $-2 \leqslant x \leqslant 6$ **6** $-5 < x \leqslant 5$

Use a number line like this for questions **7** to **12**. In each case illustrate the range on your number line.

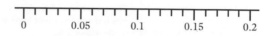

7 $0 < x < 0.1$ **10** $0.08 < x < 0.16$

8 $0.1 < x < 0.2$ **11** $0.02 < x < 0.08$

9 $0.05 < x < 0.15$ **12** $0.03 < x < 0.13$

A number is given as **3.15** correct to 2 decimal places. Illustrate on a number line the range in which this number lies.

3.15 is between 3.14 and 3.16 so we will use just that part of the number line.

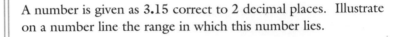

Illustrate on a number line the range of possible values for each of the following corrected numbers.

13 1.5 correct to 1 d.p.

14 0.6 correct to 1 d.p.

15 0.2 correct to 1 d.p.

16 1.3 correct to 1 d.p.

17 0.1 correct to 1 d.p.

18 6.2 correct to 1 d.p.

19 0.25 correct to 2 d.p.

20 0.52 correct to 2 d.p.

21 1.15 correct to 2 d.p.

22 6.89 correct to 2 d.p.

23 12.26 correct to 2 d.p.

24 0.05 correct to 2 d.p.

It is stated that a packet of pins contains 500 pins to the nearest 10.
Find the range in which the actual number of pins lies.

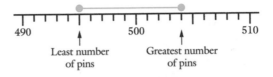

There must be a whole number of pins in the packet; 504 is the largest whole number that rounds to 500 to the nearest 10.

If n is the number of pins in the packet then n is a whole number such that

$$495 \leqslant n \leqslant 504$$

A copper tube is sold as having an internal bore (diameter) of 10 mm to the nearest millimetre. Find the range in which the actual bore lies.

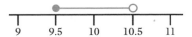

If d mm is the diameter of the tube then $9.5 \leqslant d < 10.5$

In some of the following questions you are asked to find a range of values for a quantity that can only have whole number values. When this is the case, it must be clearly stated in your answer.

25 The weight, w kg, of a bag of sand is given as 5.6 kg correct to 1 decimal place. Find the range of values in which w lies.

26 A shop is said to make a profit of £2500 a month. If this figure is given correct to the nearest £100, find the range in which the actual monthly figure, £x, lies.

27 On a certain model of bicycle, the brake pads have to be 12.5 mm thick to work efficiently. If the brake pads are x mm thick, find the range in which x lies when the figure given is correct to 1 decimal place.

28 Referring to a football match, a newspaper headline proclaimed '75 000 watch England win'. Assuming that this figure for the number of spectators is correct to the nearest 1000, find the range in which x, the number of people who actually attended the match, lies. If, in fact, the figure given was a guess, what can you say about x?

29 One of the component parts of a metal hinge is a pin. In order to work properly this pin must have a diameter of 1.25 mm correct to 2 decimal places. If d mm is the diameter, find the range in which d must lie.

30 Alan measured the width of a space between two kitchen units in metres correct to 1 decimal place, and wrote down 1.6 m.

  a Find the range within which the width of the space lies.

  b The cupboard bought to go into the space is 1.62 m wide correct to 2 decimal places. Will it fit the space?

  c Comment on Alan's measuring.

31 The length of car is given as 455 cm. If this figure is correct to the nearest 5 cm, find the range in which the actual length lies.

32 Knitting yarn is sold by weight. It is found that 10 g of double knitting pure wool has a length of 20 m correct to the nearest metre. What is the minimum length you would expect in a 50 g ball of this wool?

33 The weight of a wall tile is given as 40 g to the nearest gram. A DIY store sells these tiles in polythene-wrapped packs of ten. Ignoring the weight of the wrapping, find the range in which the weight of one of these packs lies.

34 The length of a side of a square carpet tile is 29.9 cm correct to 1 decimal place. One hundred of these carpet tiles are laid end-to-end on the floor of a shop. Find the range in which the length of this line of tiles lies. Hence give the difference between the length of the longest possible and the shortest possible line of 100 tiles.

35 Toy bricks are cubes of side 34 mm correct to the nearest millimetre. Eight of these cubes are placed side by side in a row. What is the upper bound of the length of this row?

36 A cube of side 34 mm, correct to the nearest millimetre, is to be packed in a cubical box with internal sides of 34.2 mm correct to 3 significant figures. Explain why the cube may not fit in the box.

37 John and Debbie both said that they were 15 years old. What is the greatest possible difference in their ages?

38 A wooden pole is 450 mm long and a 200 mm length is cut from it. Both of these measurements are correct to the nearest millimetre. What can you say about the length of the remaining section of the pole?

39 A rectangular sheet of stainless steel, measuring 35 mm by 23 mm correct to the nearest millimetre, needs edging with wire.

   a What length of wire is needed to make sure that there is enough to edge one of these sheets?

   b The wire costs £18.20 per metre. What is the cost of enough wire to ensure that 500 000 of these sheets can be edged? (Assume that there is no waste.)

40 A pair of digital scales gives weights correct to the nearest gram. When one nail is weighed on these scales, the reading is 8 grams. When 100 identical nails are weighed, the reading is 775 grams.

   a Explain the apparent contradiction in the readings.

   b Give a more accurate weight for one nail than the scales are capable of showing when only one nail is weighed.

FRACTIONS AND
DECIMALS

Questions from the last exercise show that an answer worked out using corrected numbers is not exact but lies within a range. If we are not aware of this, we may get errors when we use corrected numbers as the following situation shows.

A caterer was asked to provide 800 filled baguettes for a local function.

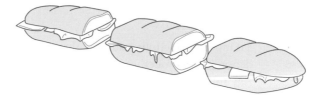

One long French loaf is enough for three filled baguettes.
The caterer reasoned that he needed 800 thirds of French loaves and worked this number out as follows.

$$800 \text{ thirds } = 800 \times \tfrac{1}{3}$$

$$= 800 \times 0.33 = 264$$

He ordered 264 loaves, made them up, then found that he was 8 baguettes short.

The caterer could have avoided this situation by checking his answer, that is, 264 loaves makes 264×3 rolls, which is 792 rolls and this is not enough.

He would not have made the error in the first place if

- he had not replaced $\tfrac{1}{3}$ with 0.33 because, although correct to 2 decimal places, $\tfrac{1}{3}$ is not exactly equal to 0.33.
  In fact $\tfrac{1}{3} = 0.333\,33\ldots$ so 0.33 is less than $\tfrac{1}{3}$; hence the shortage.

- he had worked with fractions and had calculated $800 \times \tfrac{1}{3}$ as $\tfrac{800}{1} \times \tfrac{1}{3} = \tfrac{800}{3} = 266\tfrac{2}{3}$. He would then have seen that he needed 267 loaves for 800 baguettes with 1 to spare.

This situation shows, firstly, that we need to be aware that some fractions do not have exact decimal equivalents. Secondly, if we choose to use the decimal form of a fraction, it is important to use enough decimal places to ensure the accuracy we need.
It also illustrates that if we are comfortable when working with fractions we can choose to do so when it is appropriate rather than always resorting to decimals.

Revision exercises at the front of this book can be used to practise the basic operations with fractions. In the next section we introduce a different way of interpreting division by a fraction and use fractions in less simple calculations.

**ADDITION AND
SUBTRACTION
OF FRACTIONS**

We cannot add apples to oranges unless we reclassify them both as, say, fruit. In much the same way, we cannot add tenths to quarters unless we change them both into the same kind of fraction, that is, change them so that they have a common denominator. To do this we use the following fact. The value of a fraction is unaltered if both numerator and denominator are multiplied by the same number.

**MULTIPLICATION
OF FRACTIONS**

To multiply fractions, multiply the numerators together and multiply the denominators together.
Any mixed numbers must be changed into improper fractions, and factors that are common to the numerator and denominator should be cancelled before multiplication.

RECIPROCALS

> If the product of two numbers is 1 then each number
> is called the *reciprocal* of the other.

We know that $\frac{1}{3} \times 3 = 1$ so

$\frac{1}{3}$ is the reciprocal of 3   and   3 is the reciprocal of $\frac{1}{3}$.

To find the reciprocal of $\frac{3}{4}$ we require the number which, when multiplied by $\frac{3}{4}$ gives 1.

Now $\frac{4}{3} \times \frac{3}{4} = 1$ so $\frac{4}{3}$ is the reciprocal of $\frac{3}{4}$.

In all cases the reciprocal of a fraction is obtained by turning the fraction upside down.

A number can be written as a fraction, for example, $3 = \frac{3}{1}$, $2.5 = \frac{2.5}{1}$, and so on.

so   the reciprocal of $\frac{3}{1}$ is $\frac{1}{3}$ or $1 \div 3$

and   the reciprocal of $\frac{2.5}{1}$ is $\frac{1}{2.5}$ or $1 \div 2.5 \, (= 0.4)$

> The reciprocal of a number is 1 divided by that number.

**DIVISION BY A
FRACTION**

Consider $\frac{2}{5} \div \frac{3}{7}$: this can be interpreted as $\frac{2}{5} \times 1 \div \frac{3}{7}$

Now $1 \div \frac{3}{7}$ is the reciprocal of $\frac{3}{7}$, i.e. $\frac{7}{3}$

Therefore $\frac{2}{5} \div \frac{3}{7} = \frac{2}{5} \times \frac{7}{3} = \frac{14}{15}$

i.e.   > to divide by a fraction we multiply by its reciprocal.

EXERCISE 1C Write down the reciprocals of the following numbers.

1 4   **4** 10   **7** 100   **10** 2.5

2 $\frac{1}{2}$ **5** $\frac{1}{8}$ **8** $\frac{2}{9}$ **11** 3.2

3 $\frac{2}{5}$ **6** $\frac{3}{11}$ **9** $\frac{15}{4}$ **12** 1.6

Find, without using a calculator

13 $\frac{2}{3} \div \frac{1}{2}$ **16** $5 \div \frac{4}{5}$ **19** $\frac{3}{7} \div 1\frac{3}{4}$

14 $1\frac{2}{3} \div \frac{5}{6}$ **17** $\frac{2}{9} \div 1\frac{2}{7}$ **20** $\frac{5}{9} \div 10$

15 $2\frac{1}{2} \div 4$ **18** $\frac{1}{2} \div \frac{3}{4}$ **21** $3 \div \frac{2}{3}$

Find $\;2\frac{1}{2} + \frac{3}{5} \div 1\frac{1}{2} - \frac{1}{2}\left(\frac{3}{5} + \frac{1}{3}\right)$

> Remember that brackets are worked out first, then multiplication and division and lastly addition and subtraction.

$$2\frac{1}{2} + \frac{3}{5} \div 1\frac{1}{2} - \frac{1}{2}\left(\frac{3}{5} + \frac{1}{3}\right) = 2\frac{1}{2} + \frac{3}{5} \div 1\frac{1}{2} - \frac{1}{2}\left(\frac{9+5}{15}\right)$$

$$= 2\frac{1}{2} + \frac{3}{5} \div \frac{3}{2} - \frac{1}{2} \times \frac{14}{15}$$

$$= 2\frac{1}{2} + \frac{\cancel{3}}{5} \times \frac{2}{\cancel{3}} - \frac{1}{\cancel{2}} \times \frac{\cancel{14}}{15}$$

$$= 2\frac{1}{2} + \frac{2}{5} - \frac{7}{15}$$

$$= 2 + \frac{15 + 12 - 14}{30}$$

$$= 2 + \frac{13}{30} = 2\frac{13}{30}$$

Find, without using a calculator

22 $1\frac{2}{3} \times \frac{1}{2} - \frac{2}{5}$ **25** $5\frac{1}{2} \div 3 + \frac{2}{9}$ **28** $2\frac{1}{2} \div \frac{7}{9} + 1\frac{1}{3}$

23 $\frac{3}{7} + \frac{1}{4} \div 1\frac{1}{3}$ **26** $\frac{4}{5} \div \frac{1}{6} + \frac{1}{3} \times 1\frac{1}{2}$ **29** $\frac{3}{5}\left(1\frac{1}{4} - \frac{2}{3}\right)$

24 $\frac{2}{5} \div \left(\frac{1}{2} + \frac{3}{4}\right)$ **27** $\frac{9}{11} - \frac{2}{5} \times \frac{3}{4}$ **30** $3\frac{1}{2} - \frac{2}{3} \times 6$

31 $2\frac{1}{3} + \frac{1}{2}\left(2 \div \frac{4}{5}\right)$　　**35** $\dfrac{1\frac{1}{5} - \frac{3}{4}}{2\frac{1}{2}}$　　　　**39** $\frac{2}{3} \times \frac{6}{7} - \frac{5}{8} \div 1\frac{1}{4}$

32 $\frac{3}{4}\left(5\frac{1}{3} - 2\frac{1}{5}\right) \div \frac{7}{9}$　　**36** $\dfrac{\frac{9}{10}}{\frac{5}{6}}$　　　　**40** $\dfrac{\frac{7}{8}}{3\frac{1}{2} - \frac{2}{3}}$

33 $\frac{1}{2} + \left(\frac{3}{4} \div \frac{1}{6}\right)$ of 3　　**37** $\left(\frac{2}{3} - \frac{1}{2}\right) \div \left(\frac{3}{4} - \frac{1}{3}\right)$　　**41** $\dfrac{3\frac{1}{4}}{2\frac{3}{5}}$

34 $\dfrac{\frac{1}{3} + \frac{1}{4}}{\frac{5}{6} - \frac{3}{4}}$　　　　**38** $\frac{7}{9} - \frac{1}{3}$ of $1\frac{2}{7}$　　**42** $\dfrac{\frac{2}{3} \times \frac{3}{4}}{\frac{5}{6} \times \frac{3}{10}}$

INTERCHANGING DECIMALS AND FRACTIONS

When a fraction is written in the form $\frac{2}{3}$, it is called a *common*, or vulgar, fraction.

Another way of representing fractions is by placing a point after the number of units and continuing with figures to the right.
Fractions written in the form **0.75** are called *decimal* fractions.

Usually we refer to common fractions simply as fractions and to decimal fractions simply as decimals.

EXERCISE 1D

Express **0.705** as a fraction.

$0.705 = \dfrac{7}{10} + \dfrac{5}{1000}$　　　（This step is usually omitted.）

　　　$= \dfrac{705}{1000} = \dfrac{141}{200}$

Express the following decimals as fractions.

1 0.35	**4** 1.36	**7** 0.005	**10** 2.05
2 0.216	**5** 0.03	**8** 1.01	**11** 1.104
3 0.204	**6** 0.012	**9** 0.11	**12** 0.0001

Express $\frac{7}{8}$ as a decimal.

$\frac{7}{8} = 0.875$　　　（$\frac{7}{8}$ means $7 \div 8$）　　　$\begin{array}{r} 0.875 \\ 8\overline{)7.000} \end{array}$

Express the following fractions as decimals. Do not use a calculator.

13 $\frac{3}{20}$ **15** $\frac{3}{5}$ **17** $1\frac{1}{16}$ **19** $1\frac{3}{4}$ **21** $\frac{4}{25}$ **23** $2\frac{3}{8}$

14 $\frac{1}{8}$ **16** $\frac{6}{25}$ **18** $\frac{27}{50}$ **20** $\frac{5}{32}$ **22** $\frac{5}{16}$ **24** $\frac{1}{500}$

RECURRING DECIMALS

If we try to change $\frac{1}{6}$ to a decimal,   i.e.

$$6\overline{)1.0000\ldots}\quad 0.1666\ldots$$

we discover that

- we cannot write $\frac{1}{6}$ as an exact decimal

- from the second decimal place, the 6 recurs for as long as we have the patience to continue the division.

Similarly if we convert $\frac{2}{11}$ to a decimal by dividing 2 by 11, we get $0.181\,818\,18\ldots$ and we see that

- $\frac{2}{11}$ cannot be expressed as an exact decimal

- the pair of figures '18' recurs indefinitely.

Decimals like these are called *recurring decimals*. To save time and space we place a dot over the figure that recurs. In the case of a group of figures recurring we place a dot over the first and last figure in the group.

Therefore   we write $0.166\,666\ldots$ as $0.1\dot{6}$

and   we write $0.181\,818\ldots$ as $0.1\dot{8}$

Similarly   we write $0.316\,316\,316\ldots$ as $0.\dot{3}1\dot{6}$

EXERCISE 1E

Use the dot notation to write the following fractions as decimals.

1 $\frac{1}{3}$ **3** $\frac{5}{6}$ **5** $\frac{1}{7}$ **7** $\frac{1}{11}$ **9** $\frac{5}{12}$ **11** $\frac{7}{30}$

2 $\frac{2}{9}$ **4** $\frac{1}{15}$ **6** $\frac{1}{12}$ **8** $\frac{1}{18}$ **10** $\frac{1}{14}$ **12** $\frac{1}{13}$

Some recurring decimals can be expressed as fractions by recognition,

e.g. $\frac{1}{3} = 0.\dot{3},$ $\frac{1}{9} = 0.\dot{1},$ $\frac{1}{11} = 0.\dot{0}\dot{9}$

Express as a fraction   **a** $0.0\dot{3}$ **b** $0.\dot{2}$

a $0.0\dot{3} = 0.\dot{3} \div 10$

$\qquad = \frac{1}{3} \div 10 = \frac{1}{30}$

b $0.\dot{2} = 0.\dot{1} \times 2$

$\qquad = \frac{1}{9} \times 2 = \frac{2}{9}$

13 Express as a fraction **a** $0.\dot{5}$ **b** $0.0\dot{2}$ **c** $0.0\dot{5}$ **d** $0.00\dot{4}$

14 a Express $0.\dot{9}$ as a fraction.

   b What do you deduce about the value of $0.\dot{9}$?

   c Express $0.0\dot{9}$ as a fraction.

USING NUMBERS IN STANDARD FORM

Calculations involving numbers in standard form can be done on a calculator.

For example, to enter 1.738×10^{-6}, the number 1.738 is entered normally followed by the **EXP** button and then the power of 10.

However, as with all calculations, it is important to know whether the answer is about right; this means that we need to be able to estimate results using non-calculator methods. These are illustrated in the worked example.

EXERCISE 1F

If $a = 1.2 \times 10^{-2}$ and $b = 6 \times 10^{-4}$, find **a** ab **b** $\dfrac{a}{b}$ **c** $a+b$

a $ab = (1.2 \times 10^{-2}) \times (6 \times 10^{-4}) = 7.2 \times 10^{-6}$

b $\dfrac{a}{b} = \dfrac{1.2 \times 10^{-2}}{6 \times 10^{-4}} = \dfrac{1.2}{6} \times 10^{-2-(-4)} = 0.2 \times 10^2 = 2 \times 10^1$

c (Multiplication must be done before addition, so each number must be written in full.)

   $a+b = 1.2 \times 10^{-2} + 6 \times 10^{-4}$

   $\quad\quad = 0.012 + 0.0006 = 0.0126 = 1.26 \times 10^{-2}$

1 Without using a calculator, write down the value of ab in standard form if

   a $a = 2.1 \times 10^2$, $b = 4 \times 10^3$

   b $a = 5.4 \times 10^4$, $b = 2 \times 10^5$

   c $a = 7 \times 10^{-2}$, $b = 2.2 \times 10^{-3}$

   d $a = 5 \times 10^{-4}$, $b = 2.3 \times 10^{-2}$

   e $a = 1.6 \times 10^{-2}$, $b = 2 \times 10^4$

   f $a = 6 \times 10^5$, $b = 1.3 \times 10^{-7}$

2 Without using a calculator, write down the value of $\dfrac{p}{q}$ in standard form if

   a $p = 6 \times 10^5$, $q = 3 \times 10^2$

   b $p = 9 \times 10^3$, $q = 3 \times 10^5$

   c $p = 7 \times 10^{-3}$, $q = 5 \times 10^2$

   d $p = 1.8 \times 10^{-3}$, $q = 6 \times 10^{-4}$

3 Without using a calculator, write down the value of $x+y$ in standard form if

   a $x = 2 \times 10^2$, $y = 3 \times 10^3$

   b $x = 3 \times 10^{-2}$, $y = 2 \times 10^{-3}$

   c $x = 2.1 \times 10^4$, $y = 3.1 \times 10^5$

   d $x = 1.3 \times 10^{-4}$, $y = 4 \times 10^{-3}$

For questions **4** and **5**, first estimate the answer and then use a calculator to give the answer correct to 3 significant figures.

4 The special theory of relativity states that a mass m is equivalent to a quantity of energy, E, where $E = mc^2$.
c m/s is the speed of light and $c = 2.998 \times 10^8$.
Find E when $m = 1.66 \times 10^{-27}$.

5 The quantity of nitrate in one bottle of mineral water is 1.5×10^{-3} g. The quantity of nitrate in another bottle is 7.3×10^{-4} g.
The two bottles are emptied into the same jug. How much nitrate is there in the water in the jug?

PRACTICAL WORK

You will need access to a pair of scales that give weights in kilograms and grams. They do not need to be able to measure small masses accurately; a set of kitchen scales is ideal.

You also need a bag of uncooked rice.

a Try weighing one grain of rice and report on the result.

b Describe a method by which it is possible to estimate the mass of one grain of rice in grams to 2 decimal places.

c Use your method to estimate the mass of one grain.

d Suggest a way to judge the accuracy of your estimate.

INVESTIGATION

a Express $\frac{1}{99}$ as a decimal.

b Use the result from part **a** to express $0.1\dot{5}$ as a fraction.

c Find a rule for expressing any recurring decimal with a two-digit repeating pattern as a fraction.

d Express $\frac{1}{999}$ as a decimal.

e Explain how the result from part **d** can be used to express any recurring decimal with a three-digit repeating pattern as a fraction.

f Find a rule for expressing any recurring decimal as a fraction and test your rule with examples of four-digit repeating patterns and five-digit repeating patterns.

PROBABILITY

Fruit machines provide a very popular way of losing money. You put a coin in the machine and pull the lever. This makes three drums rotate quickly, then slow down and finally stop. Each drum has pictures of several fruits. When the drums stop three fruits are shown in the centre of the display. Depending on what they are there may be a prize.

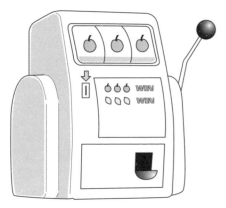

Chris has often been attracted to fruit machines but he has never played them. He would win a prize on the machine he is looking at if cherries appear on all three drums or lemons show on all three drums. Before he takes the plunge, he would like to assess the risk of losing money.

- To do this, he needs to know how to find the probability of combined events such as

  getting either a cherry or a lemon on the first drum,
  getting a cherry on both the first and the second drum.

  Both these examples involve two events, but in different ways.

- The first example involves either one or the other event occurring, whereas the second example involves both events happening.

EXERCISE 2A

Each sentence describes a situation where two events are involved. Discuss what the events are and whether they fall into the 'either ... or' category or into the 'both ... and' category.

1 An ordinary six-sided dice is rolled and scores five or six.

2 Two dice are rolled and a double six is scored.

3 Tim picks a box from a lucky dip; some boxes contain a prize and the others are empty.

4 The England cricket captain tosses a coin to find out who has the choice to bat or to field.

MUTUALLY
EXCLUSIVE
EVENTS

When an ordinary dice is rolled, it is possible to score either a five or a six. It is not possible to score both a five and a six. Such events are called *mutually exclusive*.

INDEPENDENT
EVENTS

When two ordinary dice are rolled, it is possible to score a six on the first dice and a six on the second dice Also the score obtained on the second dice is not affected in any such way by the score on the first dice. Such events, where both can happen but each has no influence on the occurrence or otherwise of the other, are called *independent events*.

Not all events are independent as this example shows. There are two green sweets and two red sweets in a bag. Mary takes one of these sweets then Tom takes one. If Mary's sweet is red there is only one out of three ways in which Tom can choose a red sweet. But if Mary's sweet is green there are two out of three ways in which Tom can choose a red sweet. So the probability that the second sweet is red depends on the colour of the first sweet taken.

EXERCISE 2B

Decide whether the events described are mutually exclusive, independent or dependent.

1 Mona and Clive each buy a ticket for a raffle and one of them wins first prize.

2 Two coins are tossed.

   a The first coin lands heads up or tails up.

   b Both coins land head up.

3 A 10 pence coin is tossed and a dice is rolled.

  a The coin lands head up and an even number is scored on the dice.

  b A three or a six is scored on the dice.

4 A blue bag and a red bag each contain a large number of coins, some of which are counterfeit. One coin is selected at random from each bag.

  a The coin taken from the blue bag is counterfeit or not counterfeit.

  b Both coins are counterfeit.

5 Hartfield Airport has 100 scheduled flights due to depart on Saturday.

  a Two or three flights are cancelled.

  b One flight is cancelled because the plane is faulty and another flight is cancelled because of a hurricane at its destination.

6 A box contains 6 blue pens and 3 red pens. One pen is removed at random.

  a The pen is put back then a pen is removed again.

  b The pen is not put back and another pen is removed.

ADDING PROBABILITIES

If we select a card at random from a pack of 52, the probability of drawing an ace is $\frac{4}{52}$ and the probability of drawing a black king is $\frac{2}{52}$.

Now drawing either an ace or a black king involves two events that are mutually exclusive since it is impossible to draw one card which is both an ace and a black king.

There are 4 aces and 2 black kings so if we want to find the probability of drawing either an ace or a black king there are 6 cards that we would count as 'successful', therefore

$$P(\text{ace or a black king}) = \frac{6}{52}$$

> Remember that the probability that an event A happens is $P(A)$ where
>
> $$P(A) = \frac{\text{the number of ways in which } A \text{ can occur}}{\text{the total number of equally likely outcomes}}$$

$$P(\text{ace}) = \frac{4}{52} \quad \text{and} \quad P(\text{black king}) = \frac{2}{52}$$

Since $$\frac{6}{52} = \frac{4}{52} + \frac{2}{52}$$ it follows that

$$P(\text{ace or black king}) = P(\text{ace}) + P(\text{black king})$$

Now consider the probability of scoring 5 or 6 when one dice is rolled.

$$P(\text{ score 5 or 6 }) = \tfrac{2}{6}$$

From one roll of a dice, a score of 5 and a score of 6 are mutually exclusive where $P(\text{ score 5 }) = \tfrac{1}{6}$ and $P(\text{ score 6 }) = \tfrac{1}{6}$

i.e.
$$P(\text{ score 5 or 6 }) = \tfrac{2}{6} = \tfrac{1}{6} + \tfrac{1}{6}$$
$$= P(\text{ score 5 }) + P(\text{ score 6 })$$

From these examples we see that

> if A and B are mutually exclusive events then
> $$P(\,A \text{ or } B\,) = P(\,A\,) + P(\,B\,)$$

EXERCISE 2C

1 A card is drawn at random from an ordinary pack of 52. What is the probability that the card is

   a a red ace   **b** a black king   **c** a red ace or a black king?

2 Gemma rolls an ordinary dice once. What is the probability the number shown is   **a** 2   **b** 3 or 4   **c** 2, 3 or 4?

3 A card is drawn at random from the 12 court cards (jacks, queens and kings). What is the probability that the card is

   a a black jack

   b a red queen

   c either a black jack or a red queen?

4 Graham is looking for his house key. The probability that it is in a pocket is $\tfrac{5}{9}$, while the probability that it is in the car is $\tfrac{1}{13}$. What is the probability that

   a the key is either in a pocket or in the car

   b the key is somewhere else?

5 When Mrs George goes shopping the probability that she returns by bus is $\tfrac{3}{7}$, in a taxi $\tfrac{1}{7}$, on foot $\tfrac{5}{14}$. What is the probability that she returns

   a by bus or taxi   **b** by bus or on foot   **c** by none of these ways?

6 Jo has a bag containing discs of four different colours. One disc is removed at random. The table shows the probabilities of choosing three of the four colours.

Colour	red	white	blue	pink
Probability	$\frac{2}{7}$	$\frac{2}{9}$	$\frac{1}{4}$	

Jo removes one disc at random. What is the probability that this disc is

a red or white   **c** red, white or blue

b white or blue   **d** pink ?

7 Maarit rolls an ordinary dice. What is the probability that the number on the dice is

a an even number   **b** a prime number   **c** either even or prime ?

Your answer to part **c** should not be the sum of the answers to parts **a** and **b**. Why not ?

MULTIPLICATION
OF
PROBABILITIES

When a coin is tossed and a dice is rolled, we can use a table to list all the possible outcomes:

Dice

		1	2	3	4	5	6
	H	H, 1	H, 2	H, 3	H, 4	H, 5	H, 6
Coin	T	T, 1	T, 2	T, 3	T, 4	T, 5	T, 6

From the table we can see that

$$P(\text{a head and an even number}) = \frac{3}{12} = \frac{1}{4}$$

Now a head from one toss of the coin and an even number from one throw of the dice are independent events, where

$$P(\text{a head}) = \frac{1}{2} \ \text{ and } \ P(\text{an even number}) = \frac{3}{6} = \frac{1}{2}$$

But

$$P(\text{a head and an even number}) = \frac{1}{4} = \frac{1}{2} \times \frac{1}{2}$$
$$= P(\text{a head}) \times P(\text{an even number})$$

This example illustrates that

if A and B are independent events then
$$P(A \text{ and } B) = P(A) \times P(B)$$

1 Two coins are tossed. What is the probability that they both land head up?

2 Two dice are tossed. Find the probability of getting a double six.

3 Peter has two tubes of Smarties. Each tube contains 10 red Smarties and 30 Smarties of other colours. Peter takes one Smartie, chosen at random, from each tube. Find

   a the probability that a red Smartie is taken from a tube

   b the probability that a Smartie other than a red one is taken from a tube

   c the probability that both Smarties removed are not red.

4 The probability that Heather will win the girls' 100 m is $\frac{2}{5}$ and the probability that Colin will win the boys' 100 m is $\frac{3}{5}$. What is the probability that

   a both of them will win their events

   b neither of them will win their event?

5 A mother has an equal chance of giving birth to a boy or a girl. Jane plans to have two children.

   a What is the probability that the first is a girl?

   b What is the probability that both are boys?

   c What is the probability that neither is a boy?

6 The probability that Eve will have to wait before she can cross Westgate Street is $\frac{1}{3}$ and the probability that she will be able to cross High Street without waiting is $\frac{1}{4}$.

What is the probability that

   a she does not have to wait to cross Westgate Street

   b she has to wait to cross High Street

   c she can cross both streets without waiting?

7 A bag contains 3 red sweets and 2 green sweets. Camilla takes one sweet at random and eats it. She then takes another sweet, also at random.

   a Make a possibility table to show the possible combinations of colours of the two sweets and use it to find the probability that both sweets removed are red.

   b Explain why, in this case, the multiplication rule does not give the correct answer to part **a**.

EXERCISE 2E Some of the events described are mutually exclusive and some are independent.

1 A red dice and a blue dice are rolled. Find the probability of getting

a a 5 or a 6 on the red dice   **c** a 2 on both dice

b a 1 or a 2 on the blue dice   **d** an even number on both dice.

2 A card is drawn at random from an ordinary pack of 52 playing cards. What is the probability that the card is

a a 2   **b** a red ace   **c** a 2 or a red ace ?

3 When Kim goes to Weightwatchers the probability that she returns on foot is $\frac{2}{3}$, by bus $\frac{1}{6}$ and in a friend's car $\frac{1}{6}$.
What is the probability that she returns

a by bus or in a friends' car   **b** on foot or by bus ?

4 The probability that Sam will complete the 5000 km race is 0.9 and the probability that Mike will complete it is 0.6. What is the probability that both Sam and Mike will complete the 5000 km race ?

5 A pack of cards is cut, reshuffled and cut again. What is the probability that

a the first card cut is an ace or a king

b the second card cut is an ace or a king

c both cards cut are aces ?

TREE DIAGRAMS When two coins are tossed, one possible outcome is a head and a tail. This outcome involves two events but they do not fit neatly into the 'either ... or' category or the 'both ... and' category.
This is because a head and a tail can be obtained by getting
either a head on the first coin and a tail on the second,
or a tail on the first coin and a head on the second.
So getting a head and a tail when two coins are tossed involves a mixture of independent and mutually exclusive events and we need an organised approach to deal with such a combination. One such approach is to draw up a table showing all the equally likely outcomes, but this method cannot be used if all the possible outcomes are not equally likely such as the possible outcomes when two people take a driving test.

Now suppose that three coins are tossed and we want the probability of getting two heads and a tail. Three events are involved here so we cannot use a table to list all the outcomes because a table can only cope with two events.

These examples show that we need a different way of listing outcomes and finding probabilities. We will illustrate this approach with a simple example.

Suppose that we have two discs, a red one marked A on one side and B on the other, and a blue one marked E on one side and F on the other.

Tossing the red disc, the probability that we get A is $\frac{1}{2}$ and the probability that we get B is also $\frac{1}{2}$. This information can be shown in the adjacent diagram.

Suppose that the red disc shows A and we go on to toss the blue disc. The probability of getting E is $\frac{1}{2}$ and the probability of getting F is $\frac{1}{2}$. We can add this information to the diagram.

We complete the diagram by considering what the probabilities are if the red disc shows a B before we toss the blue disc.

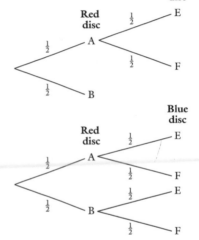

Diagrams like this are called *tree diagrams* or *probability trees*.

To use the tree diagram to find the probability that we get first an A and then an E, follow the path from left to right for an A on the first branch and an E on the second. The two probabilities we find there are $\frac{1}{2}$ and $\frac{1}{2}$. The blue disc landing showing E is independent of the letter obtained on the red disc so we multiply the probabilities together to get $\frac{1}{4}$.

To find the probability that we get a B on the red disc and an F on the blue one, follow the B and F path and multiply the probabilities,

i.e. $\quad P(\text{ B and F }) = \frac{1}{2} \times \frac{1}{2}$

$$= \frac{1}{4}$$

> In general we multiply probabilities when we follow a path along branches.

A coin is tossed and a dice is thrown. Find the probability that

a the coin lands head up and the dice does not show a six

b the coin lands tail up and the dice shows a six.

> There are only 2 possible outcomes when the coin is tossed; so we need two 'branches' to show these. There are six possible outcomes when the dice is thrown, but we only need to consider these in two groups, throwing a six or not throwing a six, and we need only two branches to show these.

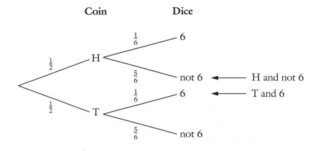

a $P(\,H \text{ and not } 6\,) = \dfrac{1}{2} \times \dfrac{5}{6} = \dfrac{5}{12}$

b $P(\,T \text{ and } 6\,) = \dfrac{1}{2} \times \dfrac{1}{6} = \dfrac{1}{12}$

1 The probability that Mark gets to work on time is $\frac{7}{8}$ and the probability that he leaves work on time is $\frac{3}{5}$.

a Find the probability that he does not leave work on time.

b Copy and complete the given probability tree.

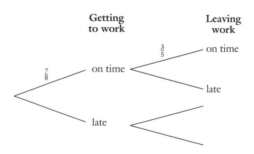

What is the probability that

c Mark gets to work on time but does not leave on time

d Mark is late for work but leaves on time ?

2 When a drawing pin falls to the ground the probability that it lands point up is 0.2.

a Find the probability that a pin does not land point up.

Two drawing pins fall one after the other.

b Copy and complete the tree diagram.

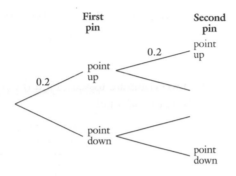

First
pin

Second
pin

point
up

0.2

point
up

0.2

point
down

point
down

Find the probability that

c both drawing pins land point up

d both drawing pins land point down.

3 The first of two boxes of tennis balls contains one white and two yellow balls; the second box contains three yellow and two lime green balls. A ball is taken at random from each box.

a Copy and complete the tree diagram.

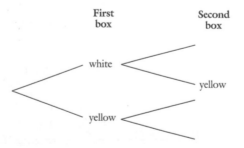

First
box

Second
box

white

yellow

yellow

Find the probability that

b both balls are yellow

c one is white and one is lime green.

4 Two soldiers fire at a target. The probability that Becker hits the target is 0.5 and the probability that Crossley does not hit the target is 0.3. Becker fires at the target first, then Crossley fires.
Draw a tree diagrams to show the possibilities and use it to find the probability that

a both Becker and Crossley hit the target

b neither hits the target

c Becker hits the target but Crossley misses.

d Crossley hits the target but Becker misses.

Two coins are tossed. Find the probability that they land showing a head and a tail.

We start by drawing a tree diagram.

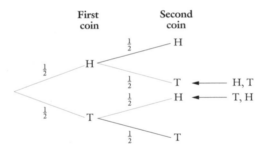

We can see that there are two paths through the tree that give a head and a tail where

$$P(\text{H on first coin and T on the second}) = P(\text{H, T}) = \frac{1}{2} \times \frac{1}{2} = \frac{1}{4}$$

$$P(\text{T on first coin and H on the second}) = P(\text{T, H}) = \frac{1}{2} \times \frac{1}{2} = \frac{1}{4}$$

Now the coins can land showing either H, T or T, H so these are mutually exclusive. Therefore we can find P(H, T or T, H) by adding the probabilities at the ends of the two paths, i.e. $P(\text{H and T}) = \frac{1}{4} + \frac{1}{4} = \frac{1}{2}$.

$$P(\text{H and T}) = \left(\frac{1}{2} \times \frac{1}{2}\right) + \left(\frac{1}{2} \times \frac{1}{2}\right)$$

$$= \frac{1}{4} + \frac{1}{4} = \frac{1}{2}$$

The worked example illustrates the general rule that

we *multiply* the probabilities when we follow a path along the branches and *add* the results of following several paths.

5 The probability that my bus has to wait at the traffic lights in the morning on the way to school is $\frac{1}{5}$.

Draw a probability tree to show the possibilities that the bus has to wait or can drive through the traffic lights on two consecutive mornings. Find the probability that, on two consecutive mornings, the bus

a has to wait at the lights on both occasions

b does not have to wait on either morning

c has to wait on just one morning.

6 a If a dice is rolled what is the probability of getting

i a six   **ii** a number other than six?

b Two dice, one red and the other blue, are rolled. Draw a tree diagram to show the possibilities of getting a six or not getting a six on each dice. Find the probability that

i   both dice show sixes

ii   the red dice gives a six but the blue dice does not

iii the blue dice gives a six but the red dice does not

iv   the probability that just one six appears.

For each of the remaining questions, draw a probability tree to illustrate the given information.

7 In a group of six girls, four are fair and two are dark. Of five boys, two are fair and three are dark. One boy and one girl are picked at random. What is the probability, that of the two pupils picked, one is fair and one is dark?

8 In a class of 20, four are left-handed. In a second class of 24, six are left-handed. One pupil is chosen at random from each class. What is the probability that one of the pupils is left-handed and one is not?

9 Derek and Alexis keep changing their minds about whether to send Christmas cards to each other. In any one year, the probability that Derek sends a card is $\frac{3}{4}$ and that Alexis sends one is $\frac{5}{6}$.

Find the probability that next year

a they both send cards

b only one of them sends a card

c neither sends a card.

What should the three answers add up to and why?

10 Copy the tree in the worked example on page 54 and, by adding branches to the right, show the following information. Three unbiased coins are tossed, one after the other.

Find the probability that

a three heads appear **b** three tails appear

c two heads and one tail appear in any order.

11 The weather forecast gives the probability that it will rain on Saturday as 0.07 and the probability that it will not rain on Sunday as 0.89.

a On which of these two days is it more likely to rain and why?

b Copy and, complete this tree diagram.

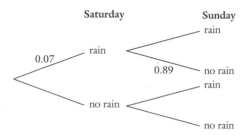

c Use your tree diagram to find the probability that it will rain on

i both days **ii** just one of the days.

The probability that it will rain on Monday is 0.3. Add more branches to your tree to include Monday.

d Use your new tree to find the probability that it will rain on

i none of the three days **ii** at least one of the three days

12 A coin is tossed three times. Use the tree diagram drawn for question **10** to find the probability of getting

a a head and two tails **c** at least one head.

b exactly one tail **d** at least two heads.

13 In a group of 120 girls, 24 have blue eyes, 48 have hazel eyes, 36 have green eyes and the remainder have brown eyes. All the girls have either long hair or short hair and the probability that a given girl has long hair is 0.25. The probability that a girl has freckles is 0.65. Assume that each attribute is independent of the others. What is the probability that a girl chosen at random from this group has

a brown eyes, freckles and short hair

b long hair, no freckles and either blue or green eyes?

The next exercise contains mixed problems on probability. Some of the questions can be answered directly from the basic definition of probability and some can be answered using the sum and product rules; draw a tree diagram or a possibility table only when you think it is needed.

1 A letter is picked at random from the word CATASTROPHE. Find the probability that

   a the letter is a vowel   **b** the letter is A or T.

2 A knitting wool sample card has 1 green, 1 black, 4 blue and 2 red samples. If one sample is picked at random, what is the probability that it is

   a yellow   **b** black, green, red or blue?

3 The scores on a four-sided spinner are 1, 2, 3 or 4. On a second four-sided spinner the scores are 5, 6, 7 or 8. If the two are spun, find the probability that

   a the score on both spinners is odd

   b the score on both spinners is even

   c the score on neither spinner is prime.

4

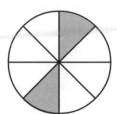 

A sector is chosen at random from each circle. What is the probability that

   a both sectors picked are green

   b both sectors picked are white

   c one is green and the other not?

5 There are two bags. The first contains 2 white and 3 black marbles and the second contains 1 red and 2 blue marbles. Two marbles are drawn, one from each bag. Find the probability that

   a a white and a blue marble are drawn

   b a black and a red marble are drawn

   c neither a white marble nor a red marble is drawn.

6   In a game of skittles the probability that Ted scores more than 5
is $\frac{2}{7}$ and the probability that George scores more than 5 is $\frac{2}{9}$.
Ted goes first followed by George. Use a probability tree to find the
probability that

a   both Ted and George score more than 5

b   Ted scores more than 5 but George does not

c   both score 5 or less

d   one scores more than 5 but the other does not.

INVESTIGATION

An agricultural society wishes to hold a two-day show in September at
one of four possible venues. The table shows the number of days it
rained each week, at the four different places, over the last four years.

Venue	A				B				C				D			
Year	1993	1994	1995	1996	1993	1994	1995	1996	1993	1994	1995	1996	1993	1994	1995	1996
Week 1	2	3	2	1	2	2	0	2	2	3	1	2	2	3	3	2
Week 2	1	2	2	0	3	1	0	2	1	2	1	1	1	2	2	2
Week 3	2	3	1	1	2	0	1	2	2	2	1	1	2	3	2	1
Week 4	2	3	0	1	2	2	1	2	1	2	2	1	1	3	2	2

a   Investigate the probability that it will be dry at venue A on any one
day during the first week of September.

b   What about a day during the second, third or fourth week?

c   Repeat parts **a** and **b** for the other three venues.

d   What is the probability that it will be dry on two consecutive days
at venue A during the first week in September?

e   Repeat part **d** for other weeks and other venues.

f   What is the probability that at venue A, on two consecutive days
during the first week in September, it will be
i dry on the first day but not on the second
ii dry on one of the two days but not on the other?

g   Repeat part **f** for other weeks and venues.

h   Which week and venue would you recommend to the organisers?
Justify your answer.
Would they be certain to get at least one dry day?

i   Compare the chance that it will be dry for two consecutive days
during the driest week at the chosen venue with the wettest week
anywhere.

PERCENTAGES

Ben Oakley is a buyer in the mens' department of a large department store. He buys a batch of shirts from a manufacturer for £11.50 each. His normal practice is to add a mark-up of 50%, and then value added tax at $17\frac{1}{2}$% has to be added to give the selling price.

His calculations are:

$$\text{Cost of shirt} = £11.50$$

$$\text{Mark-up at 50\% is } 0.5 \times £11.50 = £5.75$$

$$\text{Therefore selling price (ex VAT) is } £11.50 + £5.75 = £17.25$$

$$\text{VAT at } 17\tfrac{1}{2}\% \text{ is } 0.175 \times £17.25 = £3.02$$

$$\text{(to nearest penny)}$$

$\therefore$ selling price to the customer is £17.25 + £3.02 = £20.27

He does not want to price the shirt at such an odd amount, so decides to sell it at £19.95 including VAT.

- Ben needs to work out how much of the £19.95 is VAT, because he will have to pass this amount on to the Customs and Excise Office. This means that he must be able to work backwards to find the amount of value added tax included in the price of £19.95.

To do these calculations he needs to increase his knowledge of percentages.

EXERCISE 3A

Discuss what you need to be able to do to solve the following problems.

1 The total number of a certain species of bird in the United Kingdom is estimated to be 5000. It is believed that the number is decreasing at the rate of 5% a year. The RSPB would like to find out how many years it will be before the population drops below 2000.

2 The cost of a unit of electricity is 8.65 p including value added tax at 8%. It is rumoured that, if a particular party wins power at the next general election, the rate of VAT that applies to domestic fuel will rise from 8% to $17\frac{1}{2}$%, whereas if the other main party wins they will abolish VAT on domestic fuel altogether. Assuming that the basic cost of electricity stays the same Norman would like to know how much one unit of electricity will cost depending on which of these two parties forms the next government. He would also like to work out the percentage difference between the two.

Discussion of the above situations shows that when dealing with a variety of problems involving percentages, a clear understanding of *what the percentage is of* is very important.

PERCENTAGE INCREASE AND PERCENTAGE DECREASE

Percentage increase or decrease arises in many different areas of life today. We may read that certain workers are to receive an increase in their wages of 8%; that value added tax (VAT) may be increased from $17\frac{1}{2}$% to 18%; that the basic rate of income tax should be reduced from 23% to 20%; or that all the items in a sale are offered at a discount of 20%.

Changes are expressed in percentage terms *as a percentage of the quantity before any changes are made*. This is because this makes it easier to calculate the actual change in a particular case and to compare one change with another.

If a wage of £100 per week is increased by 8% then the new wage is

$$108\% \text{ of } £100 = 1.08 \times £100$$
$$= £108$$

If an article costs £55 plus VAT at $17\frac{1}{2}$%, then the full cost is

$$117\frac{1}{2}\% \text{ of } £55 = 1.175 \times £55$$
$$= £64.63 \text{ (correct to the nearest penny)}$$

If a woman earns £550 and has to pay tax on it at the rate of 23% she actually receives (100 − 23)% of £550, that is, 77% of £550.

$$77\% \text{ of } £550 = 0.77 \times £550$$
$$= £423.50$$

If a discount of 25% is offered in a sale, a piece of furniture, originally marked at £760, will cost 75% of £760 $= 0.75 \times £760$

$$= £570$$

Retailers buy-in goods which are usually sold at an increased price. The increase is often called 'the mark-up' and is normally given as a percentage of the buying-in price. Occasionally goods are sold at a decreased price, that is, they are sold at a loss. The loss is also given as a percentage of the buying-in price.

If a store makes a mark-up of 50% on an article it buys for £100, its mark-up is

$$50\% \text{ of } £100 = 0.5 \times £100$$
$$= £50$$

and the selling price is $150\% \text{ of } £100 = 1.5 \times £100$
$$= £150$$

EXERCISE 3B

A second-hand car dealer bought a car for £3500 and sold it for £4340. Find his percentage mark-up.

Mark-up = selling price − buying-in price
          = £4340 − £3500
          = £840

> Remember, percentage mark-up is the mark-up expressed as percentage of the buying-in price.

$$\% \text{ mark-up} = \frac{\text{mark-up}}{\text{buying-in price}} \times 100$$

$$= \frac{£840}{£3500} \times 100 = 24$$

Therefore the mark-up is 24%

Find the percentage mark-up.

1 Buying-in price £12, mark-up £3

2 Buying-in price £28, mark-up £8.40

3 Buying-in price £16, mark-up £4

4 Buying-in price £55, mark-up £5.50

A retailer bought a leather chair for £375 and sold it for £285. Find his percentage loss.

loss = buying-in price − selling price
$$= £375 − £285$$
$$= £90$$

> % loss is the loss as a percentage of the buying-in price.

$$\text{\% loss} = \frac{\text{loss}}{\text{buying-in price}} \times 100$$

$$= \frac{£90}{£375} \times 100 = 24$$

Therefore the loss is 24%

Find the percentage loss.

5   Buying-in price £20, loss £4

6   Buying-in price £125, loss £25

7   Buying-in price £64, loss £9.60

8   Buying-in price £160, loss £38.40

An article costing £30 is sold at a gain of 25%. Find the selling price.

The selling price is 100% of £30 + 25% of £30, i.e.   125% of £30

Selling price = 125% of £30
$$= 1.25 \times £30$$
$$= £37.50$$

Therefore the selling price is £37.50

> Alternatively, we can find the gain, which is 25% of the original cost, and add this to the original cost.
>
> Gain = 25% of £30
> $$= 0.25 \times £30$$
> $$= £7.50$$
> Selling price = £30 + £7.50
>
> Therefore the selling price is £37.50

Find the selling price.

9 Cost £50, gain 12%

12 Cost £36, loss 50%

10 Cost £64, gain 122%

13 Cost £75, loss 64%

11 Cost £29, gain 110%

14 Cost £128, loss $37\frac{1}{2}$%

Find the weekly cash increase for each of the following employees.

15 Ian Dickenson earning £120 per week receives a rise of 10%.

16 Nairn Williams earning £180 per week receives a rise of 12%.

17 Sylvia Smith earning £225 per week receives a rise of 8%.

18 Lyn Wyman earning £270 per week receives a rise of 9%.

19 Joe Bright earning £300 per week receives a rise of 7%.

For questions **20** to **22**, which is the better cash pay rise, and by how much?

20 a 12% on a weekly pay of £100, or

b 8% on a weekly pay of £250

21 a 7% on a weekly pay of £90, or

b $3\frac{1}{2}$% on a weekly pay of £200

22 a 4% on a weekly pay of £300, or

b 3% on a weekly pay of £400

Find the purchase price of the given object.

23 An electric fire marked £30 plus VAT at 15%

24 A food mixer marked £64 plus VAT at 18%

25 A calculator marked £8.40 plus VAT at $17\frac{1}{2}$%

26 A car tyre costing £38 plus VAT at 15%

27 A bathroom suite costing £600 plus VAT at 17%

28 A ticket for a concert costing £8 plus VAT at $18\frac{1}{2}$%

INCOME TAX

Income tax is deducted from everybody's taxable income, that is, their earnings over and above the value of their allowances. Usually people pay most of their tax at the basic rate (22% at the time of writing), although a small amount is paid at a lower rate of 10%, while those with high incomes have to pay a certain amount of higher rate tax.
(Rates of tax and allowances change from year to year in the Budget.)

EXERCISE 3C

> Peter Smith has a taxable income of £8000 per annum. How much tax will he pay if the basic rate is 23% ?
>
> Tax due = 23% of £8000
> $\phantom{\text{Tax due }}$= 0.23 × £8000
> $\phantom{\text{Tax due }}$= £1840

Assuming that the basic rate of income tax is 23%, find the yearly tax on the following taxable incomes.

1 £5000   **3** £6500   **5** £6450

2 £8000   **4** £12 500   **6** £8260

Find the yearly income tax due on a taxable income of

7 £10 000 if the basic tax rate is 33%

8 £8000 if the basic tax rate is 20%

9 £16 000 if the basic tax rate is 25%

10 £24 000 if the basic tax rate is 32%

> Edgar Brooks earns £10 000 each year. If his tax free allowances amount to £4840, how much tax will he pay when the basic rate is 23% ?
>
> Taxable income = £10 000 − £4840
> $\phantom{\text{Taxable income }}$= £5160
> $\phantom{\text{Taxab}}$Tax due = 23% of £5160
> $\phantom{\text{Taxable income }}$= 0.23 × £5160
> $\phantom{\text{Taxable income }}$= £1186.80

Use the following details to find the income tax due in each case.

	Name	Gross income	Allowances	Basic tax rate
11	Miss Deats	£8000	£2000	20%
12	Mr Evans	£10 000	£3000	23%
13	Mrs Khan	£15 000	£3200	15%
14	Mr Ames	£9000	£2600	33%
15	Miss Eyles	£20 000	£4750	28%

A dentist has an annual salary of £48 000 and can claim allowances of £10 270. The lower rate of income tax is 20% and is payable on the first £3950 of taxable income. The basic rate of income tax is 23% and is payable on the next £22 600 of taxable income. Higher rate tax of 40% is charged on any taxable income over £26 550. Find the total tax payable.

Taxable income is £48 000 − £10 270 = £37 730

Lower rate tax on £3950 at 20% = $0.20 \times £3950$
$$= £790$$

Basic rate tax due on £22 600 at 23% is 23% of £22 600
$$= 0.23 \times £22 600$$
$$= £5198$$

Higher rate tax is due on £37 730 − £3950 − £22 600
i.e. on £11 180

Higher rate tax is 40% of £11 180 = $0.40 \times £11 180$
$$= £4472$$

Total tax payable = £790 + £5198 + £4472 = £10 460

16 For each case given below, calculate the income tax payable if a lower rate of 20% is charged on the first £3950 of taxable income and the remainder of the taxable income is charged at a basic rate of 23% (assume that no higher rate of tax is due).

	Gross income	Allowances
a	£9800	£4250
b	£18 460	£5336
c	£7435	£3660

17   Repeat question **16** if the basic rate of tax is decreased from 23% to 21%.

18   For each case given below, calculate the income tax payable if a lower rate of 20% is charged on the first £3950, the basic rate is charged on the next £22 600 and the remainder of the taxable income is taxed at a higher rate.

	Yearly income	Allowances	Basic rate tax	Higher rate tax
a	£35 400	£3600	23%	40%
b	£33 000	£5740	24%	40%
c	£42 000	£4800	21%	60%
d	£253 000	£6280	25%	50%

SALE REDUCTIONS

Stores regularly encourage customers to make purchases by reducing prices. The reduced price of an article is often called the discounted price. A price reduction or *discount* is frequently expressed as a percentage of the price before the reduction is made.

EXERCISE 3D

During the January sales, a department store offers a discount of 10% off marked prices. What is the purchase price of

a   a dinner service marked £84.50

b   a pair of jeans marked £16.30 ?

a   If the discount is 10% the cash price is 90% of the marked selling price
i.e.   the purchase price of the dinner service is 90% of £84.50
$$= 0.9 \times £84.50$$
$$= £76.05$$

b   Similarly, the purchase price of the jeans is 90% of £16.30
$$= 0.9 \times £16.30$$
$$= £14.67$$

Alternatively,

a   if the discount is 10% on the marked price
$$\text{Discount} = 0.1 \times £84.50$$
$$= £8.45$$
$$\therefore \quad \text{purchase price} = £84.50 - £8.45$$
$$= £76.05$$

b   discount of 10% on £16.30 $= 0.1 \times £16.30$
$$= £1.63$$
$$\therefore \quad \text{purchase price} = £16.30 - £1.63$$
$$= £14.67$$

In a sale, a shop offers a discount of 20%. What would be the cash price for each of the following articles?

1 A dress marked £35

2 A lawn mower marked £115

3 A pair of shoes marked £32

4 A set of garden tools priced £72.50

5 Light fittings marked £42 each

In a sale, a department store offers a discount of 50% on the following articles. Find their sale price.

6 A pair of curtains marked £76.50

7 A leather football marked £32.30

8 A boy's jacket marked £28.60

9 A girl's coat marked £64.50

10 In order to clear a large quantity of woollen goods a shopkeeper puts them on sale at a discount of 33%. Find the cash price of

   a a jumper marked £18.30   **b** a skirt marked £22.20

11 A shopkeeper buys-in shirts at £12.50 each and marks them up by 50%. Some are later sold in a sale at a discount of 35%. Does the shopkeeper gain or lose on these shirts and by how much?

FINDING THE
ORIGINAL
QUANTITY

Sometimes we are given an increased or decreased quantity and we want to find the original quantity. For example, if the cost of a chair including VAT at $17\frac{1}{2}$% is £176.25, we might need to find the price of the chair before the tax was added.

EXERCISE 3E

An article is sold for £252. If this includes a mark-up of 5% find the buying-in price.

Remember that the mark-up of 5% is 5% of the buying-in price. We do not know the buying-in price, so we call it £x.

There is a mark-up of 5%

Selling price = 105% of the buying-in price
$$= 1.05 \times \text{buying-in price}$$

If the buying-in price is £x, then
$$252 = 1.05 \times x$$

$$\frac{252}{1.05} = x \qquad \text{giving} \quad 240 = x$$

Therefore the buying-in price is £240.

In questions **1** to **18**, selling price is abbreviated to SP.
Find the buying-in price.

1 SP £98, mark-up 40%   **5** SP £40, mark-up 25%

2 SP £64, mark-up 60%   **6** SP £920, mark-up 15%

3 SP £28, mark-up 75%   **7** SP £1008, mark-up 125%

4 SP £12, mark-up 100%   **8** SP £21.50, mark-up $7\frac{1}{2}$%

A book is sold for £6.30 at a loss of 30%. Find the buying-in price.

The loss of 30% is 30% of the buying-in price, so selling price is (100% − 30%)   i.e. 70% of the buying-in price.

Selling price = 70% of the buying-in price

Then if the buying-in price is x pence
$$630 = 0.7 \times x$$

∴ $630 \div 0.7 = x$ i.e. $x = 900$

The buying price of the book is 900 p, or £9.

Find the buying-in price.

9 SP £30, loss 25% **14** SP £45, loss 10%

10 SP £56, loss 30% **15** SP £120, loss 25%

11 SP £70, loss 65% **16** SP £8.50, loss 50%

12 SP £12, loss 33% **17** SP £64, loss 60%

13 SP £8.16, loss 40% **18** SP £1200, loss 40%

After a pay rise of 5% Peter's weekly pay is £126. How much did he earn before the rise?

> The 5% pay rise is 5% of Peter's original pay; so his new pay will be (100% + 5%), i.e. 105% of his original pay.

If Peter's original pay was £x

then $126 = 105\%$ of x

  $126 = 1.05 \times x$

i.e. $x = \dfrac{126}{1.05}$

  $= 120$

Peter's original weekly pay was £120.

The following table shows the weekly wage of a number of employees after percentage increases as shown. Find the original weekly wage of each employee.

	Name	% increase in pay	Weekly wage after increase
19	George Black	10%	£132
20	Anne Reed	8%	£135
21	John Rowlands	15%	£299
22	Beryl Lewis	7%	£196.88
23	Enid Jones	4%	£95.68

The purchase price of a watch is £70.50. If this includes VAT at $17\frac{1}{2}\%$, find the price before VAT was added.

> VAT is a percentage of the price before the VAT has been added,
> i.e purchase price $= \left(100\% + 17\frac{1}{2}\%\right)$ of the price before VAT is added.

If the price of the watch before VAT is added is $£\,C$

then $\quad 1.175 \times C = 70.5$

i.e. $\qquad C = 70.5 \div 1.175$

$\qquad\qquad = 60$

The price before VAT was added was £60.

24 The purchase price of a hairdryer is £13.80. If this includes VAT at 15%, find the price before VAT was added.

25 I paid £763.75 for a dining table and four chairs. If the price includes VAT at $17\frac{1}{2}\%$ find the price before VAT was added.

26 John's income last week was £112 after income tax had been deducted at 30%. Calculate his pay before the tax was deducted.

27 Water increases in volume by 4% when it is frozen. How much water is required to make $884\,\text{cm}^3$ of ice?

28 The stretched length of an elastic string is 31 cm. If this is 24% more than its unstretched length, find its unstretched length.

29 Because of a change in the exchange rate the cost of the Stone's holiday was increased by $5\frac{1}{2}\%$ to £1631.03. What was the cost of the holiday before the increase?

30 Sara's take-home pay last week was £226.46. This was after deductions that amounted to 32.4% of her gross pay.
Find　**a** her gross pay　　**b** her deductions.

MIXED PROBLEMS INVOLVING PERCENTAGE INCREASE AND DECREASE

Remember that a percentage increase or decrease is always calculated as a percentage of the quantity before the change. Questions that ask you to find the original quantity, for example question **4**, can be checked by working through the given information using your answer.

EXERCISE 3F

1 A house is bought for £68 000 and sold at a profit of 14%. Find the selling price.

2 Carpets that had been bought for £18.50 per square metre were sold at a loss of 26%. Find the selling price per square metre.

3 Potatoes bought at £4.50 per 50 kg bag are sold at 12 p per kg. Find the percentage profit.

4 An art dealer sold a picture for £1980 thereby making a profit of 65%. What did she pay for it?

5 My present average weekly grocery bill is £40.50, which is 8% more than the same goods cost me, on average, each week last year. What was my average weekly grocery bill last year?

6 What is Fred Procter's gross weekly wage if, after paying deductions of 35%, he is left with £111.80?

7 Between two elections the size of the electorate in a constituency fell by 16%. For the second election, 37 191 people were entitled to vote. How many could have voted at the first election?

8 When the rate of VAT is $17\frac{1}{2}$% a compact disc costs £7.05. What will it cost if the rate of VAT is

a increased to 20%   **b** decreased to 15%?

9 If Terry begins his journey after 9 a.m. he is allowed a discount of 30% on the cost of his rail ticket. He pays £22.75 for a ticket to London, leaving on the 9.15 a.m.

a Express the discounted price as a percentage of the full price.

b Calculate the cost of the ticket before deducting the discount.

10 A camera bought for £300 loses 40% of its value in the first year.

a What is it worth when it is one year old?

b Express its value after one year as a percentage of the amount by which it has depreciated.

11 Andrew Bullen received £955 pay last month. This sum was made up of a fixed basic wage of £245 plus commission at 2% on the value of the goods he had sold in the preceding month. Find the value of the sales Andrew made last month.

12 In a sale at a department store a tea set is sold at a discount of 15%. The sale price is £59.49.

Find   **a** the pre-sale price   **b** the discount.

13 In Speake's electrical store a Soyni video-recorder is priced £167.50.

   a Harri pays cash and so is given a discount of 8%. How much does the video-recorder cost him?

   b The ticket price of the video-recorder now is 3.5% more than it was this time last year. What was the ticket price a year ago?

14 Percy Briggs has a faulty washing machine. He calls out the service engineer, who repairs it.
   Copy and complete his bill, which is given below. Give any value that is not exact correct to the nearest penny.

	£
Fixed call-out charge	18.50
Labour: $1\frac{1}{2}$ hours at £22/h	
Parts	
Total before VAT	_____
VAT at $17\frac{1}{2}$%	
Total due	_____ 92.83

15 To the price a jeweller pays the wholesaler for a watch, she adds a mark-up of 50% to give the retail price. Value added tax at $17\frac{1}{2}$% is added to the retail price and this gives the price to the customer which is rounded up to the nearest penny. A customer buys a watch for £79.99. What did the shopkeeper pay the wholesaler for it?

16 Holidays Abroad charge $1\frac{1}{2}$% commission when selling foreign currency. How much, in pounds sterling, will Kerry have to pay for 800 French francs if the exchange rate is £1 ≡ 7.83 Ff, that is, £1 buys 7.83 Ff.

17 Ravinder is going to Canada. Her bank will exchange 1.55 Canadian dollars for each £1. It also charges 1% commission for the transaction. Ravinder wants to spend £250, at most, in buying her dollars.

   a What is the maximum whole number of dollars she can buy?

   b How much commission is she charged?

   c How much, if any, of her £250 is left over?

18 The purchase price of a diamond ring increases by £20 when the rate of VAT is increased from 12% to 20%. Find the original purchase price of the ring.

19 The table shows the original price and sale price of several items in a clothes shop.

Original price ($£P$)	24	56	65	88
Sale price ($£S$)	19.20	44.80	52	70.40

Plot these values on a graph using $2\,\text{cm} \equiv £10$ on both axes. Scale the P-axis from 0 to 100 and the S-axis from 0 to 80. Draw a straight line to pass through the four points, and use your graph to find

a the sale price of a garment originally marked **i** £36 **ii** £75

b the original price of a dress whose sale price is **i** £44 **ii** £66.

Find the gradient of the line and give a meaning to its value. Express this gradient as a percentage and use its value to find

c the sale price as a percentage of the original price

d the percentage discount.

20 Solve the problem described at the beginning of the chapter.

INTEREST

Many situations such as banks and building societies offer savings accounts. If you put money into such an account, the bank uses your money for other purposes and pays you for that use. The amount that the bank pays for the use of your money is called *interest*.

The time may come when you wish to use someone else's money to buy an expensive item such as a car or even a house. You will normally have to pay for the use of borrowed money, that is, you will have to repay more than you borrow and the extra you repay is called *interest*.

Interest on money borrowed (or lent) is usually a percentage of the sum borrowed (or lent). This percentage is often given as a charge per year (per annum, or p.a.) and it is then called the *interest rate*. For example, if £100 is put into a building society account with an interest rate of 8% p.a., then after one year, the society pays

$$8\% \text{ of } £100, \quad \text{i.e. } £8$$

EXERCISE 3G

1 Pam is given £500 for her 18th birthday. She puts the money in a savings account with an interest rate of 6% p.a. How much is added to her account after one year?

2 Ann Peters is given a loan of £650 from the bank which she agrees to repay after one year. How much does she have to repay if the interest rate is $12\frac{1}{2}$% p.a.?

Find the interest payable after one year on each of the following sums of money invested (i.e. put in a savings account) at the given interest rate.

3 £352 at 7.5% p.a.   **5** £2600 at 8.3% p.a.

4 £10 000 at 4.25% p.a.   **6** £5840 at 6.4% p.a.

7 What annual rate of interest is necessary to give interest of £238 after one year on an investment of £2800 ?

8 What is the original size of a loan that costs £45 when repaid after one year when the interest rate is 9% p.a. ?

9 Find the original sum borrowed if £287.50 has to be repaid after one year when the interest rate is 15% p.a.

10 Mr & Mrs Surefoot invest a sum of money in a deposit account with the Highway Building Society. The society quotes a gross rate of 8% but add the net interest to the account after deducting income tax at 20%. One year after the investment was made the pass book shows that net interest of £800 was added. Find

a the gross interest paid by the building society

b the amount of money invested.

COMPOUND PERCENTAGE PROBLEMS

There are many occasions when a percentage increase or decrease happens more than once. Suppose that a house is bought for £20 000 and increases in value (appreciates) by 10% of its value each year.

After one year, its value will be 110% of its initial value,

i.e.   110% of £20 000 = 1.1 × £20 000 = £22 000

The next year it will increase by 10% of the £22 000 it was worth at the beginning of the year, so its value after two years will be

$$110\% \text{ of } £22\,000 = 1.1 \times £22\,000 = £24\,200$$

While some things increase in value year after year, many things decrease in value (depreciate) each year. Should you buy a car or a motorcycle it will probably depreciate in value more quickly than anything else you buy.

If you invest money in a Building Society or Post Office Savings Account and do not spend the interest, your money will increase by larger amounts each year if the interest rate stays the same.

This kind of interest is called *compound interest*. The sum on which the interest is calculated is called the *principal* and changes each year.

EXERCISE 3H

Find the compound interest on £260.60 invested for 2 years at 8% p.a.

Interest for first year at 8% is 8% of the original principal.

New principal at end of first year
$$= 100\% \text{ of original principal} + 8\% \text{ of original principal}$$
$$= 108\% \text{ of the original principal}$$
$$= 1.08 \times \text{original principal}$$
∴ principal at end of first year
$$= 1.08 \times £260.60$$
$$= £281.448$$

> Use all available figures when an intermediate calculation does not work out exactly.

Similarly, new principal at end of second year
$$= 108\% \text{ of the principal at the beginning of the second year}$$
$$= 1.08 \times £281.448$$
∴ principal at end of second year
$$= £303.963\ldots = £303.96 \text{ correct to the nearest penny}$$

So the compound interest on £260.60 for 2 years
$$= \text{principal at end of second year} - \text{original principal}$$
$$= £303.96 - £260.60 = £43.36 \text{ correct to the nearest penny}$$

In questions **1** to **7** give all answers that are not exact correct to the nearest penny. Find the compound interest on

1 £200 for 2 years at 10% p.a.

2 £300 for 2 years at 12% p.a.

3 £400 for 3 years at 8% p.a.

4 £650 for 3 years at 9% p.a.

5 £520 for 2 years at 13% p.a.

6 £690 for 2 years at 14% p.a.

7 £624 for 3 years at 12% p.a.

8 A house is bought for £60 000 and appreciates at 8% a year. What will it be worth in 2 years' time?

9 A particular postage stamp increases in value by 15% each year. If it is bought for £50, what will it be worth in 3 years' time?

10 A motorcycle bought for £1500 depreciates in value by 20% each year. Find its value after 3 years.

a An antique silver teapot is valued at £750 and appreciates by 12% a year. Find its value after 3 years.

b By contrast a hi-fi system costing £750 depreciates by 12% a year. Find its value after 3 years.

c Express the value of the hi-fi system after 3 years as a percentage of the value of the teapot after 3 years.

a Value of teapot
$$\text{after 1 year} = 112\% \times £750$$
$$= 1.12 \times £750$$
$$= £840$$
$$\text{after 2 years} = 1.12 \times £840$$
$$= £940.80$$
$$\text{after 3 years} = 1.12 \times £940.80$$
$$= £1053.70 \text{ (to the nearest penny)}$$

b Value of hi-fi system
$$\text{after 1 year} = 88\% \times £750$$
$$= 0.88 \times £750$$
$$= £660$$
$$\text{after 2 years} = 0.88 \times £660$$
$$= £580.80$$
$$\text{after 3 years} = 0.88 \times £580.80$$
$$= £511.10 \text{ (to the nearest penny)}$$

c Value of hi-fi system after 3 years as a percentage of the value of the tea pot after 3 years

$$= \frac{\text{Value of hi-fi}}{\text{Value of teapot}} \times 100\%$$

$$= \frac{£511.10}{£1053.70} \times 100\%$$

$$= 48.5\% \text{ (correct to 3 s.f.)}$$

Remember to use all the available figures when an intermediate calculation does not work out exactly.

11 Three years ago David and Charles each invested £30 000. David put his money into shares in a pharmaceutical company while Charles invested his money in a really spectacular car. The value of the car depreciated by 20% a year while shares in the pharmaceutical company appreciated by the same percentage. Find the value of each investment now.

12

A new car costing £15 000 depreciated in value each year by 18% of its value at the beginning of that year.

a Find its value   **i** after 1 year   **ii** after 4 years.

b Calculate the percentage decrease in the value of the car over 4 years.
(Give your answer correct to the nearest tenth of a percent.)

13 A new motorbike costing £8000 depreciates each year by 18% of its value at the beginning of the year. Find

a its value   **i** after 1 year   **ii** after 3 years

b the percentage decrease in value after 3 years.
(Give your answer correct to the nearest tenth of a percent.)

14 The present toll for a car crossing the Midford suspension bridge is £3.80. This charge is set to rise by 8% this year and by a further 6% for each of the following 5 years. Each increase is rounded down to the nearest 5 p. How much will it cost to cross Midford bridge

a in 2 years' time   **b** in 4 years' time ?

15 In the state of Necka the current amount that a person can earn without paying any income tax is £3500. This amount, rounded up to the nearest £10, is set to increase in line with inflation, the projected rates of which are given in the table.

Number of years from now	1	2	3	4	5
Expected rate of inflation for that year	3%	2.5%	4.5%	6.8%	8.3%

Use these values to find the tax-free amount a single person can earn in

a 3 years' time   **b** 5 years' time

16 The rabbit population of Ditcher's Heath has increased by 20% a year for the last 3 years. The estimated population is now 400 rabbits.

    a Estimate, to the nearest 10, the number of rabbits on the heath

       i 1 year ago   **ii** 2 years ago   **iii** 3 years ago.

    b State whether each of the following statements is true or false?

       A The rabbit population of Ditcher's Heath has increased by the same number each year.

       B Every year the increase in the number of rabbits is more than the increase was the year before.

       C The increase in the number of rabbits is the same every year.

       D At the present rate of increase there will be more than twice as many rabbits on the heath within 10 years.

17 The population of Roxley has increased by **5.5**% a year for the last 5 years and is expected to grow at the same rate for the foreseeable future. The present population of Roxley is 20 000.

    a Find the expected population   **i** next year   **ii** in 2 years' time.

    b Find the population of Roxley   **i** 1 year ago   **ii** 2 years ago.

Peter buys a fixed rate bond for £5000 at the start of 1996. Interest is added at the end of each year at 10% of the value of the bond at the start of the year. For how many years will the bond need to be kept before its value is at least £8000?

> We need to find the value at start of each year from 1996 and continue until the value exceeds £8000.
> The working is easier to follow when it is laid out in a table.

Date	Elapsed time (years)	Value
1996	0	£5000
1997	1	110% of £5000 = $1.1 \times$ £5000 = £5500
1998	2	110% of £5500 = $1.1 \times$ £5500 = £6050
1999	3	$1.1 \times$ £6050 = £6655
2000	4	$1.1 \times$ £6655 = £7320.50
2001	5	$1.1 \times$ £7320.50 = £8052.55

Peter will need to keep the bond for 5 years.

18 Mr James grows prize marrows. When growing conditions are ideal, his marrows increase in weight by 10% each day. How many days will it take a marrow whose weight is now 700 g to

   a increase in weight to at least 800 g

   b increase in weight to at least 1000 g

   c double in weight?

19 A new town of 2000 houses is to be built in the year 2010 and is then planned to increase by 20% each year. At the start of which year will the number of houses be at least doubled?

20 A pest eradication scheme for a railway system aims to decrease the number of rats by 30% a month. At the start of July it is estimated that there are 10 000 rats. After how many months will the number of rats be below   **a** 5000   **b** 500?

21 A classic car is bought for £ P. Its value appreciates by 8% each year. Find, in terms of P, an expression for the value of the car after

   a 1 year   **b** 2 years   **c** 6 years   **d** n years.

   Use the formula you found in part **d** to find the value after 10 years of a classic car bought for £500.

MIXED EXERCISE

EXERCISE 3I

1 Tim earns £210 per week and received a rise of 4%. What is his new weekly wage?

2 In a sale a store gives a reduction of 20% off the ticket price. A winter coat is marked £99. How much must I pay for it?

3 Enid May earned £18 500 last year. No tax was due on the first £3900 of her income but she had to pay tax at 20% on the next £3750 and 23% on the remainder. How much income tax did she pay altogether?

4 A shopkeeper buys 80 articles for £200 and sells them for £3.50 each. Find his percentage profit.

5 Engine modifications were made to a particular model of car. As a result the number of kilometres it travels on one litre of petrol increased by 8%. If the new petrol consumption is 16.2 km/litre what was the previous value?

6 At the end of 1996 the number of a rare species of animal in North America was 6000. It is predicted that their numbers will reduce by 14% each year.

   a How many will be left at the end of 1999?

   b By the end of which year will the number first be less than 3000?

PUZZLE

Pauline manages an Art Gallery. She buys a picture for £2000 and sells it at a profit of 50%. Six months later she buys it back for 50% less than she sold it for, but soon sells it again at a profit of 50%. This process of buying back at a loss of 50% of her selling price, and selling the picture again soon after, at a profit of 50% is repeated on a further two occasions. How much profit (or loss) did she make altogether?

PRACTICAL WORK

You have been left £10 000 by a deceased relative whom you did not know existed. You decide you would like to invest the money in an Instant Access account with a building society. This means that you can withdraw the whole amount at a moment's notice. Collect leaflets that give you the current interest rates paid on Instant Access accounts by the different building societies in your area. Which society pays the best rate? How much interest will you receive at the end of one year? How much more is this than if the money were invested in the building society that pays the lowest rate? Is the choice affected by how often the interest is paid?
Are there other accounts that pay higher rates of interest? What are the disadvantages of these accounts?

RATIO
AND
PROPORTION

Richard has a new scooter.

The instructions state that the tank must be filled with fuel mixed from oil and petrol in the ratio 1 : 50. If Richard puts 20 ml of oil in the tank, the ratio 1 : 50 tells him he needs to add 50 times as much petrol, that is, 1000 ml which is 1 litre. The ratio also tells him that 100 ml of oil has to be mixed with 100×50 ml of petrol, and so on.

- The relationship between the quantity of oil and the quantity of petrol is always the same, that is, 1 : 50. Any two quantities that vary so that they are always in the same ratio behave in the same way, that is, if one quantity is doubled, so is the other; if one quantity is increased by a factor of 4, so is the other, and so on. Quantities that are related in this way are said to be *directly proportional*.

Richard uses his scooter to travel to work. The distance is 10 miles. If he leaves home at 8 a.m. the journey takes him 40 minutes. Using average speed $= \dfrac{\text{distance}}{\text{time}}$ gives his average speed as 15 mph.
If he leaves at 7.30 a.m. he avoids the rush hour and the journey only takes 20 minutes. In this case his average speed is 30 mph.

- As the distance is fixed, there is a relationship between the average speed and the time the journey takes. In this case however, when the speed doubles, the time halves. So the relationship between the speed and the time is not one where they are in the same ratio. Quantities that are related this way, that is, when one increases by becoming, say, 4 times larger then the other decreases to $\frac{1}{4}$ of its original size, are said to be *inversely proportional*.

In this chapter we are going to work with these two forms of relationship. First, however, you need to be able to recognise when quantities are related in one of these two ways, and equally importantly, when they are not.

EXERCISE 4A   In each question discuss how the quantities are related. Some may be directly proportional, some may be inversely proportional. Some may be related in a different way from either of these and others may not be related in any way.

1 Jane's pay for working an eight-hour day and her pay for working a 37-hour week assuming the rate of pay per hour is constant

2 Simon's pay for working an eight-hour day and his pay for working a 45-hour week when this includes overtime pay at a higher rate than the standard rate of pay per hour

3 The age of a woman and her weight

4 The number of £1 coins in a pile and the height of the pile

5 The time it takes to fill a swimming pool and the number of hoses used to fill it, assuming that the rate of flow of water from each hose is the same

6 The number of sweets that can be bought for £1 and the cost per kilogram of those sweets

7 The number of towels in a washing machine and the time it takes to wash them

8 The size of a telephone bill and the number of calls made from the telephone

9 The size of an electricity bill and the number of units used

10 The area of a square and the length of its side

11 The size of an interior angle of a polygon and the number of sides.

RATIOS   Ratio and direct proportion are introduced in Book 8A. Revision Exercise 1.2 at the front of this book can be used to revise this work. The next exercise also revises ratios but in different contexts.

EXERCISE 4B

In a jar of 357 mixed raisins and peanuts there are 153 raisins.
Find the ratio of the number of raisins to the number of peanuts.

The number of peanuts is $357 - 153 = 204$
Ratio of raisins to peanuts (by number)
$$= 153 : 204$$
$$= 3 : 4$$

> Dividing both
> numbers by 51.

1 In a school of 1029 pupils, 504 are girls. What is the ratio of the
number of boys to the number of girls?

2 I spend £3.60 on groceries and £2.40 on vegetables.
What is the ratio of the cost of

a groceries to vegetables

b vegetables to groceries

c groceries to the total?

3 One rectangle has a length of 6 cm and a width of 4.5 cm. A second
rectangle has a length of 9 cm and a width of 2.5 cm.
Find the ratios of

a their lengths   **c** their perimeters
b their widths   **d** their areas.

4 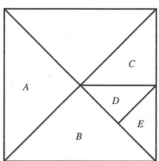   Find the ratios of the following areas

a $B : A$ **d** $E : D$
b $C : B$ **e** $E : C + D$
c $E : A + B$ **f** C : whole square

5 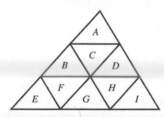   The areas of the small triangles are equal.
Find the ratios of the following areas.

a A : whole figure

b $A : A + B + C + D$

c $B + E + F + G$: whole figure

Express the ratio $5 : 7$ in the form $1 : n$

$5 : 7 = 1 : \frac{7}{5}$

$= 1 : 1.4$

> Remember that a ratio is unaltered when each number is multiplied (or divided) by the same amount. In this case we divide both numbers by 5, to give 1 as the first number in the ratio.

Express the following ratios in the form $1 : n$, giving n correct to 3 significant figures where necessary.

6 $2 : 3$	**9** $11 : 30$	**12** $3 : 4$	**15** $\frac{1}{3} : \frac{1}{4}$
7 $5 : 12$	**10** $5 : 3$	**13** $4 : 3$	**16** $0.75 : 0.25$
8 $7 : 6$	**11** $8 : 21$	**14** $7 : 10$	**17** $1\frac{1}{2} : 2\frac{2}{3}$

For one brand of emulsion paint, the information given is that 5 litres covers $45 \, \text{m}^2$. Another brand of emulsion paint has an average coverage of $20 \, \text{m}^2$ for 2 litres.
Compare the ratios of the volume of paint to area covered for the two brands.

First brand;   volume of paint : area covered $= 5 : 45$
Second brand;   volume of paint : area covered $= 2 : 20$

> The ratios are easier to compare if they are both expressed in the form $1 : n$

First brand;   ratio $= 1 : 9$
Second brand;   ratio $= 1 : 10$

Hence the second brand of paint covers a greater area per litre than the first.

For questions **18** to **22**, compare the ratio of quantity to price and hence state which is cheaper.

18 Gravel at 4 p per kilogram or at £38 per tonne

19 Eggs at 6 p each or 70 p per dozen

20 Gold-coloured chain at £16.20 per metre or 15 p per centimetre

21 Screws at 72 p for twenty or 4 p each

22 A 500 ml bottle of liquid fertiliser that covers $90 \, \text{m}^2$ and costs £2.56 or a 5 litre bag of granular fertiliser that covers $300 \, \text{m}^2$ and costs £28.40.

DIVISION IN A
GIVEN RATIO

The tank on Richard's scooter holds 20 litres of fuel. To fill the tank Richard needs to work out how much oil to put in. The tank can then be filled with petrol from the pump.

The fuel has to be 1 part oil to 50 parts petrol,

therefore the 20 litres of fuel to fill the tank has to comprise 51 parts.

Hence the quantity of oil required is $\frac{1}{51}$ of 20 litres

$$= \frac{1}{51} \times 20\,000\,\text{ml}$$

$$= 392\,\text{ml}\ (\text{correct to 3 s.f. })$$

> We have given 392 to the nearest millilitre. This is probably more accurate than necessary in this context for several reasons, one of which is that it is not easy or necessary to measure a quantity of oil as accurately as that.

EXERCISE 4C

Give your answer to each question as accurately as you consider to be appropriate in the context of the problem.

1 Concentrated orange juice has to be diluted with water in the ratio 2 : 5 by volume. How many millilitres of concentrated juice are needed to make up 2 litres of juice to drink?

2 The tank on a chemical spray holds 5 litres. For removing moss on hard surfaces, the instructions on a bottle of moss killer recommend dilution in the ratio 3 : 50.

a If the tank is to be filled, how much moss killer should be put in?

b If only 3 litres of spray is to be made up, how much moss killer is required?

3 Bronze is a metal alloy which, for one purpose, contains copper and tin in the ratio 3 : 22 by mass. What mass of copper is needed to make 5 kg of this bronze?

4 Mr Brown, Mrs Smith and Mr Shah work for AB Engineering plc. Their salaries are £12 000, £15 000 and £14 000 respectively. A bonus of £3000 is to be divided between these three employees in the ratio of their salaries.

a In what ratio are their salaries?

b What bonus is paid to each employee?

5 Two students were presented with a telephone bill for £70.22 which they had to pay between them. They decided to share the cost in the ratio of the number of calls they had each made.

a James had made 42 calls and Sarah had made 75 calls. How much should they each pay?

b After they had paid the bill, Sarah remembered that she had in fact made 6 more calls. How much should she pay James?

SIMPLE DIRECT PROPORTION

If we know the cost of one article, we can easily find the cost of ten similar articles, or if we know what someone is paid for one hour's work, we can find what the pay is for five hours.

EXERCISE 4D

> If 1 cm³ of lead weighs 11.3 g, what is the weight of
>
> **a** 6 cm³   **b** 0.8 cm³?
>
> 1 cm³ weighs 11.3 g
> **a** 6 cm³ weighs 11.3 × 6 g = 67.8 g
> **b** 0.8 cm³ weighs 11.3 × 0.8 g = 9.04 g

1 The cost of 1 kg of sugar is 90 p. What is the cost of
  a 3 kg   **b** 12 kg?

2 In one hour an electric fire uses $1\frac{1}{2}$ units. Find how much it uses in
  a four hours   **b** $\frac{1}{2}$ hour

3 One litre of petrol takes a car 18 km. At the same rate, how far does it travel on

    a four litres   **b** 6.6 litres?

4 The cost of 1 kg of mushrooms is £3.30. Find the cost of

    a 500 g   **b** 2.4 kg

We can reverse the process and, for instance, find the cost of one article if we know the cost of three similar articles.

---

18 cm^3 of copper weighs 162 g. What is the weight of 1 cm^3?

18 cm^3 weighs 162 g

1 cm^3 weighs $\frac{162}{18}$ g $= 9$ g

---

5 Six pens cost £7.20. What is the cost of one pen?

6 A car uses eight litres of petrol to travel 124 km. At the same rate, how far can it travel on one litre?

7 A man walks steadily for three hours and covers 13 km. How far does he walk in one hour?

8 A carpet costs £117.60. Its area is 12 m^2. What is the cost of 1 m^2?

We can use the same process even if the quantities given are not whole numbers of units.

The mass of 0.6 cm^3 of a metal is 3 g. What is the mass of 1 cm^3?

The mass of 0.6 cm^3 is 3 g

The mass of 1 cm^3 is $\frac{3}{0.6}$ g $= 5$ g

---

9 8.6 m^2 of carpet cost £71.38. What is the cost of 1 m^2?

10 The cost of running a refrigerator for 3.2 hours is 4.8 p. What is the cost of running the refrigerator for one hour?

11 A bricklayer takes 0.8 hours to build a wall 1.2 m high. How high a wall (of the same length) could he build in one hour?

12 A piece of webbing is 12.4 cm long and its area is 68.2 cm². What is the area of a piece of this webbing that is 1 cm long?

DIRECT
PROPORTION

If two varying quantities are always in the same ratio, they are said to be *directly proportional* to one another (or sometimes simply *proportional*).

For example, when buying pads of paper which each cost the same amount, the total cost is proportional to the number of pads. The ratio of the cost of 11 pads to the cost of 14 pads is 11 : 14, and if we know the cost of 11 pads, we can find the cost of 14 pads.

One method for solving problems involving direct proportion uses ratio. Another uses the ideas in the last exercise and is called the unitary method because it makes use of the cost of one article or the time taken by one man to complete a piece of work, and so on.

EXERCISE 4E

The mass of 16 cm³ of a metal alloy is 24 g. What is the mass of 20 cm³ of the same alloy?

First method (using ratios)
Let the mass of 20 cm³ be x grams.

Then $x : 20 = 24 : 16$

> The ratio of mass to volume stays the same. Write it starting with x.

i.e. $\dfrac{x}{20} = \dfrac{24}{16}$

$$20 \times \frac{x}{20} = \cancel{20}^{5} \times \frac{\cancel{24}^{6}}{\cancel{16}_{4}}$$

so $x = 30$

The mass of 20 cm³ is 30 grams.

Second method (unitary method)

> Rewrite the first sentence so that it *ends* with the quantity you want, i.e. the mass.

16 cm³ has a mass of 24 g

∴   1 cm³ has a mass of $\dfrac{24}{16}$ g

> There is no need to work out the value of $\frac{24}{16}$ yet.

so   20 cm³ has a mass of $\cancel{20}^{5} \times \dfrac{\cancel{24}^{6}}{\cancel{16}_{4}}$ g $= 30$ g

1 At a steady speed a car uses four litres of petrol to travel 75 km. At the same speed how much petrol is needed to travel 60 km?

2

A hiker walked steadily for four hours, covering 16 km. How long did he take to cover 12 km?

3 An electric fire uses $7\frac{1}{2}$ units in three hours.

a How many units does it use in five hours?

b How long does the fire take to use 9 units?

4 A rail journey of 300 miles costs £84. At the same rate per mile

a what would be the cost of travelling 250 miles

b how far could you travel for £63?

5 It costs £362 to turf a lawn of area 63 m². How much would it cost to turf a lawn of area 56 m²?

6 A machine in a soft drinks factory fills 840 bottles in six hours. How many could it fill in five hours?

7 A 6 kg bag of sprouts costs 396 p. At the same rate, what would an 8 kg bag cost?

8 The instructions for setting the tension on a knitting machine are 55 rows to measure 10 cm. How many rows should be knitted to give 12 cm?

9 A scale model of a ship is such that the mast is 9 cm high and the mast of the original ship is 12 m high. The length of the original ship is 27 m. How long is the model ship?

Either method will work, whether the numbers are complicated or simple. Even if the problem is about something unfamiliar, it is sufficient to know that the quantities are proportional.

In a spring balance the extension in the spring is proportional to the load. If the extension is 2.5 cm when the load is 8 newtons, what is the extension when the load is 3.6 newtons?

Ratio method

Let the extension be x cm.

$$x : 3.6 = 2.5 : 8$$

i.e. $$\frac{x}{3.6} = \frac{2.5}{8}$$

so $$3.6 \times \frac{x}{3.6} = 3.6 \times \frac{2.5}{8} \qquad \text{giving} \qquad x = 1.125$$

The extension is 1.125 cm.

Unitary method

If a load of 8 newtons gives an extension of 2.5 cm,

then a load of 1 newton gives an extension of $\dfrac{2.5}{8}$ cm,

so a load of 3.6 newtons gives an extension of

$$3.6 \times \frac{2.5}{8} \text{ cm} = 1.125 \text{ cm}$$

---

10 The rates of currency exchange published in the newspapers on a certain day showed that 14 kroner could be exchanged for 210 pesos. How many pesos could be obtained for 32 kroner?

11 At a steady speed, a car uses 15 litres of petrol to travel 164 km. At the same speed, what distance could be travelled on six litres?

12 If a 2 kg bag of sugar contains 9×10^6 crystals, how many crystals are there in   **a** 5 kg   **b** 8 kg   **c** 0.03 kg?

13 The current flowing through a lamp is proportional to the voltage across the lamp. If the voltage across the lamp is 10 volts the current is 0.6 amps. What voltage is required to make a current of 0.9 amps flow?

14 The amount of energy carried by an electric current is proportional to the number of coulombs. If five coulombs carry 19 joules of energy, how many joules are carried by 6.5 coulombs?

15 Two varying quantities, x and y, are directly proportional.
Copy and complete this table.

x	2	4	6	8
y	10			

16 A recipe for date squares lists the following quantities:

Ingredients	Costs
125 g of brown sugar	500 g cost 76 p
75 g of oats	750 g cost 102 p
75 g of flour	1.5 kg cost 88 p
100 g of margarine	250 g cost 36 p
100 g of dates	250 g cost 84 p
Pinch of bicarbonate of soda	—
Squeeze of lemon juice	1 p

Find the cost of making these date squares as accurately as possible,
then give your answer correct to the nearest penny.

INVERSE PROPORTION

At the start of this chapter we found that some quantities are not directly
proportional to one another, although there is a connection between
them. When we considered the relationship between Richard's average
speed and the time taken to travel 10 miles, we saw that when the
average speed doubled, the time halved. This means that the reciprocal,
or inverse, of the time is proportional to the average speed.
Further, when the speed changes from 20 mph to 40 mph,

the time changes from 30 minutes to 15 minutes.

If we multiply the speed and the corresponding time together, the result
is 600 in both cases.

As another example, suppose that a fixed amount of food is available for
several days. If each person eats the same amount each day, the more
people there are, the shorter is the time that the food will last. In fact, if
the food will last 2 people for 6 days, it will last 6 people for 2 days. So
when we treble the number of people, the number of days the food will
last decreases to a third. Hence the number of days the food will last is
inversely proportional to the number of people eating it.

Again, we see that the product of the number of days and the number of
people it will feed is 12 in both cases.

These examples illustrate the fact that

> when two quantities are inversely proportional,
> their product remains constant.

EXERCISE 4F In this exercise, assume that the rates are constant.

Four bricklayers can build a certain wall in ten days. How long would it take five bricklayers to build it?

Constant product method
Suppose it takes five bricklayers x days to build the wall,

then $\qquad 5 \times x = 4 \times 10$

i.e. $\qquad 5x = 40 \qquad$ giving $\qquad x = 8$

It would take them 8 days.

Unitary method
Four bricklayers take 10 days.
One bricklayer would take 40 days.
Five bricklayers would take $\frac{40}{5}$ days = 8 days.

---

1 Eleven taps fill a tank in three hours. How long would it take to fill the tank if only six taps are working?

2 The length of an essay is 174 lines with an average of 14 words per line. If it is rewritten with an average of 12 words per line, how many lines will be needed?

3 Nine children share out equally the chocolates in a large tin and get eight each. If there were only six children, how many would each get?

4 A field of grass feeds 24 cows for six days. How long would the same field feed 18 cows?

5 The dimensions of a block of stamps are 30 cm wide by 20 cm high. The same number of stamps could also have been arranged in a block 24 cm wide. How high would this second block be?

6 A batch of bottles were packed in 25 boxes taking 12 bottles each. If the same batch had been packed in boxes taking 15 each, how many boxes would be filled?

7 When used to knit a scarf 48 stitches wide, one ball of wool produces a length of 18 cm. If there had been 54 stitches instead, how long a piece would the same ball produce?

8 In a school, 33 classrooms are required if each class has 32 pupils. How many classrooms would be required if the class size was reduced to 22?

9 A factory requires 42 machines to produce a given number of articles in 63 days. How many machines would be required to produce the same number of articles in 54 days?

10 Two quantities, p and q, that can vary in value are inversely proportional.
Copy and complete the following table.

p	20	5	0.5	0.01
q	0.5			

EXERCISE 4G   This exercise contains a mixture of questions, some of which cannot be answered because the quantities are in neither direct nor inverse proportion. In these cases give a reason why there is no answer. For those questions that can be solved, give answers correct to three significant figures where necessary.

1 The list of exchange rates states that £1 = 7 French francs and £1 = 2300 lira, so that 7 francs = 2300 lira.

   a How many lira can 54 francs be exchanged for?

   b How many francs can be exchanged for 1000 lira?

2 A man earned £30.60 for an eight-hour day. How much would he earn at the same rate for a 38-hour week?

3 A typist types 3690 words in $4\frac{1}{2}$ hours. How long would it take to type 2870 words at the same rate?

4 At the age of twelve, a boy is 1.6 m tall. How tall will he be at the age of eighteen?

5 A ream of paper (500 sheets) is 6.2 cm thick. How thick is a pile of 360 sheets of the same paper?

6 If I buy balloons at 14 p each, I can buy 63 of them. If the price of a balloon increases to 18 p, how many can I buy for the same amount of money?

7 A boy's mark for a test is 18 out of a total of 30 marks. If the test had been marked out of 40 what would the boy's mark have been?

8 Twenty-four identical mathematics text books occupy 60 cm of shelf space. How many books will fit into 85 cm?

9 A lamp post 4 m high has a shadow 3.2 m long cast by the Sun. A man 1.8 m high is standing by the lamp post. At the same moment, what is the length of his shadow?

10 A contractor decides that he can build a barn in nine weeks using four men. If he employs two more men, how long will the job take? Assume that all the men work at the same rate.

11 A twelve-year-old girl gained 27 marks in a competition. How many marks did her six-year-old sister gain?

12 For a given voltage, the current flowing is inversely proportional to the resistance. When the current flowing is 2.5 amps the resistance is 0.9 ohms. What is the current when the resistance is 1.5 ohms?

13 The tables give some corresponding values of two variables. Decide whether the variables are directly proportional, inversely proportional or neither.

a

x	2	4	7	8	9	12
y	6	12	21	24	30	36

b

p	2	3	6	7	9	10
q	20	10	5	6	6.5	10

c

w	1	2	4	8	10	20
p	18	9	4.5	2.25	1.8	0.9

MIXED EXERCISE

EXERCISE 4H

1 Complete the ratio $\square : 9 = 2 : 5$

2 A car uses seven litres of petrol for a 100 km journey. At the same rate, how far could it go on eight litres?

3 Eight typists together could complete a task in five hours. If all the typists work at the same rate, how long would six typists take?

4 The ratio of zinc to copper in one type of brass is 3 : 7. What is the weight of zinc in 10 kg of this brass?

5 a Express the ratio $3 : 5$ in the form $n : 1$

b Give the ratio $5 : 6$ in the form $1 : n$

6 The ratio of two sums of money is $4 : 5$. The first sum is £6. What is the second?

7 A typist charges £25 for work which took her six hours. How much would she charge for nine hours' work at the same rate?

8 James Bond takes $4\frac{1}{2}$ minutes to complete a Grand Prix circuit when driving at an average speed of 97 mph. How long would it take to complete one circuit if he increases his average speed to 112 mph?

INVESTIGATION

Sweets at a 'Pick and Mix' counter are sold by weight at £0.56 per 100 grams.

a If x grams costs y pence, copy and complete this table giving values of y corresponding to values of x.

x	20	50	100	200	500	1000
y						

b Use a scale of 1 cm for 50 units on the x-axis and a scale of 2 cm for 50 units on the y-axis to plot these points on a graph.

c What do you notice about these points? Can you use your graph to find the cost of 162 grams?

d The cost and weight of these sweets are directly proportional. Investigate the graphical relationship between other quantities that are directly proportional. What do you notice? Is this always true?

e Extend your work to investigate the graphical relationship between two quantities that are inversely proportional. Use some of the questions in **Exercise 4F** as examples of quantities that are inversely proportional.

SUMMARY 2

NUMBERS

Rounded numbers

When a number has been rounded, its true value lies within a range that can be shown on a number line.

For example, if a nail is 23.5 mm long correct to 1 decimal place, then the length, x mm, is from 23.45 mm up to, but not including 23.55 mm as shown on this number line.

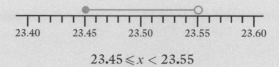

$$23.45 \leqslant x < 23.55$$

23.45 is the *lower bound* and 23.55 is the *upper bound* of x.

Reciprocals

The reciprocal of a number is 1 divided by that number,

for example, the reciprocal of 4 is $\frac{1}{4} = 0.25$

and   the reciprocal of 0.8 is $1 \div 0.8 = 1.25$

Recurring decimals

When some fractions are changed to decimals, the result is a recurring pattern of figures that repeats indefinitely. To save space, we place a dot over the first and last numbers in the group of figures that recur.

For example, $\frac{1}{11} = 1 \div 11 = 0.090909\ldots = 0.\dot{0}\dot{9}$

Division in a given ratio

To divide £200 into three amounts of money in the ratio 2 : 5 : 3, means that £200 has to be divided into $2 + 5 + 3 = 10$ equal parts; so the first amount is 2 of these parts, that is, $\frac{2}{10}$ of £200,

the second amount is $\frac{5}{10}$ of £200 and the third amount is $\frac{3}{10}$ of £200.

Direct proportion

When two quantities are related so that when one of them trebles, say, the other also trebles, the quantities are directly proportional (that is, they are always in the same ratio).

Inverse proportion

When two quantities are related so that when one of them trebles, say, the other becomes a third of its original size, the quantities are inversely proportional and their product is constant.

PERCENTAGE

Percentage change

A percentage change in a quantity is expressed as a percentage of that quantity before any changes were made.

For example, if a shirt is sold for £25 after a discount of 10% has been made, the discount is 10% of the price before it is reduced,

i.e.   £25 = original price − 10% of original price
          = 90% × original price.

Interest

When a sum of money is borrowed or lent, interest is usually charged on a yearly basis and is given as a percentage of the sum borrowed, for example, 3% p.a.

Compound percentage change

Compound percentage change is an accumulating change. If, for example, the value of a house increases by 5% of its value at the start of each year; its value after one year is 105% of its initial value;
after another year, its value is 105% of its value at the start of that year, that is, 105% of its increased value, and so on.

PROBABILITY

Adding probabilities

We add probabilities when we want the probability that one or other of two (or more) events will happen, provided that only one of the events can happen at a time.

Events such that only one of them can happen on any one occasion are called *mutually exclusive*.

For example, when one dice is rolled,

$$P(\text{scoring 5 or 6}) = P(\text{scoring 5}) + P(\text{scoring 6})$$

Multiplying probabilities

We multiply probabilities when we want the probability that two (or more) events both happen, provided that each event has no influence on whether or not the other occurs.

Events such that each has no influence on whether the other occurs are called *independent events*.

For example, when two dice A and B are rolled,

$$P(\text{scoring 6 on A and B}) = P(\text{scoring 6 on A}) \times P(\text{scoring 6 on B})$$

Tree diagrams

Tree diagrams can be used to illustrate the outcomes when two or more events occur.

This tree, for example, shows the possible outcomes when two coins are tossed.

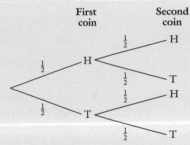

We *multiply* the probabilities when we follow a path along the branches and *add* the results of following several paths.

REVISION EXERCISE 2.1 (Chapters 1 and 2)

1 Find   **a** $4 \div \frac{2}{3}$ **b** $2\frac{1}{3} \div 1\frac{5}{9}$ **c** $3\frac{2}{3} \div \left(\frac{2}{3} + \frac{4}{5}\right)$

2 a Express each fraction as a decimal.

   i $\frac{7}{20}$ **ii** $\frac{5}{16}$ **iii** $\frac{1}{400}$ **iv** $\frac{15}{32}$

   b Express each decimal as a fraction in its lowest terms.

   i 0.55   **ii** 0.875   **iii** 0.015   **iv** 2.208

3 Use dot notation to write the following fractions as decimals.

   a $\frac{2}{30}$ **b** $\frac{4}{7}$ **c** $\frac{11}{12}$

4 Write down the reciprocal of   **a** $\frac{1}{5}$ **b** 1.8   **c** $\frac{2}{9}$

5 a It is stated that the number of screws in a box is 200 to the nearest 10. Find the range in which the actual number of screws lies.

   b A number is given as 2.47 correct to 2 decimal places. Illustrate on a number line the range in which this number lies.

6 A card is drawn at random from an ordinary pack of 52. A second card is drawn at random from another ordinary pack of 52 cards. What is the probability that

   a the first card is a red 2

   b the second card is a black queen

   c the first card is a red 2 and the second card is a black queen?

7 The probability that Simon completes his maths homework is $\frac{7}{8}$ and that Hilary completes her maths homework is $\frac{3}{4}$.
What is the probability that

a Simon does not complete his maths homework

b both Simon and Hilary complete their maths homework

c neither Simon nor Hilary complete their maths homework?

8 Kim rolls an ordinary dice and tosses a coin.
What is the probability that

a the dice shows a six

b the coin lands head up

c the dice shows a six and the coin lands head up?

9 When the post arrived Val had 3 first-class letters and 4 second-class letters while Dirk had 5 first-class letters and 2 second-class letters. A letter is taken at random from each batch.

a Copy and complete the tree diagram by writing in the probabilities on each branch.

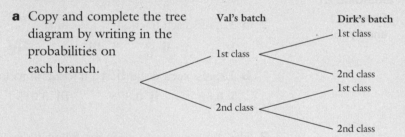

Find the probability that

b both letters are first class

c one is first class and one is second class.

10 Vicki has a box of 36 biscuits, 12 of which are square biscuits and the remainder are round biscuits. Joe also has a box of 36 biscuits but 20 of his are square and the remainder are round. A biscuit is taken at random from each box.

a Copy and complete the tree diagram by writing in the probabilities on each branch.

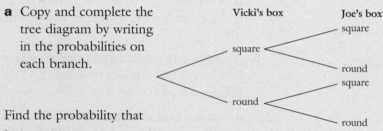

Find the probability that

b both biscuits are round

c one biscuit is square and the other is round.

**REVISION
EXERCISE 2.2
(Chapters 3
and 4)**

1 A chair costing £85 is sold at a gain of 60%. Find the selling price.

2 Find the cost of an electric fire marked £34 + VAT at $17\frac{1}{2}\%$.

3 The selling price, excluding VAT, of a CD player is £127.50. If this includes a mark-up of 50% find the buying-in price.

4 Mrs Denham has a gross annual income of £12750 and allowances of £3925. She pays tax at 23% on her taxable income.

   a How much tax must she pay?

   b Express the tax as a percentage of her gross pay. Give your answer correct to the nearest whole number.

5 Find the compound interest on £800 invested for 2 years at 5%.

6 a Express in the form $1 : n$

    i $5 : 9$ **ii** $\frac{1}{4} : \frac{1}{5}$ **iii** $3 : 7$

    (giving n correct to 3 significant figures)

  b Express in the form $n : 1$

    i $7 : 4$ **ii** $\frac{1}{4} : 2$ **iii** $11 : 3$

    (giving n correct to 3 significant figures)

7 Kim spends £1.20 on a pen and 90 p on a pencil. What is the ratio of the cost of

   a the pen to the pencil

   b the pencil to the pen

   c the pencil to the total?

8 A bus journey of 120 miles costs £15.60. At the same rate per mile

   a what would be the cost of travelling 150 miles

   b how far you can travel for £26?

9 If it takes 21 men 4 days to mark out the field in a stadium for an athletics meeting, how long would it take 12 men to do the same job?

10 For a youth camp they take enough stores to support 42 campers for 14 days. In the event 49 campers attend. How long will the stores last?

**REVISION
EXERCISE 2.3
(Chapters 1 to 4)**

1 a Give 34.678 cm correct to the nearest tenth of a centimetre.

  b The acceptable thickness of a metal plate is 4 mm, to the nearest mm. Find the range in which the actual thickness lies.

2 Find   **a** $\frac{3}{4} \div \frac{1}{2}$ **b** $4\frac{1}{7} \div 2\frac{5}{12}$ **c** $\frac{2}{5} \times \left(\frac{5}{6} - \frac{5}{12} \right) + \frac{1}{2}$

3 Write down   **a** the reciprocal of $\frac{3}{7}$

  b $\frac{19}{11}$ as a decimal using dot notation

  c 0.275 as a fraction in its lowest terms.

4 When Mrs Hussain needs to buy bread, the probability that she buys it at the supermarket is $\frac{4}{9}$, while the probability that she buys it at the local bakery is $\frac{1}{3}$. What is the probability that

  a she buys bread either at the supermarket or at the local bakery

  b she buys bread somewhere else?

5 a Sam Nolan has a taxable income of £7600 a year. How much tax must he pay if the rate of tax is 23%?

  b In a sale, a shirt is marked at £16.50 which includes a reduction of 25%. What was the pre-sale price?

6 Because they are emigrating Mr and Mrs Thomson are anxious to sell their house quickly. They put the house on the market at £100 000 but reduce the price every week by 4% of its price at the beginning of that week. It is sold the first week the price drops below £90 000.

  a How many weeks did it take to sell?

  b What was the selling price? (Give your answer correct to the nearest £1000.)

7 A car bought for £12 000 depreciates in value by 20% each year. Find its value after 3 years.

8 A hotel charges £259 per person per seven-day week. What would be the charge for 16 days at the same rate?

9 Under normal conditions a school boiler consumes 0.75 t of fuel a day and the stock of fuel is sufficient to last for 12 days. A cold spell causes the consumption rate to rise to 0.9 t per day. How long will the fuel last?

**REVISION
EXERCISE 2.4
(Chapters 1 to 4)**

1 a Express $\frac{9}{11}$ as a decimal correct to 3 decimal places.

 b Use your answer to part **a** to find $\frac{9}{11}$ of £50. Use fractions to find $\frac{9}{11}$ of £50 correct to the nearest penny. Explain your two answers.

2 Express as a fraction   **a** $0.\dot{6}$ **b** $0.0\dot{6}$ **c** $0.60\dot{6}$

3 Find   **a** $\frac{2}{3} \div 1\frac{2}{7}$ **b** $\left(3 - 1\frac{7}{8}\right) \times 1\frac{1}{3}$ **c** $\left(1\frac{2}{3} + 2\frac{1}{5}\right) \div 5\frac{4}{5}$

4 Carl draws one disc at random from a bag containing 2 red discs and 3 blue discs. He then draws another disc from a second bag containing 3 red discs and 5 blue discs. What is the probability that

 a both discs are red   **b** both discs are blue

 c either both discs are red or both discs are blue?

5 Celia has two bags. Bag A contains 4 sweet apples and 5 sour apples while bag B contains 6 sweet apples and 4 sour apples. Ian takes one apple from each bag.

 a Copy and complete the tree diagram by filling in the probabilities on the branches.

Find the probability that

 b both apples are sweet

 c one apple is sweet and the other is sour.

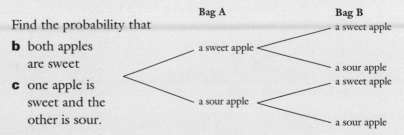

Bag A   Bag B

a sweet apple

a sweet apple

a sour apple

a sweet apple

a sour apple

a sour apple

6 A dress costing £55 is sold at a loss of 20%. Find the sale price.

7 When a fireside rug is sold for £39 the retailer suffers a loss of 40%. Find the buying-in price.

8 a Find the interest payable after one year if £8500 is invested at 5.25% a year.

 b If the original sum, plus the interest, remains invested at the same rate, find the total value of the investment at the end of the second year.

9 A bookshelf will hold 360 books of thickness 2 cm. How many copies of *National Geographic* will it hold if each copy is 6 mm thick?

ALGEBRAIC PRODUCTS

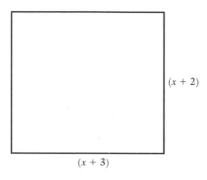

(x + 2)

(x + 3)

The area of this rectangle can be found by multiplying its length by its width,

i.e.   the area $= (x + 3)(x + 2)$.

The area can also be found by dividing the rectangle into sections as shown, giving the area as $x^2 + 3x + 2x + 6$.

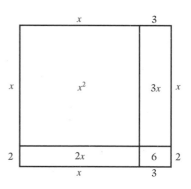

- This means that $(x + 3)(x + 2)$ and $x^2 + 3x + 2x + 6$ are two forms of the same expression. Being able to write $(x + 3)(x + 2)$ as $x^2 + 3x + 2x + 6$ *without* having to draw a diagram is clearly an advantage. This chapter shows how this can be done.

We start, however, with a reminder about multiplying out simpler brackets.

BRACKETS   Remember that $5(x + 1) = 5x + 5$

and that $4x(y + z) = 4xy + 4xz$

EXERCISE 5A Expand, i.e. multiply out

1 $2(x+1)$ **5** $5x(3y+z)$ **9** $8r(2t-s)$

2 $3(x-1)$ **6** $4y(4x+3z)$ **10** $3a(b-5c)$

3 $4(x+3)$ **7** $2n(3p-5q)$ **11** $4x(3y+2z)$

4 $-3(x+2)$ **8** $-3b(a-c)$ **12** $-3x(y-2z)$

THE PRODUCT OF TWO BRACKETS

Frequently, we wish to find the product of two brackets, each of which contains two terms, for example, $(a+b)(c+d)$.

$(a+b)(c+d)$ means $a\times(c+d)+b\times(c+d)$, that is, each term in the second brackets has to be multiplied by each term in the first bracket.

The multiplication could be done in any order but it is easier if you always follow the same pattern.

Always multiply the brackets together in the following order:

1   the first terms in the brackets
2   the outside terms
3   the inside terms
4   the second terms in the brackets.

Thus

$$(a+b)(c+d) = ac + ad + bc + bd$$

EXERCISE 5B

Expand $(x+2y)(2y-z)$

$$(x+2y)(2y-z) = 2xy - xz + 4y^2 - 2yz$$

Remember that the product of two numbers with the same sign gives a positive result and the product of two numbers with different signs gives a negative result.

Expand

1 $(a+b)(c+d)$ **6** $(a-b)(c+d)$

2 $(p+q)(s+t)$ **7** $(x+y)(y+z)$

3 $(2a+b)(c+2d)$ **8** $(2a+b)(3c+d)$

4 $(5x+2y)(z+3)$ **9** $(5x+4y)(z+2)$

5 $(x+y)(z-4)$ **10** $(3x-2y)(5-z)$

11 $(p+q)(2s-3t)$

12 $(a-2b)(c-d)$

13 $(6u-5v)(w-5r)$

14 $(3a+4b)(2c-3d)$

15 $(3x+2y)(3z+2)$

16 $(3p-q)(4r-3s)$

17 $(3a-4b)(3c+4d)$

18 $(7x-2y)(3-2z)$

19 $(2a+b)(5c-2)$

20 $(5a-4b)(3-2d)$

We get a slightly simpler form when we find the product of two brackets such as $(x+2)$ and $(x+3)$,
i.e. using the order we chose earlier

$$(x+2)(x+3) = x^2 + 3x + 2x + 6$$
$$= x^2 + 5x + 6$$

i.e. $(x+2)(x+3) = x^2 + 5x + 6$

> Since $2x$ and $3x$ are like terms they can be collected.

EXERCISE 5C

Expand

1 $(x+3)(x+4)$

2 $(x+2)(x+4)$

3 $(x+1)(x+6)$

4 $(x+5)(x+2)$

5 $(x+8)(x+3)$

6 $(a+4)(a+5)$

7 $(b+2)(b+7)$

8 $(c+4)(c+6)$

9 $(p+3)(p+12)$

10 $(q+7)(q+10)$

Expand $(x-4)(x-6)$

$$(x-4)(x-6) = x^2 - 6x - 4x + 24$$
$$= x^2 - 10x + 24$$

Expand

11 $(x-2)(x-3)$

12 $(x-5)(x-7)$

13 $(a-2)(a-8)$

14 $(x-10)(x-3)$

15 $(b-5)(b-5)$

16 $(x-3)(x-4)$

17 $(x-4)(x-8)$

18 $(b-4)(b-2)$

19 $(a-4)(a-4)$

20 $(p-7)(p-8)$

Expand $(x+3)(x-6)$

$$(x+3)(x-6) = x^2 - 6x + 3x - 18$$
$$= x^2 - 3x - 18$$

Expand

21 $(x+3)(x-2)$

22 $(x-4)(x+5)$

23 $(x-7)(x+4)$

24 $(a+3)(a-10)$

25 $(p+5)(p-5)$

26 $(x+7)(x-2)$

27 $(x-5)(x+6)$

28 $(x+10)(x-1)$

29 $(b-8)(b+7)$

30 $(z+1)(z-12)$

FINDING THE PATTERN

You may have noticed in the previous exercise, that when you expanded the brackets and simplified the answers, there was a definite pattern.

e.g. $(x+5)(x+9) = x^2 + 9x + 5x + 45$
$$= x^2 + 14x + 45$$

We could have written it

$$(x+5)(x+9) = x^2 + (9+5)x + (5) \times (9)$$
$$= x^2 + 14x + 45$$

Similarly $(x+4)(x-7) = x^2 + (-7+4)x + (4) \times (-7)$
$$= x^2 - 3x - 28$$

and $(x-3)(x-8) = x^2 + (-8-3)x + (-3) \times (-8)$
$$= x^2 - 11x + 24$$

In each case there is a pattern:
the *product* of the two numbers in the brackets gives the number term in the expansion, while *collecting* them gives the number of *x*s.

EXERCISE 5D

Use the pattern given above to expand the following products.

1 $(x+4)(x+5)$

2 $(a+2)(a+5)$

3 $(x-4)(x-5)$

4 $(a-2)(a-5)$

5 $(x+8)(x+6)$

6 $(a+10)(a+7)$

7 $(x-8)(x-6)$

8 $(a-10)(a-7)$

9 $(a+2)(a-5)$

10 $(y-6)(y+3)$

11 $(z+4)(z-10)$

12 $(p+5)(p-8)$

13 $(a-10)(a+7)$

14 $(y+10)(y-2)$

15 $(z-12)(z+1)$

16 $(p+2)(p-13)$

The pattern is similar when the brackets are slightly more complicated.

EXERCISE 5E

Expand the product $(2x+3)(x+2)$

$(2x+3)(x+2) = 2x^2 + 4x + 3x + 6$

$\qquad\qquad\quad = 2x^2 + 7x + 6$

Expand the following products.

1 $(2x+1)(x+1)$

2 $(x+2)(5x+2)$

3 $(5x+2)(x+3)$

4 $(3x+4)(x+5)$

5 $(3x+2)(x+1)$

6 $(x+3)(3x+2)$

7 $(4x+3)(x+1)$

8 $(7x+2)(x+3)$

Expand the product $(3x-2)(2x+5)$

$(3x-2)(2x+5) = 6x^2 + 15x - 4x - 10$

$\qquad\qquad\quad = 6x^2 + 11x - 10$

Expand

9 $(3x+2)(2x+3)$

10 $(4x-3)(3x-4)$

11 $(5x+6)(2x-3)$

12 $(7a-3)(3a-7)$

13 $(5x+3)(2x+5)$

14 $(7x-2)(3x-2)$

15 $(3x-2)(4x+1)$

16 $(3b+5)(2b-5)$

17 $(2a+3)(2a-3)$

21 $(4x+3)(4x-3)$

18 $(3b-7)(3b+7)$

22 $(5y-2)(5y+2)$

19 $(7y-5)(7y+5)$

23 $(3x-1)(3x+1)$

20 $(5a+4)(4a-3)$

24 $(4x-7)(4x+5)$

Expand $(4+x)(2-x)$

$$(4+x)(2-x) = 8 - 4x + 2x - x^2$$
$$= 8 - 2x - x^2$$

Expand

25 $(2-x)(5+x)$

29 $(7+x)(3-x)$

26 $(4+3x)(2-x)$

30 $(1+4x)(2-x)$

27 $(x-1)(1-x)$

31 $(x-3)(2-x)$

28 $(5-y)(4+y)$

32 $(4-2p)(5+2p)$

Expand $(3x-2)(5-2x)$

$$(3x-2)(5-2x) = 15x - 6x^2 - 10 + 4x$$
$$= 19x - 6x^2 - 10$$
$$= -6x^2 + 19x - 10$$

Expand

33 $(2x+1)(1+3x)$

38 $(5x+2)(4+3x)$

34 $(5x+2)(2-x)$

39 $(7x+4)(3-2x)$

35 $(6x-1)(3-x)$

40 $(4x-3)(3-5x)$

36 $(5a-2)(3-7a)$

41 $(3-p)(4+p)$

37 $(3x+2)(4-x)$

42 $(x-5)(2+x)$

IMPORTANT
PRODUCTS

Three very important products are:

$$(x+a)^2 = (x+a)(x+a)$$
$$= x^2 + xa + ax + a^2$$
$$= x^2 + 2ax + a^2 \quad (\text{since } xa \text{ is the same as } ax)$$

i.e.

$$(x+a)^2 = x^2 + 2ax + a^2$$

so

$$(x+3)^2 = x^2 + 6x + 9$$

$$(x-a)^2 = (x-a)(x-a)$$
$$= x^2 - xa - ax + a^2$$

i.e.

$$(x-a)^2 = x^2 - 2ax + a^2$$

so

$$(x-4)^2 = x^2 - 8x + 16$$

$$(x+a)(x-a) = x^2 - xa + ax - a^2$$
$$= x^2 - a^2$$

i.e.

$$(x+a)(x-a) = x^2 - a^2$$

and

$$(x-a)(x+a) = x^2 - a^2$$

so

$$(x+5)(x-5) = x^2 - 25$$

and

$$(x-3)(x+3) = x^2 - 9$$

You should learn these three results thoroughly, for they will appear time and time again. Given the left-hand side you should know the right-hand side and vice versa.

EXERCISE 5F

Expand $(x+5)^2$

> Comparing $(x+5)^2$ with $(x+a)^2$ tells us that $a = 5$ in this case.
> So $x^2 + 2ax + a^2$ becomes $x^2 + 2(5)x + (5)^2$

$$(x+5)^2 = x^2 + 10x + 25$$

> Alternatively, write $(x+5)^2$ as $(x+5)(x+5)$ and multiply out in the usual way.

Expand, by comparing with $(x+a)^2$

1 $(x+1)^2$ **4** $(b+4)^2$ <u>**7**</u> $(c+d)^2$

2 $(x+2)^2$ **5** $(x+z)^2$ **8** $(m+n)^2$

3 $(a+3)^2$ **6** $(y+x)^2$ **9** $(a+9)^2$

Expand by writing as the product of two brackets.

10 $(t+10)^2$ **13** $(p+7)^2$ <u>**16**</u> $(e+f)^2$

11 $(x+12)^2$ **14** $(p+q)^2$ **17** $(u+v)^2$

12 $(x+8)^2$ **15** $(a+b)^2$ <u>**18**</u> $(M+m)^2$

Expand $(2x+3)^2$

$$(2x+3)^2 = (2x)^2 + 2(2x)(3) + (3)^2$$

i.e. $(2x+3)^2 = 4x^2 + 12x + 9$

Using $(x+a)^2$ and replacing x by $2x$ and a by 3.

Expand

19 $(2x+1)^2$ **22** $(6c+1)^2$ <u>**25**</u> $(3a+4)^2$

20 $(4b+1)^2$ **23** $(3a+1)^2$ **26** $(4y+3)^2$

21 $(5x+2)^2$ **24** $(2x+5)^2$ <u>**27**</u> $(3W+2)^2$

Expand $(2x+3y)^2$

$$(2x+3y)^2 = (2x)^2 + 2(2x)(3y) + (3y)^2$$

$$(2x+3y)^2 = 4x^2 + 12xy + 9y^2$$

Expand

28 $(x+2y)^2$ **31** $(3a+2b)^2$ <u>**34**</u> $(7x+2y)^2$

29 $(3x+y)^2$ **32** $(3a+b)^2$ **35** $(3s+4t)^2$

30 $(2x+5y)^2$ **33** $(p+4q)^2$ <u>**36**</u> $(3s+t)^2$

Expand $(x-5)^2$

$(x-5)^2 = x^2 - 10x + 25$

Expand

37 $(x-2)^2$ <u>**41**</u> $(x-3)^2$

38 $(x-6)^2$ <u>**42**</u> $(x-7)^2$

39 $(a-10)^2$ <u>**43**</u> $(a-b)^2$

40 $(x-y)^2$ <u>**44**</u> $(u-v)^2$

Expand $(2x-7)^2$

$(2x-7)^2 = (2x)^2 + 2(2x)(-7) + (-7)^2$

$(2x-7)^2 = 4x^2 - 28x + 49$

Expand

45 $(3x-1)^2$ <u>**49**</u> $(2a-1)^2$

46 $(5z-1)^2$ <u>**50**</u> $(4y-1)^2$

47 $(10a-9)^2$ <u>**51**</u> $(7b-2)^2$

48 $(4x-3)^2$ <u>**52**</u> $(5x-3)^2$

Expand $(7a-4b)^2$

$(7a-4b)^2 = (7a)^2 + 2(7a)(-4b) + (-4b)^2$

$(7a-4b)^2 = 49a^2 - 56ab + 16b^2$

Expand

53 $(2y-x)^2$ <u>**57**</u> $(a-3b)^2$

54 $(5x-y)^2$ <u>**58**</u> $(m-8n)^2$

55 $(3m-2n)^2$ <u>**59**</u> $(5a-2b)^2$

56 $(7x-3y)^2$ <u>**60**</u> $(3p-5q)^2$

EXERCISE 5G

> Expand   **a** $(a+2)(a-2)$ **b** $(2x+3)(2x-3)$
>
> **a** $(a+2)(a-2) = a^2 - 4$
> **b** $(2x+3)(2x-3) = 4x^2 - 9$

Expand

1 $(x+4)(x-4)$ **5** $(x+5)(x-5)$

2 $(b+6)(b-6)$ **6** $(a-7)(a+7)$

3 $(c-3)(c+3)$ **7** $(q+10)(q-10)$

4 $(x+12)(x-12)$ **8** $(x-8)(x+8)$

9 $(2x-1)(2x+1)$ **13** $(5x+1)(5x-1)$

10 $(3x+1)(3x-1)$ **14** $(2a-3)(2a+3)$

11 $(7a+2)(7a-2)$ **15** $(10m-1)(10m+1)$

12 $(5a-4)(5a+4)$ **16** $(6a+5)(6a-5)$

> Expand $(3x+2y)(3x-2y)$
>
> $$(3x+2y)(3x-2y) = (3x)^2 - (2y)^2$$
> $$= 9x^2 - 4y^2$$

Expand

17 $(3x+4y)(3x-4y)$ **21** $(10a-9b)(10a+9b)$

18 $(2a-5b)(2a+5b)$ **22** $(5a-4b)(5a+4b)$

19 $(1-2a)(1+2a)$ **23** $(1+3x)(1-3x)$

20 $(7y+3z)(7y-3z)$ **24** $(3-5x)(3+5x)$

The results from this exercise are very important when written the other way around,

i.e. $a^2 - b^2 = (a+b)(a-b)$

We refer to this as 'factorising the difference between two squares' and we will deal with it in detail in Chapter 7.

HARDER EXPANSIONS

EXERCISE 5H

Simplify $(x+2)(x+5)+2x(x+7)$

$$(x+2)(x+5)+2x(x+7)=x^2+5x+2x+10+2x^2+14x$$
$$=3x^2+21x+10$$

Simplify

1 $(x+3)(x+4)+x(x+2)$

2 $x(x+6)+(x+1)(x+2)$

3 $(x+4)(x+5)+6(x+2)$

4 $(a-6)(a-5)+2(a+3)$

5 $(a-5)(2a+3)-3(a-4)$

6 $(x+3)(x+5)+5(x+2)$

7 $(x-3)(x+4)-3(x+3)$

8 $(x+7)(x-5)-4(x-3)$

9 $(2x+1)(3x-4)+(2x+3)(5x-2)$

10 $(5x-2)(3x+5)-(3x+5)(x+2)$

Expand $(xy-z)^2$

$$(xy-z)^2=(xy)^2-2(xy)(z)+z^2$$

$$(xy-z)^2=x^2y^2-2xyz+z^2$$

Expand

11 $(xy-3)^2$

12 $(5-yz)^2$

13 $(xy+4)^2$

14 $(3pq+8)^2$

15 $(a-bc)^2$

16 $(ab-2)^2$

<u>17</u> $(6-pq)^2$

<u>18</u> $(mn+3)^2$

<u>19</u> $(uv-2w)^2$

SUMMARY

The following is a summary of the most important types of examples considered in this chapter that will be required in future work.

1 $2(3x + 4) = 6x + 8$

2 $(x + 2)(x + 3) = x^2 + 5x + 6$

3 $(x - 2)(x - 3) = x^2 - 5x + 6$

4 $(x - 2)(x + 3) = x^2 + x - 6$

5 $(2x + 1)(3x + 2) = 6x^2 + 7x + 2$

6 $(2x - 1)(3x - 2) = 6x^2 - 7x + 2$

7 $(2x + 1)(3x - 2) = 6x^2 - x - 2$

8 $(2 + x)(3 - x) = 6 + x - x^2$

Note that a) if the signs in the brackets are the same, i.e. both $+$ or $-$, then the number term is $+$
(examples 2, 3, 5 and 6)

whereas b) if the signs in the brackets are different, i.e. one $+$ and one $-$, then the number term is $-$
(examples 4 and 7)

c) the middle term is given by collecting the product of the outside terms in the brackets and the product of the inside terms in the brackets,

i.e. in 2 the middle term is $3x + 2x$ or $5x$

in 3 the middle term is $-3x - 2x$ or $-5x$

in 4 the middle term is $3x - 2x$ or x

in 5 the middle term is $4x + 3x$ or $7x$

in 6 the middle term is $-4x - 3x$ or $-7x$

in 7 the middle term is $-4x + 3x$ or $-x$

in 8 the middle term is $-2x + 3x$ or x.

Most important of all we must remember the general expansions:

$$(x + a)^2 = x^2 + 2ax + a^2$$
$$(x - a)^2 = x^2 - 2ax + a^2$$
$$(x + a)(x - a) = x^2 - a^2$$

MIXED EXERCISE

EXERCISE 5I   Expand

1 $5(x+2)$

2 $8p(3q-2r)$

3 $(3a+b)(2a-5b)$

4 $(4x+1)(3x-5)$

5 $(x+6)(x+10)$

6 $(x-8)(x-12)$

7 $(4y+3)(4y-7)$

8 $(4y-9)(4y+9)$

9 $(5x+2)^2$

10 $(2a-7b)^2$

11 $4(2-5x)$

12 $8a(2-3a)$

13 $(4a+3)(3a-11)$

14 $(x+11)(9-x)$

15 $(2x+5)(1-10x)$

16 $(y+2z)^2$

17 $(6y-z)(6y+5z)$

18 $(4a+1)^2$

19 $(5a-7)^2$

20 $(6z-13y)^2$

21 $3(2-a)$

22 $4a(2b+c)$

23 $(5a+2b)(2c+5d)$

24 $(x-7)(x-12)$

25 $(a+7)(a+9)$

26 $(a+4)(a-5)$

27 $(3x+1)(2x+3)$

28 $(5x-2)(5x+2)$

29 $(3x-7)^2$

30 $(5x+2y)(5x-2y)$

EXPRESSIONS, IDENTITIES AND EQUATIONS

$2x$ is an expression. When we write $x+x=2x$ we are showing the equivalence between two forms of the *same* expression. The equality between two forms of the same expression is called an *identity*; in an identity the equality is true for any value of x. For example when $x=6$, $x+x=12$ and $2x=12$, and so on for any value of x.

If, on the other hand, we have $x+2=2x$
then when $x=6$, $x+2=8$ and $2x=12$,
i.e. $x+2$ and $2x$ are *not* equal for all values of x.
$x+2=2x$ is called an *equation*; we can find the values of x (if any) for which the equality is true by solving the equation.

EXERCISE 5J

Determine which of these are expressions, which are identities and which are equations.

1 $2(x+1)=2x+2$

2 $2(x+1)=4$

3 $3x+6=3(x+2)$

4 $x(x+1)=x^2+x$

5 $5(x+2)-3(x-4)$

6 $(x-1)(x+1)=x^2-1$

7 $3x+4(x-2)$

8 $(x+1)(x+4)=(x+2)^2$

**PRACTICAL
WORK**

In this chapter we have shown algebraically that
$$(x+a)^2 = x^2 + 2ax + a^2$$
and $$(x+a)(x-a) = x^2 - a^2$$

These identities are easy to prove geometrically as you will find if you observe the following instructions.

1 To show geometrically that $(x+a)^2 = x^2 + 2ax + a^2$

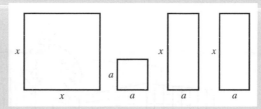

a Draw a square of side x cm and another square of side a cm. Choose any values you wish for x and a. Draw two rectangles each x cm long and a cm wide.

b Write down the area of each of the four shapes in terms of x and/or a. Find their total area in terms of x and a expressing this area as simply as possible.

c Cut out the four shapes and rearrange them to form a large square. What is the length of a side of this square in terms of x and a? Write down the area of this square as the square of the length of its edge.

d Compare your answers to parts **b** and **c**. This shows geometrically that
$$(x+a)^2 = x^2 + 2ax + a^2$$

2 Draw a square of side x cm and then cut from one corner a square of side a cm.

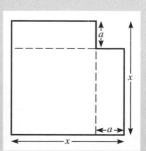

Write down an expression for the area of the remaining shape.
Cut the shape into three pieces and rearrange them to form a rectangle.
Write down an expression for the area of this rectangle.
Explain how your two expressions show that
$$(x-a)(x+a) = x^2 - a^2$$

INEQUALITIES

Some problems involve relationships that can be represented by inequalities. In this chapter we will discover methods for dealing with them, and find solutions to problems like those given in the following exercise.

EXERCISE 6A

1 Kingfisher Travel arrange a mystery tour for the Finley Social Club using one of their 44-seater coaches. For this they will charge £200. The club secretary charges £9 per adult and £5 per child and each adult brings at least one child but not more than three. The tour will be cancelled if the cost of hiring the coach cannot be covered. If there is a profit it will be used to give the children a treat.
Discuss how you could try to find the possible numbers of adults and children necessary for the trip to run.

2 A factory producing motor spares employs skilled workers who earn £300 a week and unskilled workers who earn £200 a week. The machinery requires that there must be a minimum of 110 operators of whom at least 45 must be skilled. An agreement with the union states that at least one third of the workforce must be skilled.
Discuss

a how many unknown quantities there are and how they are related

b whether there is just one possible set of values for the unknown quantities or if there are several.

Discussion arising from the situations in **Exercise 6A** shows that there are problems where relationships between unknowns involve inequalities, and that some problems can have several solutions. We have seen in Book 8A how we can solve equations involving two unknowns. To do this we need two distinct equations, but for the problems in the exercise above some of our information is in the form of inequalities and so we need to find ways of dealing with inequalities in two unknowns. For an inequality such as $0 < x + y < 10$, there are several values of x and y that satisfy it. This means that the problems we are now considering may have several possible solutions.

One way of finding solutions to these and similar problems is to form inequalities and then represent them on a graph. In Book 8A we discussed inequalities in a purely algebraic way.
Now we look at them more visually, using graphs.

USING TWO-DIMENSIONAL SPACE

If we have the inequality $x \geqslant 2$, x can take any value greater than, or equal to, 2. This can be represented by the following diagram.

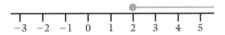

On this number line, x can take any value on the green line including 2 itself, as indicated by the solid circle at 2.

If $x > 2$ then the diagram is as shown below.

In this case x cannot take the value 2 and this is shown by the open circle at 2.

It is sometimes more useful to use two-dimensional space with x- and y-axes, rather than a one-dimensional line. We represent $x \geqslant 2$ by the set of points whose x-coordinates are greater than or equal to 2. (y is not mentioned in the inequality so y can take any value.)

The boundary line represents all the points for which $x = 2$ and the region to the right contains all points with x-coordinates greater than 2.

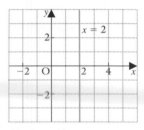

To indicate this, and to make future work easier, we use a continuous line for the boundary when it is included and we shade the region we do *not* want.

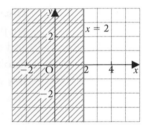

The inequality $x > 2$ tells us that x may not take the value 2. In this case we use a broken line for the boundary.

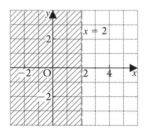

We can draw a similar diagram for $y > -1$.

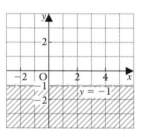

EXERCISE 6B

Draw diagrams to represent the inequalities

a $x \leqslant 1$ **b** $2 < y$

a $x \leqslant 1$

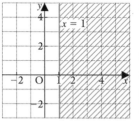

> The boundary line is $x = 1$ and is included.

The unshaded region represents $x \leqslant 1$.

b $2 < y$

> The boundary line is $y = 2$ and is not included.

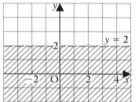

The unshaded region represents $2 < y$.

Use squared paper throughout this exercise.
Draw diagrams to represent the following inequalities.

1 $x \geqslant 2$ **4** $y < 4$ **7** $x \leqslant -4$

2 $y \leqslant 3$ **5** $x \geqslant 0$ **8** $y > -3$

3 $x > -1$ **6** $0 > y$ **9** $2 < x$

Draw a diagram to represent $-3 < x < 2$ and state whether or not the points $(1, 1)$ and $(-4, 2)$ lie in the given region.

> $-3 < x < 2$ gives two inequalities, $-3 < x$ and $x < 2$.
> Boundary lines are $x = -3$ and $x = 2$; neither are included.

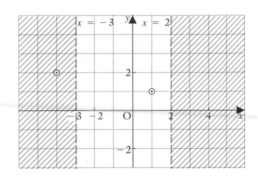

The unshaded region represents $-3 < x < 2$
$(-4, 2)$ does not lie in the given region.
$(1, 1)$ lies in the given region.

Draw diagrams to represent the following pairs of inequalities.

10 $2 \leqslant x \leqslant 4$ **13** $4 < y < 5$ **16** $-\frac{1}{2} \leqslant x \leqslant 1\frac{1}{2}$

11 $-3 < x < 1$ **14** $0 \leqslant x < 4$ **17** $-2 \leqslant y < -1$

12 $-1 \leqslant y \leqslant 2$ **15** $-2 < y \leqslant 3$ **18** $3 \leqslant x < 5$

19 In each of the questions **10** to **12**, state whether or not the point $(1, 4)$ lies in the unshaded region.

Give the inequality that defines the unshaded region.

The boundary line is $x = 3$ and is included.

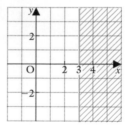

The inequality is $x \leqslant 3$

Give the inequalities that define the unshaded region.

Boundary lines $y = 4$; not included
and $y = -1$; included.

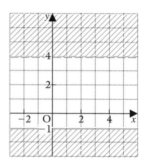

The inequalities are $-1 \leqslant y < 4$

Give the inequalities that define the unshaded regions.

20

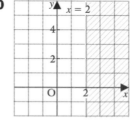

22

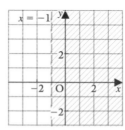

21

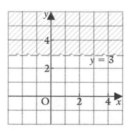

23

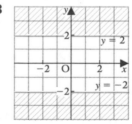

24 **25**

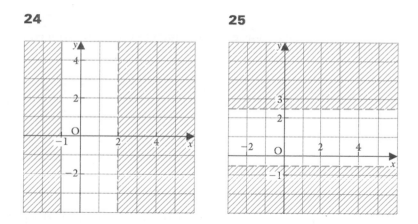

26 In each of the questions **20** to **25** state whether or not the point $(2, -1)$ is in the unshaded region.

Give the inequalities that define the *shaded* regions.

27 **29**

28 **30**

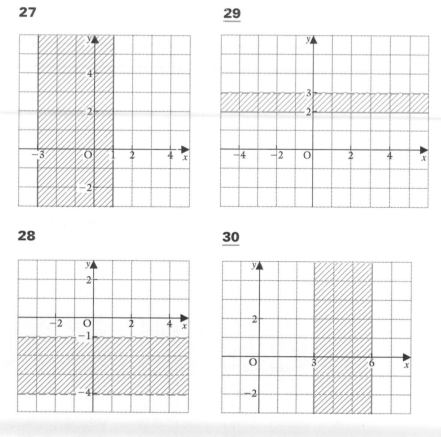

31 In each of the questions **27** to **30** state whether or not the point $(0, 2)$ is in the shaded region.

EXERCISE 6C

> Draw a diagram to represent the region defined by the set of inequalities $-1 \leqslant x \leqslant 2$ and $-5 \leqslant y \leqslant 0$.
>
> The boundary lines are
>
> $x = -1$
> $x = 2$
> $y = -5$
> $y = 0$
>
>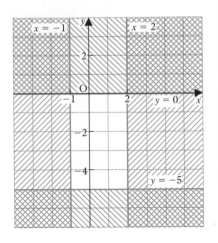
>
> The unshaded region represents the inequalities.

Draw diagrams to represent the regions described by the following sets of inequalities. In each case, draw axes for values of x and y from -5 to 5.

1 $2 \leqslant x \leqslant 4,\ -1 \leqslant y \leqslant 3$

5 $-4 < x < 0,\ -2 < y < 2$

2 $-2 < x < 2,\ -2 < y < 2$

6 $-1 < x < 1,\ -3 < y < 1$

3 $-3 < x \leqslant 2,\ -1 \leqslant y$

7 $x \geqslant 0,\ y \geqslant 0$

4 $0 \leqslant x \leqslant 4,\ 0 \leqslant y \leqslant 3$

8 $x \geqslant 1,\ -1 \leqslant y \leqslant 2$

Give the sets of inequalities that describe the unshaded regions.

9

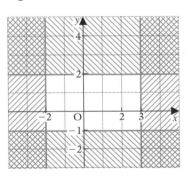

10

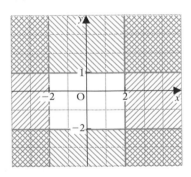

11 Is the point $\left(2\tfrac{1}{2},0\right)$ in either of the unshaded regions in questions **9** and **10**?

Give the sets of inequalities that describe the unshaded regions.

12

15

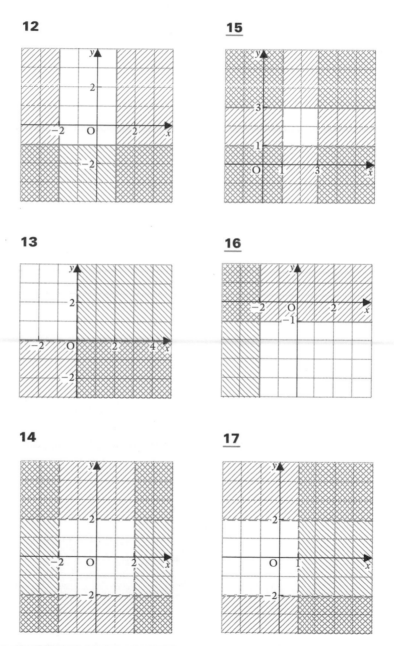

13

16

14

17

18 Is the point (2, −1) in the unshaded region in either of questions **16** and **17**?

INEQUALITIES
INVOLVING TWO
VARIABLES

The boundary lines for inequalities are parallel to the x- or y-axis when the inequalities contain either x or y, but not both.

Now we will consider some inequalities involving *both x and y* and we will find that the boundary lines are no longer parallel to an axis.

Consider $x + y \geqslant 4$.

In this case the boundary line is $x + y = 4$; as it is included in the region it is drawn as a solid line.

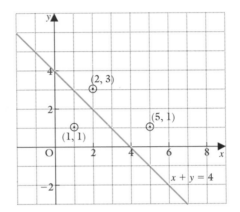

The boundary line divides the space into two regions, one on each side of the line. We need to decide which of the two regions is the one that we want.

Test a point such as $(2, 3)$.

When $x = 2$ and $y = 3$, $x + y = 5$ which is greater than 4, so the point $(2, 3)$ is in the required region.

The point $(5, 1)$ is also in the region, but the point $(1, 1)$ is not.

We can now see which region is required.

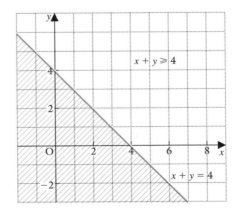

EXERCISE 6D

Leave unshaded the region defined by the inequality
$2x + 3y < 12$

The boundary line is $2x + 3y = 12$ (not included in the inequality).

> A table for values of x and y helps with drawing the line.

x	0	6	3
y	4	0	2

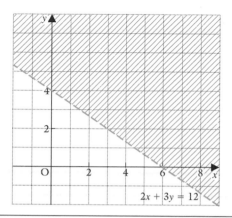

$2x + 3y = 12$

> To decide which of the two regions is wanted, we test an easy point such as $(0, 0)$.

When $x = 0$ and $y = 0$, $2x + 3y = 0$ which is less than 12.
Therefore $(0, 0)$ is in the required region.

> Shade the region on the other side of the line.

The unshaded region represents the inequality $2x + 3y < 12$

Draw diagrams to represent the regions defined by the following inequalities (draw axes for values of x and y from -6 to 6). Leave the required region unshaded.

1 $x + y \leqslant 3$

2 $x + 4y \leqslant 8$

3 $x + y > 1$

4 $x + y \leqslant 2$

5 $2x + 5y \geqslant -6$

6 $3x + 4y \geqslant 12$

7 $4x + y < 4$

8 $2x + 5y > 10$

9 $2x + y \leqslant 6$

10 $3x + 2y > 5$

Sometimes the x and y terms are not on the same side of the inequality.

Leave unshaded the region defined by the inequality $y \geqslant 2x + 1$

The boundary line is $y = 2x + 1$

x	−2	0	2
y	−3	1	5

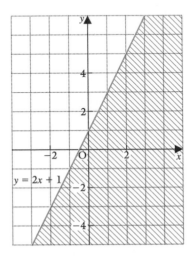

Test $(0, 0)$

$$y = 0$$
$$2x + 1 = 1$$

$0 < 1$ so $(0, 0)$ is not in the required region.
The unshaded region represents the inequality $y \geqslant 2x + 1$

Draw diagrams to represent the regions defined by the following inequalities. Leave the required region unshaded.

11 $y \leqslant x + 1$

12 $y > 2x - 1$

13 $y \geqslant \frac{1}{2}x + 1$

14 $y < 2 - 2x$

15 $y < 4 - x$

16 $y \geqslant 2x - 2$

17 $y < 2x + 3$

18 $y < 5 + 3x$

19 $y > 3 + x$

20 $y \leqslant 5 - 2x$

21 $y \leqslant x - 4$

22 $y \geqslant 1 - x$

EXERCISE 6E

Find the inequality defining the unshaded region.

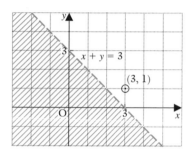

The boundary line is $x + y = 3$ and is not included.
Test the point $(3, 1)$, which is in the required region.
When $x = 3$ and $y = 1$, $x + y = 4$
$4 > 3$, so the inequality is $x + y > 3$

Find the inequalities that define the unshaded regions.

1

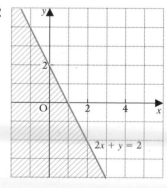

3

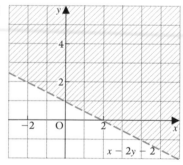

2

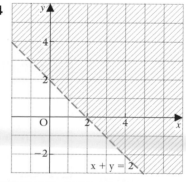

4

5

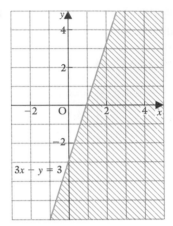

6

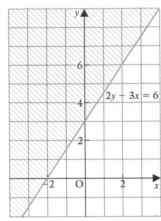

Find the inequality that defines the unshaded region.

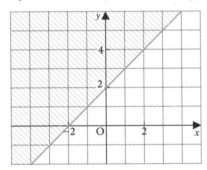

Remember that the equation of a straight line is $y = mx + c$ where m is the gradient and c is the intercept on the y-axis. We can use the diagram to work out the value of m and to write down the value of c.

The gradient of the boundary line is

$$\frac{2-0}{0-(-2)} = \frac{2}{2} = 1$$

The line cuts the y-axis when $y = 2$ so the equation of the boundary line is $y = x + 2$

The boundary line is included in the region.

Test the point $(0, 0)$, which is in the required region.

At $(0, 0)$, $y = 0$ and $x + 2 = 2$

$0 < 2$ so the inequality is $y \leqslant x + 2$

Find the inequalities that define the unshaded regions.

7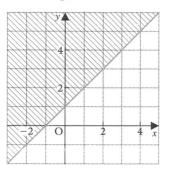

8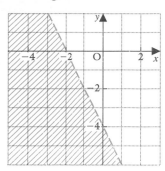

9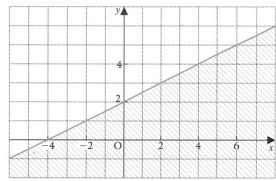

10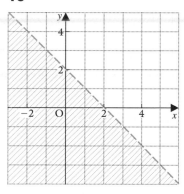

12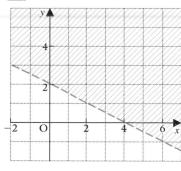

11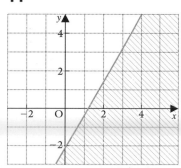

13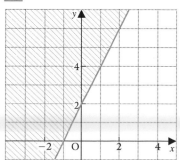

EXERCISE 6F

Leave unshaded the region defined by the set of inequalities
$x + y < 4$, $x \geqslant 0$ and $x + 2y \geqslant 2$

1st boundary line (not included) $x + y = 4$

x	4	0	2
y	0	4	2

2nd boundary line (included) $x = 0$

3rd boundary line (included) $x + 2y = 2$

x	0	2	4
y	1	0	−1

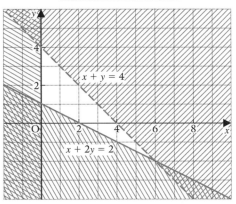

The unshaded region is defined by the given inequalities.

Leave unshaded the regions defined by the following sets of inequalities.

1 $x \geqslant -3$, $y \geqslant -2$, $x + y \leqslant 3$

2 $x > -1$, $-2 < y < 4$, $x + y < 4$

3 $y > 0$, $y \leqslant \frac{1}{2}x$, $x + y \geqslant 1$, $x + y \leqslant 5$

4 $y < 3$, $2x + 3y \geqslant 6$, $y > x - 2$

5 $y \leqslant 0$, $x \leqslant 0$, $x + y \geqslant -4$

6 $y > x$, $y < 4x$, $x + y < 5$

7 $y \geqslant 1$, $x \leqslant 0$, $y \leqslant x + 2$

8 $x + y \leqslant 6$, $3x + y \geqslant 3$, $y \geqslant -1$

9 $x \geqslant 0$, $y \geqslant x - 1$, $2y + x < 4$

10 $x > 0$, $y \geqslant \frac{1}{2}x$, $x + y \geqslant 1$, $x + y \leqslant 5$

11 $y \geqslant 0, \; x \geqslant 0, \; x + y \leqslant 1$

12 What can you say about the region defined by $x + y > 4$, $x + y < 1$, $x > 0$ and $y > 0$?

13 Do the regions defined by the following sets of inequalities exist?

    a $x + y \geqslant 3, \; y \leqslant 2, \; y \geqslant 2x$

    b $x + y > 3, \; y > 2, \; y > 2x$

SHADING THE REQUIRED REGION

In some simple cases you might be asked to shade the region defined by the inequality, instead of leaving it unshaded.

Occasionally, you may be asked to shade the required region when it is defined by several inequalities. If you try to do it by shading the required side of each boundary line, you will find yourself with overlapping shadings, resulting in a confused diagram.

For instance if $y \geqslant 0, \; x \geqslant 0$ and $x + y \leqslant 3$, the diagram looks like this and the required region disappears in a muddle.

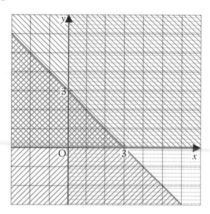

A better method is to do the shading as before so that the required region is left unshaded, then draw a second diagram on which you shade the required area.

1st diagram

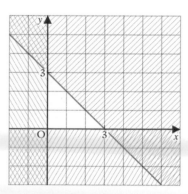

2nd diagram
The required region is shaded.

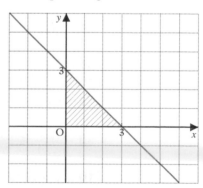

EXERCISE 6G

Shade the region defined by the set of inequalities
$x < 3$, $y < 4$, $x + y > 0$

Boundary lines are $x = 3$, $y = 4$ and $x + y = 0$

1st diagram

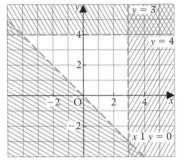

2nd diagram
The required region is shaded.

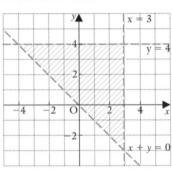

Shade the regions defined by the sets of inequalities.

1 $x \geqslant 0$, $y \geqslant 0$, $x + y \leqslant 3$ **3** $x \leqslant 4$, $y \leqslant 3$, $x + y \geqslant 0$

2 $x > -2$, $y \geqslant 2x$, $x + y < 4$ **4** $2x + y \geqslant 4$, $y \leqslant 0$, $x \leqslant 4$

5 $y > \frac{1}{2}x$, $0 < x < 2$, $x + y < 4$, $y < 3$

6 $\frac{x}{2} + \frac{y}{3} \geqslant 1$, $\frac{x}{2} - \frac{y}{3} \geqslant 1$, $x \leqslant 8$

7 $3x + 4y \leqslant 12$, $y \leqslant 2x + 1$, $y \geqslant -1$

8 $5x + 2y \leqslant -10$, $y \geqslant -1$, $x \geqslant -6$

9 $\frac{x}{5} + \frac{y}{4} \leqslant 1$, $y \leqslant 3x$, $y \geqslant \frac{1}{2}x$

10 $3x + 2y \geqslant -6$, $y \geqslant -3$, $x \geqslant -1$

If you wish to use diagrams for solving problems, it is best to leave the required regions unshaded.

EXERCISE 6H

Give the inequalities that define the unshaded region.

The first two inequalities are $x \geqslant 0$, $y \geqslant 0$
The 3rd boundary line is $5x + 2y = 10$

Test the point $(1, 1)$
When $x = 1$ and $y = 1$,
$5x + 2y = 7$
$7 < 10$ so the 3rd inequality is $5x + 2y \leqslant 10$

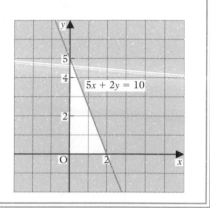

Give the sets of inequalities that define the unshaded regions:

1

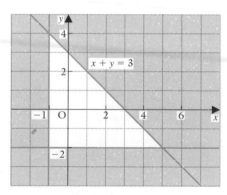

3

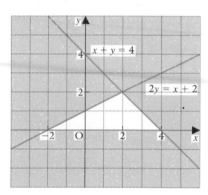

2

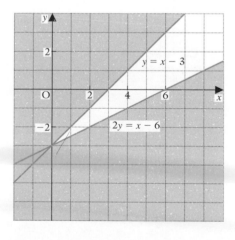

4

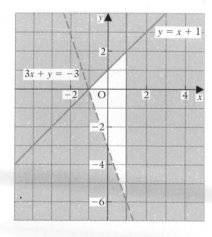

In the next four questions it is necessary to find the equations of the boundary lines first.

5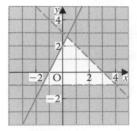

7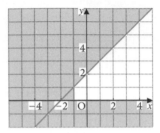

6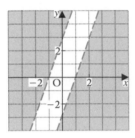

8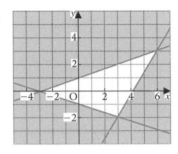

9 Use inequalities to describe the regions

a **A**   d **D**
b **B**   e **E**
c **C**   f **A + C**

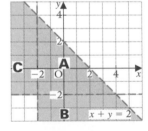

10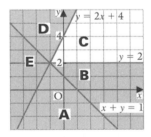

Use inequalities to describe the regions

a **A**   c **C**   e **E**
b **B**   d **D**   f **A + B**

(The axes are not boundary lines.)

11

In the diagram, which region (**A**, **B** or **C**) does each set of inequalities refer to?

a $x + y < 2$, $x < -2$, $y > -2$
b $x + y < 2$, $x > -2$, $y > -2$
c $x + y < 2$, $x > -2$, $y < -2$

EXERCISE 6I

The total number of cats and dogs kept in an animal boarding home is limited by regulations to a maximum of 15. There are 4 permanently boarded cats. A cat requires 2 units of accommodation and a dog requires 5. For owners of the home to make the business worthwhile, the total number of units occupied must not be less than 40.

a If there are x cats and y dogs write down as many inequalities as possible to express the relationship between x and y.

b Represent these inequalities graphically on squared paper.

c Give one pair of values for x and y that
  i satisfies all the inequalities
  ii does not satisfy all the inequalities.

a The regulations state that there are to be at most 15 animals
Therefore $x + y \leqslant 15$
Since there are always at least 4 cats, $x \geqslant 4$
The accommodation units required by x cats is $2x$ units and by y dogs is $5y$ units. At least 40 accommodation units must be used.
Therefore $2x + 5y \geqslant 40$
It is also obvious that $y \geqslant 0$

b The region that satisfies all the inequalities listed in part **a** is shown unshaded.

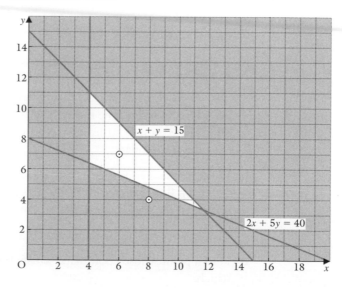

c **i** If $x = 6$ and $y = 7$ all the inequalities are satisfied.

   ii If $x = 8$ and $y = 4$
      the inequality $2x + 5y \geqslant 40$
      is not satisfied.

> We cannot have half a dog (or cat) so x and y are positive integers.

Use squared paper for this exercise.

1 Duncan wants to buy some pencils and some ball-point pens. He wants more than twice as many pencils as ball-point pens, but must have at least 2 pens.

    a If he decides to buy x pencils and y pens write down two inequalities involving x and y.

    b A pen costs 50 p and a pencil costs 40 p. Duncan has £6.50 to spend on pens and pencils. Copy the inequality $\Box x + \Box y \leqslant 650$, fill in the blanks (marked with $\Box$), and simplify it.

    c Illustrate the three inequalities graphically. Take 2 squares for 1 unit on both axes for $0 \leqslant x \leqslant 14$ and $0 \leqslant y \leqslant 14$.

    d If he buys the maximum number of pens how many pens and pencils could he have altogether?

    e If he buys the maximum total number of pens and pencils, how many of each does he buy?

2 A youth group of 60 are going camping. They wish to hire tents and find that the two sizes that are available are: small tents that take 2 people and large tents that take 5. The hire company supplies x small tents and y large tents and all the tents will be filled.

    a For various reasons several conditions must be borne in mind. Express these conditions as inequalities.
       i There must be at least 5 large tents.
      ii There must not be more than 20 small tents.
     iii There must be at least twice as many of the smaller tents as the larger ones.

    b Illustrate the inequalities in part **a** on a graph. Use 2 squares for 1 unit on both axes for $0 \leqslant x \leqslant 21$ and $0 \leqslant y \leqslant 12$. Leave unshaded the area that satisfies all the conditions.

    c Copy and complete the equation $\Box x + \Box y = 60$.

    d Find a point whose coordinates satisfy the equation in part **c** and which
       i satisfies all the inequalities
      ii does not satisfy all the inequalities.

    e Explain why some pairs of values of x and y satisfy all the inequalities and the equation but do not give a solution to the problem. Write down a pair of such values.

3 Sally is going to sell plants at the Summer Fayre and is allocated an area of 60 square feet. She decides to sell trays of plants which need an area of 1.5 square feet each and pot plants which need 0.5 square feet each.

a If she has x trays and y pot plants, copy and complete the inequality $\Box x + \Box y \leqslant 60$ and simplify it.

b After 1 hour she has sold 20 trays. Write another inequality involving x and/or y.

c The organisers stipulate that there must be at least twice as many pots as trays. Write another inequality involving x and y.

d Using 1 square for 1 unit on both axes for $0 \leqslant x \leqslant 36$ and $20 \leqslant y \leqslant 64$, illustrate the inequalities on a graph. Leave unshaded the area that satisfies all the conditions.

e Write down the coordinates of two points within the unshaded area. Are all three inequalities satisfied by each pair of values? What do the coordinates tell us about the number of trays and pots?

4 For a concert 500 seats are laid out in the school hall. It is agreed that there are x seats at £8 and y seats at £5, and that the number of £5 seats is not less than the number of £8 seats. The Parents Association have agreed to buy 200 of the cheaper seats. This information gives the relationships $x + y = a$ and $y \geqslant bx$ where a and b stand for numbers.

a Write down the values of a and b.

b Use the given information to write down another inequality involving x and/or y.

c Illustrate these three relationships on a graph and leave unshaded the area in which all the inequalities are satisfied. Use 1 square for 10 units on both axes for $0 \leqslant x \leqslant 320$, $0 \leqslant y \leqslant 520$.

d If there are 150 expensive seats how many cheaper seats are there? In this case are the inequalities satisfied?

e Is it possible to have 260 expensive seats? Explain your answer.

f If equal numbers of cheap and expensive seats are sold, what is
i the lowest possible income   **ii** the highest possible income

5

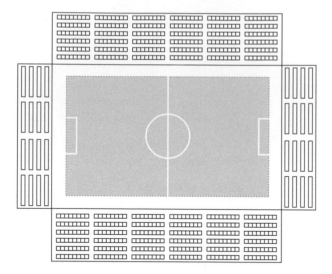

Hansford County Borough plan to build a new sports complex around a rectangular playing field. The perimeter of the field is to be more than 500 metres but less than 700 metres. It is proposed to build grandstands (which must be at least 120 m in length) to fill the longer sides of the field, which will seat 40 spectators for each metre length, and to put seating (which must be at least 80 m long) to fill the shorter sides which will hold 30 spectators for each metre length. The capacity of the ground is to be 24 000 and the length of the field must be no more than twice its width. Eventually they decide on a field that is x m long and y m wide.

a Write down all the inequalities that can be extracted from the given information.

b Show that $4x + 3y = 1200$.

c Illustrate the equation and the inequalities graphically. Leave unshaded the area that satisfies all the inequalities. Use graph paper and a scale of 1 cm to 10 units on both axes. Scale the x-axis from $x = 100$ and the y-axis from $y = 70$.

d Use the equation to find x when $y = 108$. Do these values of x and y satisfy all the inequalities?

e Write down a pair of values of x and y that satisfy the equation and all the inequalities.

f Use the equation to find y when $x = 213$. How many of the inequalities are satisfied by this pair of values of x and y?

g Indicate the different solutions that are possible to the problem as posed assuming that x and y are integers.

INVESTIGATION

Dana Ceramics produce a new commemorative plate and mug. Russell runs a china shop and has to decide how many of each to order. Plates will cost him £15 each and mugs will cost him £8 each. After going through the various options he decides that his order must

- contain at least 10 plates and 25 mugs
- contain at least twice as many mugs as plates
- not exceed 40 pieces
- have a value of at least £360 to qualify for a 5% discount off his next order

a Investigate the possible orders he can make if all four conditions are met.

b His profit on a plate is £5 and his profit on a mug is £4. On the assumption that he can sell all that he buys, he needs to know how many of each he should order to maximise his profit. Can you help him? Write a report on the strategies you use to search for an answer.

ALGEBRAIC FACTORS

In Chapter 5 we saw that we could expand algebraic expressions involving brackets, for example, $(x+3)(x+2)$ can be multiplied out to give $x^2 + 5x + 6$.

Later in this book we meet situations where we need to reverse this process, that is, we need to change from the form $x^2 + 5x + 6$ to the form $(x+3)(x+2)$.

In this chapter we build up the skills necessary to do this.

FINDING FACTORS

When we reverse the process of expanding expressions we are finding the factors of an expression. This is called *factorising*.

COMMON FACTORS

In the expression $7a + 14b$ we could write the first term as $7 \times a$ and the second term as $7 \times 2b$,

i.e. $$7a + 14b = 7 \times a + 7 \times 2b$$

The 7 is a common factor.
However we already know that $7(a + 2b) = 7 \times a + 7 \times 2b$

$\therefore$ $$7a + 14b = 7 \times a + 7 \times 2b = 7(a + 2b)$$

EXERCISE 7A

> Factorise $3x - 12$
>
> $3x - 12 = 3 \times x - 3 \times 4$
> $\qquad = 3(x - 4)$
>
> > Expand your answer to check that it is correct.

Expand your answer, in your head, to check that it is correct.

Factorise

1 $4x + 4$ **2** $12x - 3$ **3** $6a + 2$

4 $5a - 10b$ **6** $10a - 5$ **8** $2a + 4b$

5 $3t - 9$ **7** $12a + 4$ **9** $14x - 7$

Factorise $x^2 - 7x$

$x^2 - 7x = x \times x - 7 \times x$
$\qquad\quad = x(x - 7)$

Factorise

10 $x^2 + 2x$ **13** $2x^2 + x$ **16** $x^2 - 4x$

11 $x^2 - 7x$ **14** $4t - 2t^2$ **17** $b^2 + 4b$

12 $a^2 + 6a$ **15** $x^2 + 5x$ **18** $4a^2 - a$

Factorise $9ab + 12bc$

$9ab + 12bc = 3b \times 3a + 3b \times 4c$
$\qquad\qquad\; = 3b(3a + 4c)$

Factorise

19 $2x^2 - 6x$ **22** $12x^2 + 16x$ **25** $2a^2 - 12a$

20 $2z^3 + 4z$ **23** $5ab - 10bc$ **26** $6p^2 + 2p$

21 $25a^2 - 5a$ **24** $3y^2 - 27y$ **27** $9y^2 - 6y$

Factorise $ab + 2bc + bd$

$ab + 2bc + bd = b(a + 2c + d)$

Factorise

28 $2x^2 + 4x + 6$ **33** $3x^2 - 6x + 9$

29 $10a^2 - 5a + 20$ **34** $4a^2 + 8a - 4$

30 $ab + 4bc - 3bd$ **35** $5xy + 4xz + 3x$

31 $8x - 4y + 12z$ **36** $5ab + 10bc + 5bd$

32 $9ab - 6ac - 3ad$ **37** $2xy - 4yz + 8yw$

Factorise $8x^3 - 4x^2$

$8x^3 - 4x^2 = 4x^2 (2x - 1)$

> Sometimes we do not 'see' all the common factors to begin with. In this case, we may spot that 4 is a common factor and not 'see' the x^2, giving
>
> $$8x^3 - 4x^2 = 4(2x^3 - x^2)$$
>
> A check on the terms inside the bracket shows that there is another common factor, namely x^2.
>
> so $\qquad 8x^3 - 4x^2 = 4x^2(2x - 1)$

Remember to check that *all* the common factors have been removed from inside the bracket.

Factorise

38 $x^3 + x^2$

39 $x^2 - x^3$

40 $20a^2 - 5a^3$

41 $12x^3 - 16x^2$

42 $4x^4 + 12x^2$

43 $a^2 + a^3$

44 $b^3 - b^2$

45 $4x^3 - 2x^2$

46 $27a^2 - 18a^3$

47 $10x^2 - 15x^4$

48 $12x + 8$

49 $8x^2 + 12x$

50 $9x^2 - 6x + 12$

51 $5x^3 - 10x$

52 $8pq + 4qr$

53 $x^2 - 8x$

54 $12 + 9y^2$

55 $12xy + 16xz + 8x$

56 $4x^3 + 6x$

57 $12abc - 8bcd$

Factorise **a** $2\pi r^2 + 2\pi rh$ **b** $\frac{1}{2}Mu^2 - \frac{1}{2}mu^2$

a $2\pi r^2 + 2\pi rh = 2\pi r(r + h)$

b $\frac{1}{2}Mu^2 - \frac{1}{2}mu^2 = \frac{1}{2}u^2(M - m)$

Factorise

58 $\frac{1}{2}ah + \frac{1}{2}bh$

59 $mg - ma$

60 $\frac{1}{2}mv^2 - \frac{1}{2}mu^2$

65 $\frac{1}{2}mv^2 - mgh$

61 $P + \dfrac{PRT}{100}$

66 $\frac{4}{3}\pi r^3 - \frac{1}{3}\pi r^2 h$

62 $2\pi r^2 + \pi rh$

67 $3\pi r^2 + 2\pi rh$

63 $\pi R^2 + \pi r^2$

68 $\frac{1}{2}mu^2 + \frac{1}{2}mv^2$

64 $2gh_1 - 2gh_2$

69 $\frac{1}{2}bc - \frac{1}{4}ca$

FACTORISING QUADRATIC EXPRESSIONS

The type of expression we are most likely to want to factorise is one such as $x^2 + 7x + 10$.

To factorise such an expression, called a quadratic expression, we look for two brackets whose product is the original expression.

When we expanded $(x+2)(x+4)$ we had

$$(x+2)(x+4) = x^2 + 6x + 8$$

If we write $x^2 + 6x + 8 = (x+2)(x+4)$ we say we have factorised $x^2 + 6x + 8$,

i.e.   just as 10 is 2×5 so $x^2 + 6x + 8$ is $(x+2) \times (x+4)$.

To factorise an expression of the form $x^2 + 7x + 10$, that is, where all the terms are positive, we remind ourselves of the patterns we observed in Chapter 5 and summarised on page 115.

We found when expanding brackets that

- if the sign in each bracket is $+$ then the number term in the expansion is $+$
- the x^2 term comes from $x \times x$
- the number term in the expansion comes from multiplying the numbers in the brackets together
- the middle term, or x term in the expansion, comes from collecting the product of the outside terms in the brackets and the product of the inside terms in the brackets.

Using these ideas in reverse order

$$x^2 + 7x + 10 = (x + \quad)(x + \quad)$$
$$= (x+2)(x+5)$$

Choosing two numbers whose product is 10 and whose sum is 7.
The other pair of numbers whose product is 10 is 1 and 10 but the sum of 1 and 10 is 11.

Factorise $x^2 + 8x + 15$

> The product of 3 and 5 is 15, and their sum is 8.
> The other possible pair is 1 and 15 but $1 + 15 = 16$ not 8.

$x^2 + 8x + 15 = (x + 3)(x + 5)$

> Remember that 2×3 is the same as 3×2
> so that $(x + 3)(x + 5)$ is the same as $(x + 5)(x + 3)$
> i.e. the order in which the brackets are written does not matter.

Factorise   *+ and + at end*

1 $x^2 + 3x + 2$ **6** $x^2 + 8x + 7$

2 $x^2 + 6x + 5$ **7** $x^2 + 8x + 12$

3 $x^2 + 7x + 12$ **8** $x^2 + 13x + 12$

4 $x^2 + 8x + 15$ **9** $x^2 + 16x + 15$

5 $x^2 + 21x + 20$ **10** $x^2 + 12x + 20$

11 $x^2 + 8x + 16$ **16** $x^2 + 6x + 9$

12 $x^2 + 15x + 36$ **17** $x^2 + 20x + 36$

13 $x^2 + 19x + 18$ **18** $x^2 + 9x + 18$

14 $x^2 + 22x + 40$ **19** $x^2 + 11x + 30$

15 $x^2 + 9x + 8$ **20** $x^2 + 14x + 40$

To factorise an expression of the form $x^2 - 6x + 8$ remember the pattern

- the numbers in the brackets must multiply to give $+8$, that is, they must have the same sign. Since the middle term in the expression is − they must both be −
- the x^2 term comes from $x \times x$
- the middle term, or x term, comes from collecting the product of the outside terms and the product of the inside terms.

Thus $x^2 - 6x + 8 = (x - 2)(x - 4)$

Since $(-2) \times (-4) = +8$

and $x \times (-4) + (-2) \times x = -4x - 2x = -6x$

EXERCISE 7C

Factorise $x^2 - 7x + 12$

> The product of -3 and -4 is $+12$.
> The outside product is $-4x$ and the inside product is $-3x$.
> Collecting these gives $-7x$.
>
> Other pairs looked at and discarded are -2 and -6, and -1 and -12.

$x^2 - 7x + 12 = (x - 3)(x - 4)$

— and — at end

Factorise

1 $x^2 - 9x + 8$　　　　　　**6** $x^2 - 5x + 6$

2 $x^2 - 7x + 12$　　　　　　**7** $x^2 - 16x + 15$

3 $x^2 - 17x + 30$　　　　　**8** $x^2 - 6x + 9$

4 $x^2 - 11x + 28$　　　　　**9** $x^2 - 18x + 32$

5 $x^2 - 13x + 42$　　　　**10** $x^2 - 16x + 63$

Similarly　　　　　$x^2 + x - 12 = (x + 4)(x - 3)$

If the number term in the expansion is negative the signs in the brackets are different.

Thus　　　　　　　$(+4) \times (-3) = -12$

Working as before, the product of the outside terms is $-3x$
　　　　　　　　and the product of the inside terms is $+4x$

Therefore the total is $+x$.

Similarly　　　　$x^2 + 2x - 15 = (x + 5)(x - 3)$

or　　　　　　　$x^2 + 2x - 15 = (x - 3)(x + 5)$

+ and — at the end

EXERCISE 7D　　Factorise

1 $x^2 - x - 6$　　　　　　　**4** $x^2 + 3x - 28$

2 $x^2 + x - 20$　　　　　　**5** $x^2 + 2x - 15$

3 $x^2 - x - 12$　　　　　　**6** $x^2 - 2x - 24$

7 $x^2 + 6x - 27$

8 $x^2 - 9x - 22$

9 $x^2 - 2x - 35$

10 $x^2 - 8x - 20$

Most of the values in the previous three exercises have been easy to spot. Should you have difficulty, set out all possible pairs of numbers, as shown below, until you find the pair that gives the original expression when you multiply back.

Factorise

a $x^2 - 11x + 24$

(Because the number term is $+$
the two numbers in the brackets
must have the same sign)

Possible numbers		Sum
−1	−24	−25
−2	−12	−14
−3	−8	−11

$\therefore \quad x^2 - 11x + 24 = (x - 3)(x - 8)$

b $x^2 + 5x - 24$

(Because the number term is $-$
the two numbers in the brackets
have different signs)

Possible numbers		Sum
−1	+24	+23
−2	+12	+10
−3	+8	+5

$\therefore \quad x^2 + 5x - 24 = (x - 3)(x + 8)$

Remember that a '+' before the number term means that the signs in the brackets are the same, whereas a '−' before the number term means that they are different.

EXERCISE 7E

Factorise $x^2 + 13x + 36$

> The possible pairs of numbers whose product is 36 are 1×36, 2×18, 3×12, 4×9 and 6×6; 4 and 9 is the only pair that gives a sum of 13.

$x^2 + 13x + 36 = (x + 4)(x + 9)$

Factorise

1 $x^2 + 9x + 14$

2 $x^2 - 10x + 21$

3 $x^2 + 5x - 14$

4 $x^2 + x - 30$

5 $x^2 + 9x + 8$

6 $x^2 - 10x + 25$

7 $x^2 + 8x - 9$

8 $x^2 - 15x + 26$

9 $x^2 + x - 56$

10 $x^2 + 32x + 60$

11 $x^2 - 6x - 27$

12 $x^2 + 16x - 80$

13 $x^2 + 14x + 13$

14 $x^2 + 12x - 28$

15 $x^2 + 2x - 80$

16 $x^2 - 11x + 30$

17 $x^2 + 8x - 48$

18 $x^2 + 18x + 72$

19 $x^2 + 17x + 52$

20 $x^2 - 12x - 28$

21 $x^2 + 11x + 24$

22 $x^2 - 11x - 42$

23 $x^2 - 18x + 32$

24 $x^2 - 7x - 60$

Factorise $6 + x^2 - 5x$

> This needs to be rearranged into the familiar form,
> i.e. x^2 term first, then the x term and finally the number.

$$6 + x^2 - 5x = x^2 - 5x + 6$$
$$= (x - 2)(x - 3)$$

> Possible pairs:
> 1, 6, sum 7, reject
> 2, 3, sum 5, correct.

Factorise

25 $8 + x^2 + 9x$

26 $9 + x^2 - 6x$

27 $11x + 28 + x^2$

28 $x - 20 + x^2$

29 $9 + x^2 + 6x$

30 $8 + x^2 - 9x$

31 $17x + 30 + x^2$

32 $6x - 27 + x^2$

33 $x^2 + 22 + 13x$

34 $x^2 - 11x - 26$

35 $7 + x^2 - 8x$

36 $x + x^2 - 42$

37 $x^2 - 5x - 24$

38 $14 + x^2 - 9x$

39 $28x + 27 + x^2$

40 $2x - 63 + x^2$

Factorise $x^2 + 6x + 9$

> If you cannot see the numbers required, write down all the pairs whose product is 9.

$$x^2 + 6x + 9 = (x+3)(x+3)$$
$$= (x+3)^2$$

> 3×3 or
> 1×9

Factorise

41 $x^2 + 10x + 25$

45 $x^2 + 12x + 36$

42 $x^2 - 10x + 25$

46 $x^2 - 12x + 36$

43 $x^2 + 4x + 4$

47 $x^2 - 4x + 4$

44 $x^2 - 14x + 49$

48 $x^2 + 16x + 64$

EXERCISE 7F

Factorise $6 - 5x - x^2$

> When the x^2 term is negative, the terms should be arranged: number term, then the x term and finally the x^2 term. This means that the x term appears at the end of each bracket.

> 2×3 or
> 6×1

$$6 - 5x - x^2 = (6+x)(1-x)$$

Factorise

1 $2 - x - x^2$

5 $6 - x - x^2$

2 $6 + x - x^2$

6 $2 + x - x^2$

3 $4 - 3x - x^2$

7 $8 - 2x - x^2$

4 $8 + 2x - x^2$

8 $5 - 4x - x^2$

9 $10 - 3x - x^2$

13 $6 + 5x - x^2$

10 $12 + 4x - x^2$

14 $20 - x - x^2$

11 $5 + 4x - x^2$

15 $15 - 2x - x^2$

12 $14 - 5x - x^2$

16 $12 + x - x^2$

In Chapter 5, one of the expansions we listed was

$$(x + a)(x - a) = x^2 - a^2$$

If we reverse this we have

$$x^2 - a^2 = (x + a)(x - a)$$
$$\text{or } x^2 - a^2 = (x - a)(x + a)$$

(the order of multiplication of two brackets makes no difference to the result).

This result is known as *factorising the difference between two squares* and is very important.

When factorising do not confuse $x^2 - 4$ with $x^2 - 4x$.

$$x^2 - 4 = (x + 2)(x - 2)$$

whereas $\qquad x^2 - 4x = x(x - 4) \qquad$ (4x is *not* a square)

EXERCISE 7G

Factorise $x^2 - 9$

$$x^2 - 9 = x^2 - 3^2$$
$$= (x + 3)(x - 3)$$

$\boxed{\text{or } (x - 3)(x + 3)}$

Factorise

1 $x^2 - 25$ **4** $x^2 - 1$ **7** $x^2 - 36$

2 $x^2 - 4$ **5** $x^2 - 64$ **8** $x^2 - 81$

3 $x^2 - 100$ **6** $x^2 - 16$ **9** $x^2 - 49$

Factorise $4 - x^2$

$$4 - x^2 = 2^2 - x^2$$
$$= (2 + x)(2 - x)$$

$\boxed{\text{or } (2 - x)(2 + x)}$

Factorise

10 $9 - x^2$ **13** $a^2 - b^2$ **16** $25 - x^2$

11 $36 - x^2$ **14** $9y^2 - z^2$ **17** $81 - x^2$

12 $100 - x^2$ **15** $16 - x^2$ **18** $x^2 - y^2$

We began this chapter by considering common factors. A little revision is now necessary followed by factorising expressions such as $2x^2 - 8x - 10$ where 2 is a common factor.

EXERCISE 7H

Factorise $12x - 6$

$$12x - 6 = 6(2x - 1)$$

Factorise

1 $3x + 12$ **4** $14x + 21$ **7** $9x^2 - 18x$

2 $25x^2 + 10x$ **5** $4x^2 + 2$ **8** $20x + 12$

3 $12x^2 - 8$ **6** $21x - 7$ **9** $8x^2 - 4x$

Factorise $2x^2 - 8x - 10$

$$2x^2 - 8x - 10 = 2(x^2 - 4x - 5)$$
$$= 2(x - 5)(x + 1)$$

Now check to see if the quadratic expression factorises.

Factorise

10 $3x^2 + 12x + 9$ **15** $4x^2 - 24x + 20$

11 $5x^2 - 15x - 50$ **16** $3x^2 + 18x + 24$

12 $4x^2 + 8x - 32$ **17** $5x^2 - 45$

13 $3x^2 - 12$ **18** $3x^2 - 12x - 63$

14 $2x^2 - 18x + 28$ **19** $18 - 3x - 3x^2$

CALCULATIONS USING FACTORISING

EXERCISE 7I

Find $1.7^2 + 0.3 \times 1.7$

$$1.7^2 + 0.3 \times 1.7 = 1.7(1.7 + 0.3)$$
$$= 1.7 \times 2$$
$$= 3.4$$

Find, without using a calculator

1 $2.5^2 + 0.5 \times 2.5$ **5** $5.2^2 + 0.8 \times 5.2$

2 $1.3 \times 3.7 + 3.7^2$ **6** $2.6 \times 3.4 + 3.4^2$

3 $5.9^2 - 2.9 \times 5.9$ **7** $4.3^2 - 1.3 \times 4.3$

4 $8.76^2 - 4.76 \times 8.76$ **8** $16.27^2 - 5.27 \times 16.27$

Find $100^2 - 98^2$

$$100^2 - 98^2 = (100 + 98)(100 - 98)$$
$$= 198 \times 2$$
$$= 396$$

Find, without using a calculator

9 $55^2 - 45^2$ **13** $10.2^2 - 9.8^2$

10 $20.6^2 - 9.4^2$ **14** $13.5^2 - 6.5^2$

11 $7.82^2 - 2.82^2$ **15** $8.79^2 - 1.21^2$

12 $2.667^2 - 1.333^2$ **16** $0.763^2 - 0.237^2$

MIXED QUADRATIC EXPRESSIONS

Some quadratic expressions such as $x^2 + 9$ and $x^2 + 3x + 1$ will not factorise. The next exercise in this chapter includes some expressions that will not factorise.

EXERCISE 7J

Factorise where possible

1 $x^2 + 13x + 40$ **5** $x^2 - 8x + 12$

2 $x^2 - 11x + 18$ **6** $x^2 - 11x - 10$

3 $x^2 - 36$ **7** $x^2 + 6x - 7$

4 $x^2 + 4$ **8** $x^2 + 13x - 30$

9 $x^2 - 11x + 24$

13 $x^2 + 8x + 12$

10 $x^2 + 11x + 12$

14 $x^2 - x - 30$

11 $x^2 + 14x - 15$

15 $x^2 - 49$

12 $28 - 12x - x^2$

16 $x^2 - 7x + 2$

17 $x^2 - 7x - 10$

21 $x^2 + 13x - 68$

18 $x^2 + 13x + 42$

22 $x^2 + 11x - 26$

19 $x^2 - 9$

23 $a^2 - 16a + 63$

20 $x^2 - 10x + 24$

24 $28 + 3x - x^2$

MIXED EXERCISES

EXERCISE 7K

1 Factorise **a** $10a + 20$ **b** $15p^2 - 10p$

2 Factorise **a** $4ab - 8bc$ **b** $5b^2 + 15b - 5$

3 Factorise **a** $a^2 + 9a + 18$ **b** $x^2 - 7x - 8$

4 Factorise **a** $21 + 10x + x^2$ **b** $10 - 7x + x^2$

5 Factorise **a** $a^2 - 36$ **b** $16 - x^2$

6 Find, without using a calculator
a $3.7^2 + 1.3 \times 3.7$ **b** $7.7 \times 2.3 + 2.3^2$

7 Find, without using a calculator
a $101^2 - 99^2$ **b** $5.21^2 - 4.79^2$

EXERCISE 7L

1 Factorise **a** $8z^3 - 4z^2$ **b** $5xy - 20yz$

2 Factorise **a** $7a - a^2$ **b** $x^2 - 6x - 27$

3 Factorise **a** $x^2 + 12x + 35$ **b** $2a^2 - 6a - 8$

4 Factorise **a** $100 - x^2$ **b** $x^2 - 9$

5 Factorise   **a** $7x - 8 + x^2$ **b** $a^2 - 14a + 49$

6 Find, without using a calculator

     a $4.92^2 + 5.08 \times 4.92$ **b** $6.09^2 - 1.09 \times 6.09$

7 Find, without using a calculator

     a $10.3^2 - 9.7^2$ **b** $0.643^2 - 0.357^2$

EXERCISE 7M

1 Factorise   **a** $12z^2 - 6z$ **b** $8xy - 12yz$

2 Factorise   **a** $x^2 + 10x + 25$ **b** $x^2 - 2x - 24$

3 Factorise   **a** $a^2 + a - 6$ **b** $x^2 + 7x - 44$

4 Factorise   **a** $b^2 - 49$ **b** $16p^2 - p$

5 Factorise   **a** $30 - 17x + x^2$ **b** $12 - x - x^2$

6 Find, without using a calculator

     a $13.2 \times 6.8 + 13.2^2$ **b** $7.38^2 - 2.38 \times 7.38$

7 Find, without using a calculator

     a $0.553^2 - 0.447^2$ **b** $997^2 - 797^2$

INVESTIGATION

Using the digits 3 and 6 it is possible to make two two-digit numbers, namely 36 and 63.

The difference between the squares of these two numbers is

$$63^2 - 36^2 = 2673$$
$$= 99 \times 27$$
$$= 99 \times 9 \times 3$$
$$= 99 \times (\text{the sum of the original digits})$$
$$\times (\text{the difference between the original digits})$$

Investigate whether or not this is true for other pairs of digits.

If you cannot find a pair of digits for which the above result is not true, then start again letting the two digits be x and y. Write the two numbers in terms of x and y. (They are not xy and yx). Use the knowledge you have gained from this chapter to prove that

$$(10x + y)^2 - (10y + x)^2 = 99 \times (x + y) \times (x - y)$$

ORGANISING AND SUMMARISING DATA

The local health authority wants some information about the heights of five-year-old child in its area. The information is obtained from the first school medical examination, when the height of each child is recorded.

- There are 1256 five-year-olds in infant schools in the area covered by the health authority, so the data needs organising and summarising before it can yield useful information.
- When the health authority has found the information it needs for its own purposes, it may want to compare the distribution of heights in its area with those in other health authority areas.

Data can be organised by grouping it. The data can be summarised by giving the range, and one or more measures of central tendency, that is, the mean, the median or the mode. Summary 1 and Revision Exercise 1.7 at the front of this book can be used to remind you what these measures are and how to calculate them.

EXERCISE 8A   Use these questions for discussion.

1 In Area A, the height of the shortest child is 92 cm and that of the tallest child is 112 cm; the range of heights is therefore 20 cm. In Area B, the range of heights is also 20 cm.

    a Does this mean that the shortest child in Area B is 92 cm tall?

    b What can you say about the height of the tallest child in Area B?

    c Is the range on its own a good way of describing a set of data?

    d An unusually tall five-year-old joined an infant school in Area B. As a consequence, the range of heights in this area increased to 30 cm. Is it reasonable to use the ranges alone to compare the heights in Area B with those in Area A?

2 The first three children entering a classroom had with them 2 books, 2 books and 8 books respectively. The next three children entering the room had with them 3 books, 4 books and 5 books respectively.

a For the first group of children, write down the mean, median and the range of the number of books.

b Repeat part **a** for the second group of children.

c Is the mean number of books, on its own, a satisfactory way of describing either set?

d If the mean and the range are used, is this a better way to describe a set of data?

e Is there any advantage in using the median instead of the mean?

f The numbers of books brought in by the first three children lie in the interval shown on this number line.

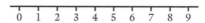

For a third set of three children who enter the room, the range is 4 books. Try drawing this range on a number line. What problems do you have?

g For the third set of children, the median number of books is 4. Does this help to locate the range on a number line?

3 The health authority wanted to investigate the possible long-term harm that pupils may suffer by carrying heavy loads to and from school in unsuitable bags. A start was made by gathering information about the weights carried by 800 secondary school pupils.
The data collected made it possible to ask questions such as

What is the range of weights?
What is the mean weight?
What is the range of the lowest 90% of these weights?

Discuss how you could go about answering these questions and what problems you might have.

ANALYSING LARGE SETS OF INFORMATION

Discussion arising from the questions in the last exercise shows that we need to use either the mean or the median together with the range, to give a reasonable summary of a set of data. It also shows that the range has disadvantages as a measure of spread, and other ways of describing the spread of a distribution are discussed later in this chapter.
Discussion about question **3** may have produced the conclusion that a large set of data presents problems in carrying out the calculations needed to analyse it unless the data is entered into a spreadsheet, or other computer statistical package.

Consider these figures, collected on behalf of the local health authority.

Heights (rounded down to the nearest centimetre) of ninety five-year-olds from one infant school

99	107	102	98	115	95	106	110	108	105	118	102	114	108	94
104	113	102	105	95	105	110	109	101	106	108	107	107	101	109
105	116	109	114	110	97	110	113	116	112	101	92	105	104	115
111	103	110	99	93	104	103	113	107	94	102	117	116	104	99
114	106	114	98	109	107	114	106	107	109	113	112	100	109	113
118	104	94	114	107	96	108	103	112	106	115	111	115	101	108

These figures were written on record cards in the same order as the children came into the medical examination, so the heights are listed in a random order. Disorganised figures like these are called *raw data*. Some form of summary, such as the mean, the median, the mode and the range, are needed to describe these figures.

- If the mean height of five-year-olds in one school is required, it can be found from the raw figures. For the 90 heights given, the mean can be calculated by adding up the heights and then dividing by 90; this is a tedious job and, even with the help of a calculator, mistakes are likely. With 1256 heights, it is not sensible to use the raw figures unless they have been entered into, say, a spreadsheet that can do the calculations.
- Grouping the data not only helps to give a 'picture' of the distribution of heights, it also reduces the complexity by replacing hundreds of individual figures with a much smaller number of groups of figures; it does, though, reduce the amount of detail given by the individual figures.

First we will organise the data into groups. If we use a number line to represent the height of these children there is no point on the line which could not represent someone's height, that is, heights are continuous data. So the grouping we choose must not have any 'gaps' between the values included in consecutive groups.

Taking h cm to represent the height of any child, a suitable grouping is

$90 \leqslant h < 95, 95 \leqslant h < 100, 100 \leqslant h < 105, 105 \leqslant h < 110, 110 \leqslant h < 115, 115 \leqslant h < 120$

and these are used to construct the following frequency table.

Height, h cm	Tally	Frequency, f
$90 \leqslant h < 95$	~~////~~	5
$95 \leqslant h < 100$	~~////~~ ////	9
$100 \leqslant h < 105$	~~////~~ ~~////~~ ~~////~~ //	17
$105 \leqslant h < 110$	~~////~~ ~~////~~ ~~////~~ ~~////~~ ~~////~~ ///	28
$110 \leqslant h < 115$	~~////~~ ~~////~~ ~~////~~ ~~////~~ /	21
$115 \leqslant h < 120$	~~////~~ ~~////~~	10
	Total	90

When you make a frequency table from raw data, work down the columns, making a tally mark in the appropriate row for each value. Do not go through the data looking for values that fit into the first group and the second group and so on.

Now we can see that the modal group is 105 cm to 110 cm, and we can estimate the range of height as

(upper bound of last group − lower bound of first group) = (120 − 90) cm = 30 cm

STEM-AND-LEAF DIAGRAMS

Placing the data in groups loses some of the detail; for example, we can give the modal group but not the mode from the table on page 158. Another way of organising the data is to draw a stem-and-leaf diagram. This also groups the data but preserves the detail of individual figures. The groups form the *stem*: the heights given are all between 90 cm and 120 cm, so we will use the number of tens as the numbers for the stem, i.e. 9, 10 and 11. We write these down the left-hand side. The leaves are the corresponding units, and these are written on the right, next to the appropriate stem so 102 is represented by a 2 in the leaf next to 10 in the stem. Start by working across the rows marking each unit in the appropriate place but do not attempt to order them.

Stem	Leaves
9	9 8 5 4 5 7 2 9 3 4 9 8 4 6
10	7 2 6 8 5 2 8 4 2 5 5 9 1 6 8 7 7 1 9 8 5 9 1 5 4 3 4 3 7 2 4 6 9 7 6 7 9 0 9 4 7 8 3 6 1
11	5 0 8 4 3 0 6 4 0 0 3 6 2 5 1 0 3 7 6 4 4 4 3 2 3 8 4 2 5 1 5

Then 9|2 means 92, 10|1 means 101, 11|3 means 113, etc.

Next redraw the diagram with the numbers in order of size and give a key.

Stem	Leaves	9|2 means 92
9	2 3 4 4 4 5 5 6 7 8 8 9 9 9	
10	0 1 1 1 1 2 2 2 2 3 3 3 4 4 4 4 4 5 5 5 5 5 6 6 6 6 6 7 7 7 7 7 7 7 8 8 8 8 8 9 9 9 9 9 9	
11	0 0 0 0 0 1 1 2 2 2 3 3 3 3 3 4 4 4 4 4 4 5 5 5 5 6 6 6 7 8 8	

Now we can see that the mode is 107 cm. We can also find the median − there are 90 heights so the median is the average of the 45th and 46th height − these are both 107 cm, so the median height is 107 cm.

The range can also be found without having to estimate it. What is the range?

Questions such as 'How many of these five-year-olds are less than 103 cm tall?' cannot be answered exactly from the grouped frequency table but they can from the stem-and-leaf diagram. What is the exact answer?

This stem-and-leaf diagram illustrates the number of pages in twenty-five text books.

Number of pages	1 | 62 means 162
1	62 74 80 88
2	01 20 34 41 58 75 92
3	09 10 25 36 71 77 79 80 86
4	10 24 63
5	06 10

Use this diagram to answer these questions.

a How many pages are there in the shortest book?

b What is the range of the number of pages?

c How many books have more than 250 pages?

d How many books have fewer than 350 pages?

Next we need to develop methods for finding the mean and the median from grouped data. Firstly, we will remind ourselves how to find the mean of an ungrouped frequency distribution.

FINDING THE MEAN OF AN UNGROUPED FREQUENCY DISTRIBUTION

In Book 7A, we found the mean value of a frequency distribution by multiplying each value by its frequency, adding these products and dividing the result by the total number of values, that is, the sum of the frequencies. This example (from Book 7A) summarises the process.

The symbol 'Σ' means 'the sum of all items such as'.

Test marks

Mark, x	Frequency, f	Frequency × mark, fx
0	1	0
1	1	1
2	8	16
3	11	33
4	5	20
5	4	20
	Total, $\Sigma f, = 30$	Total, $\Sigma fx, = 90$

The mean mark is given by $\dfrac{\Sigma fx}{\Sigma f} = \dfrac{90}{30} = 3$.

This method can be used to find the mean of any ungrouped frequency distribution, i.e.

> the mean value of a frequency distribution is given by
> $$\frac{\Sigma fx}{\Sigma f}$$
> where x is the value of an item and f is its frequency.

FINDING THE MEAN OF A GROUPED FREQUENCY DISTRIBUTION

From the table below we can see that 5 children had heights, h cm, in the range $90 \leqslant h < 95$.

Height, h cm	Frequency, f
$90 \leqslant h < 95$	5
$95 \leqslant h < 100$	9
$100 \leqslant h < 105$	17
$105 \leqslant h < 110$	28
$110 \leqslant h < 115$	21
$115 \leqslant h < 120$	10

If we *assume* that the mean height of these five children is halfway between 90 cm and 95 cm, i.e. 92.5 cm, then we can estimate the total heights of the 5 children as 92.5 × 5 cm = 462.5 cm.

The middle value of a group is called the *midclass value*.

Using the midclass value as an estimate for the mean value in each group, we can find (approximately) the total height of the children in each group and hence the total height of all the 90 five-year-olds.

It is easier to keep track of the calculations if we add another two columns to the frequency table.

Height, h cm	Frequency, f	Midclass value, x	fx
$90 \leqslant h < 95$	5	92.5	462.5
$95 \leqslant h < 100$	9	97.5	877.5
$100 \leqslant h < 105$	17	102.5	1742.5
$105 \leqslant h < 110$	28	107.5	3010
$110 \leqslant h < 115$	21	112.5	2362.5
$115 \leqslant h < 120$	10	117.5	1175
Totals	$\Sigma f = 90$		$\Sigma fx = 9630$

The total height of all 50 children is estimated as 9630 cm, so the mean height is approximately $\frac{9630}{90}$ cm $= 107$ cm

Remember that this calculation is based on the assumption that the average height in each group is half-way through the group so what we have found is an estimate for the mean.

This process can be used with any grouped frequency distribution, so

> the estimated mean value of a grouped frequencydistribution is given by
> $$\frac{\Sigma f x}{\Sigma f}$$
> where x is the midclass value and f is the frequency of items in the group.

EXERCISE 8B

1 Fifty boxes of peaches were examined and the number of bad peaches in each box was recorded, with the following result. Estimate the mean number of bad peaches per box.

No. of bad peaches per box	0–4	5–9	10–14	15–19
Frequency	34	11	4	1

2   7   12   17

2 Twenty tomato seeds were planted in a seed tray. Four weeks later, the heights of the resulting plants were measured and the following frequency table was made. Estimate the mean height of the seedlings.

Height, h cm	$1 \leqslant h < 4$	$4 \leqslant h < 7$	$7 \leqslant h < 10$	$10 \leqslant h < 13$
Frequency	2	5	10	3

2.5   5.5   8.5   11.5

3 The table shows the result of a survey among 100 pupils on the amount of money each of them spent in the school tuck shop on one particular day. Find an estimate for the mean amount of money spent.

Amount (pence)	0–24	25–49	50–74	75–99
Frequency	26	15	38	21

12   37   62   87

4 The bar chart shows the result of an examination of 20 boxes of screws.

Make a frequency table and estimate the mean number of defective screws per box.

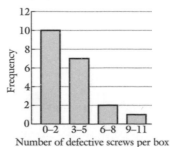

Number of defective screws per box

5 The table shows the distribution of heights of 100 adult females, rounded down to the nearest centimetre. Find the mean height.

Height, h cm	$145 \leqslant h < 150$	$150 \leqslant h < 155$	$155 \leqslant h < 160$	$160 \leqslant h < 165$	$165 \leqslant h < 170$	$170 \leqslant h < 175$
Frequency	2	6	42	36	10	4

6 A new income-tax form was trialled by asking some people to complete it. The time each person took was recorded and the frequency polygon summarises the results.

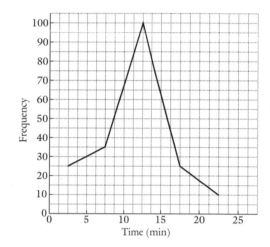

a How many people were asked to complete the form?

b Estimate the range of times taken.

c Copy and complete this table.

Midclass value, t minutes	2.5				
Frequency	25				

d Estimate the mean time taken to complete this form.

THE MEDIAN

The table on page 158 shows the heights of 90 five-year-olds collected from one infant school in the area covered by a local health authority. These are some of the 1256 heights collected from all infant schools in the area.

- The authority may want to use these figures to find the median height, that is, the height matched or exceeded by half the five-year-old.
- It may also want answers to questions such as 'What proportion of five-year-olds are at least 105 cm tall?' or 'Below what height are the shortest 25% of the children?'

For these questions to be answered, the heights need to be arranged in ascending order. Without the help of a computer this is a time-consuming task for 90 figures; for 1256 figures it is daunting.

RUNNING
TOTALS

The proportion of the 90 children in one school who are at least 105 cm tall can be found from the grouped frequency table by working out a 'running total', that is, by finding the total number of heights below 95 cm, the total number of heights below 100 cm, and so on. We add another column to the table given on page 158 to show the running totals.

Height, h cm	Frequency, f	Running total
$90 \leqslant h < 95$	5	5 heights below 95 cm
$95 \leqslant h < 100$	9	$5 + 9$, i.e. 14, heights are below 100 cm
$100 \leqslant h < 105$	17	$14 + 17$, i.e. 31, heights are below 105 cm
$105 \leqslant h < 110$	28	$31 + 28$, i.e. 59, heights are below 110 cm
$110 \leqslant h < 115$	21	...
$115 \leqslant h < 120$	10	

The last column shows that 31 out of 90 children are less than 105 cm tall,

i.e. $\frac{31}{90} = 34.4\%$ are less than 105 cm tall,

so **65.6%** are taller than 105 cm.

EXERCISE 8C

1 The following table shows the separate subject results achieved by a certain pupil, with the running total given in the fourth column.

Lesson	Subject	Mark	Running total
1	Physics	54	54
2	French	72	126
3	Biology	62	
4	Chemistry	45	
5	History	78	
6	Mathematics	64	
7	English	45	
8	Geography	82	

Copy and complete the table.

2 The table shows the running totals of pupils having school lunch each day during a certain school week. Complete the table to find out how many had lunch on each day.

Weekday	Number of lunches served each day	Running total of lunches served
Monday		126
Tuesday		280
Wednesday		424
Thursday		599
Friday		717

3 The mile-posts along the M4 motorway show that the distances, in miles, between various places are as follows.

Cardiff to Newport 10.   Newport to Severn Bridge 16.
Severn Bridge to Leigh Delamere 28.   Leigh Delamere to Swindon 18.
Swindon to Reading 39.   Reading to Heathrow Airport 28.
Heathrow to Central London 15.

a Make a running total of the distances along the motorway from Cardiff to Central London.

b Use your table to find the distance from
   i Swindon to Heathrow Airport   ii Newport to Reading.

4 During a week's holiday a family spent the following amounts.

	Amount spent	Running total of expenditure
Monday	£25	
Tuesday	£48	
Wednesday	£8	
Thursday	£55	
Friday	£34	
Saturday	£15	
Sunday	£5	

a Copy and complete the table.

b How much in total did the family spend in the week?

c By which day had they spent over half of their total expenditure?

CUMULATIVE FREQUENCY

The running total of frequencies is called the *cumulative frequency*.

A *cumulative frequency table* is constructed by adding each frequency to the sum of all those that have gone before it.

The cumulative frequency table for the 90 heights of five-year-old children can be constructed as follows:

Height, h cm	Frequency, f	Height, h cm	Cumulative frequency
$90 \leqslant h < 95$	5	$h < 95$	5
$95 \leqslant h < 100$	9	$h < 100$	$(5 + 9 =\,)\,14$
$100 \leqslant h < 105$	17	$h < 105$	$(14 + 17 =\,)\,31$
$105 \leqslant h < 110$	28	$h < 110$	$(31 + 28 =\,)\,59$
$110 \leqslant h < 115$	21	$h < 115$	$(59 + 21 =\,)\,80$
$115 \leqslant h < 120$	10	$h < 120$	$(80 + 10 =\,)\,90$

Even if the second column is omitted, the frequency of heights in any group can be found from the cumulative frequencies.

For example, the number of heights in the group $105 \leqslant h < 110$ is given by the cumulative frequency up to 110 cm minus the cumulative frequency up to 105 cm, that is, $59 - 31 = 28$.

Notice that the last number in the cumulative frequency column can be used as a check on accuracy. It confirms that the total number of heights is 90.

One of the questions the health authority may want to answer is 'What proportion of the children are shorter than 115 cm?'

From the table above we can see that 80 of these 90 children are less than 115 cm tall,

that is, $\frac{80}{90} = \frac{8}{9} = 88\%$ (correct to 2 s.f.) are shorter than 115 cm.

If the health authority needs to know the percentage of children that are taller than 115 cm, then, since 88% are shorter than 115 cm, ($100 - 88$), i.e. 12%, are taller than 115 cm.

EXERCISE 8D Keep your tables for questions **1** to **3** because you will need them for the next exercises.

1 Complete the following table which shows the distribution of goals scored by the home sides in a football league one Saturday.

Score	Frequency	Score	Cumulative frequency
0	3	$\leqslant 0$	3
1	8	$\leqslant 1$	$3 + 8 = 11$
2	4	$\leqslant 2$	
3	3	$\leqslant 3$	
4	5	$\leqslant 4$	
5	2	$\leqslant 5$	
6	1	$\leqslant 6$	

a How many matches were played?

b In how many matches were 3 or more goals scored by the home side?

2 Complete the following table which shows the distribution of the marks scored by Year 7 pupils in their English test.

Mark	Frequency (no. of pupils' scores within each range)	Mark	Cumulative frequency
1–10	7	$\leqslant 10$	
11–20	14	$\leqslant 20$	
21–30	18	$\leqslant 30$	
31–40	33	$\leqslant 40$	
41–50	36	$\leqslant 50$	
51–60	43	$\leqslant 60$	
61–70	21	$\leqslant 70$	
71–80	15	$\leqslant 80$	
81–90	8	$\leqslant 90$	
91–100	5	$\leqslant 100$	

a How many Year 7 pupils are there?

b How many scored 50 or less?

c How many scored more than 60?

d Can you say how many pupils scored 75? Explain your answer.

e If you were asked to give the number of pupils who scored less than 55, what difficulties would you have in providing an answer?

3 The table is based on a cricketer's scores in one season. Complete the table to show the cumulative frequencies.

Score	0–19	20–39	40–59	60–79	80–99	100–119	120–139
Frequency	8	14	33	6	5	3	1
Score	⩽ 19	⩽ 39	⩽ 59	⩽ 79	⩽ 99	⩽ 119	⩽ 139
Cumulative frequency							

a How many innings did he play?

b In how many innings did he score less than 60?

c In how many innings did he score at least 40?

4 A school organises a Grand Prize Draw to raise money to buy a minibus. Tickets are sold at 50 p per book and pupils are encouraged to sell as many books as possible by the award of inducement prizes, including a first prize of £20 to the pupil who sells the most books. The table shows the distribution of the numbers of books sold by the pupils in the school.

Number of books sold	0–5	6–10	11–15	16–20	21–25	26–30	31–35	36–40	41–45	46–50
Frequency	77	124					73	32	22	9
Number of books sold	⩽ 5	⩽ 10	⩽ 15	⩽ 20	⩽ 25	⩽ 30	⩽ 35	⩽ 40	⩽ 45	⩽ 50
Cumulative frequency			383	611	775	867				

Complete the table and hence find

a the number of pupils who sold more than 30 books

b the number of pupils who sold fewer than 21 books

c the number of pupils who sold more than 10 books but fewer than 31 books.

d Was the £20 inducement prize won by one pupil or could it have been shared?

5 The table is based on a golfer's scores on the professional circuit one summer.

Score	67	68	69	70	71	72	73	74	75	76	77	78
Frequency	2	4	9									
Score	$\leqslant 67$	$\leqslant 68$	$\leqslant 69$	$\leqslant 70$	$\leqslant 71$	$\leqslant 72$	$\leqslant 73$	$\leqslant 74$	$\leqslant 75$	$\leqslant 76$	$\leqslant 77$	$\leqslant 78$
Cumulative frequency	2	6	15	24	36	51	64	72	77	85	91	95

Complete this table and hence find

a the number of rounds in which he scored 73

b the number of rounds in which he scored 75 or more.

CUMULATIVE FREQUENCY DIAGRAMS

We used this cumulative frequency table for the height of 90 five-year-old children to find the number of children who were less than 105 cm tall.

Height, h cm	Frequency, f	Height, h cm	Cumulative frequency
$90 \leqslant h < 95$	5	$h < 95$	5
$95 \leqslant h < 100$	9	$h < 100$	14
$100 \leqslant h < 105$	17	$h < 105$	31
$105 \leqslant h < 110$	28	$h < 110$	59
$110 \leqslant h < 115$	21	$h < 115$	80
$115 \leqslant h < 120$	10	$h < 120$	90

It is not possible to find the number of children who are less than 102 cm tall, because 102 cm is in the middle of a group and we do not know the individual heights in this group; they may all be less than 102 cm, or all greater than 102 cm, or any distribution between 100 cm and 105 cm.

We can make an estimate of this number as follows:

102 cm is just under halfway through the group 100 cm to 105 cm,

there are 17 heights in this group so *assume* that just under half of them (8) are less than 102 cm, that is, assume that the heights are evenly spread throughout the group.

This gives $14 + 8 = 22$ as an estimate of the total number of heights less than 102 cm.

This estimate can be found more easily from a graph drawn by plotting cumulative frequencies against the upper ends of the groups. When the points are joined with straight lines, the graph is called a *cumulative frequency polygon*. If we draw a smooth curve through the points, we have a *cumulative frequency curve*.

This graph shows the cumulative frequency polygon (green) and the curve (black) for the distribution of heights given above.

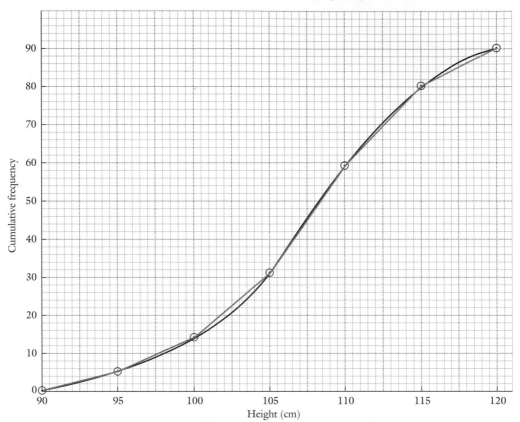

Notice that the graph starts at the point where the cumulative frequency is zero at the lower end of the first group; this is because there are no heights less than 90 cm.

When we join the points with straight lines, we are assuming that the items in any one group are evenly spread throughout that group. For example, we are assuming that the five heights in the group $90 \leqslant h < 95$ and the nine heights in the group $95 \leqslant h < 100$ are spread as shown on this number line:

They are more likely to be spread as shown on this number line.

Drawing a smooth curve through the points assumes this more likely distribution of the heights.

We can now use the curve to estimate the number of children whose heights are less than 102 cm.

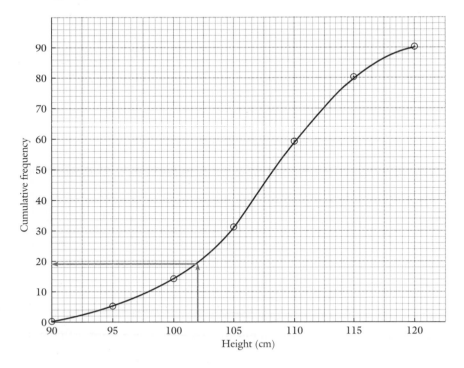

We draw a line up from 102 cm to the curve and then draw a line across to the cumulative frequency axis; the reading is 19. So we estimate that 19 children are shorter than 102 cm.

EXERCISE 8E

Keep your curves; you will need them for **Exercise 8F** and **Exercise 8G**.

1 Draw a cumulative frequency curve for question **1** in **Exercise 8D**.

2 a Draw a cumulative frequency curve to illustrate the data given in question **2** of **Exercise 8D**.

 b Use your curve to estimate the number of pupils who scored 75 or more.

 c The pass mark for the examination was 45%. How many pupils passed ?

3 a Draw a cumulative frequency curve for question **3** in **Exercise 8D**.

 b Estimate the number of innings in which the batsman scored 90 or more.

FINDING THE MEDIAN OF A GROUPED FREQUENCY DISTRIBUTION

For an ungrouped frequency distribution, we can find the median directly from the cumulative frequency table.

For example, this table gives the scores from shots in a shooting competition.

Score	Cumulative frequency (i.e. number of shots)
$\leqslant 1$	3
$\leqslant 2$	7
$\leqslant 3$	25
$\leqslant 4$	41
$\leqslant 5$	50

The median of these 50 scores is the $\dfrac{50+1}{2}$ th score, that is, the average of the 25th and 26th score.

From the cumulative frequencies, we see that the 25th score is 3 and the 26th score is 4, so the median score is $\frac{1}{2}(3+4) = 3.5$

Now consider again the grouped distribution of 90 heights.

Height, h cm	Cumulative frequency
$h < 95$	5
$h < 100$	14
$h < 105$	31
$h < 110$	59
$h < 115$	80
$h < 120$	90

The median of these 90 heights is the average of the 45th and 46th heights when they have been arranged in order of size. The table shows that these both lie between 105 cm and 110 cm.

Therefore the median is in the interval $105 \leqslant h < 110$.

While we cannot locate the median exactly, we can estimate its value from the cumulative frequency curve.

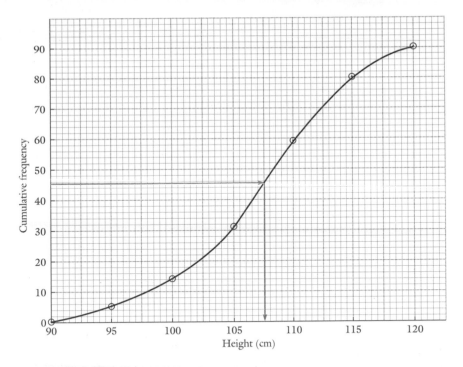

The median is halfway through the 90 heights; 45 is halfway between 0 and 90 on the cumulative frequency axis, so we read across from 45 and then down to the horizontal axis.

The reading here is 107.5 cm, and this is used as an estimate for the median.

We can also use the curve to estimate answers to questions such as 'Below what height are the shortest 25% of the children?'

Now 25% of 90 is 22.5; reading across from 22.5 and then down to the horizontal axis gives 103 cm, so we estimate that 25% of these children are shorter than 103 cm.

EXERCISE 8F

1 Find the median from each of the graphs drawn for **Exercise 8E**.

2 Use the cumulative frequency table that follows to draw the cumulative frequency curve for the prices of all the houses advertised in a property magazine one weekend.

Price (thousands of £s)	⩽ 30	⩽ 40	⩽ 50	⩽ 60	⩽ 70	⩽ 80	⩽ 90	⩽ 100	⩽ 110	⩽ 120
Cumulative frequency	10	22	60	128	170	187	197	203	208	210

Use your graph to estimate the median 'asking price' for a house.

3 This is the cumulative frequency curve for the marks awarded in a maths exam.

   a What is the median mark?

   b The pass mark is to be set so that 25% of candidates fail. What should this mark be?

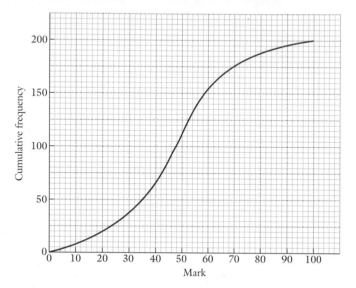

4 The cumulative frequency curve for the marks in an English test is given below. Use the graph to find

   a the number of pupils sitting the test   **b** the median mark

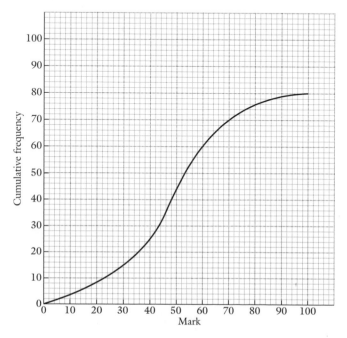

5 Use the following cumulative frequency table to draw the corresponding cumulative frequency curve for the marks obtained by candidates in an examination.

Marks	⩽ 9	⩽ 19	⩽ 29	⩽ 39	⩽ 49	⩽ 59	⩽ 69	⩽ 79	⩽ 89	⩽ 99
Cumulative frequency	7	16	28	47	80	125	174	202	212	220

Use your graph to estimate the median mark.

6 The number of cars using a cross-channel ferry on each trip during a particular month was noted and the results are given in the following table.

Number of cars	40	41	42	43	44	45	46	47	48	49	50
Number of crossings	2	4	6	10	10	12	8	6	2	1	1

Give the corresponding cumulative frequency table and hence find the median number of cars making the crossing. How many crossings did the ferry make during the month?

7 A traffic survey counted the number of cars per hour passing Southwood Post Office each hour of the day from 8 a.m. to 6 p.m. for a week. The results are given in the table.

	8 a.m.–9 a.m.	9 a.m.–10 a.m.	10 a.m.–11 a.m.	11 a.m.–12 noon	12 noon–1 p.m.	1 p.m.–2 p.m.	2 p.m.–3 p.m.	3 p.m.–4 p.m.	4 p.m.–5 p.m.	5 p.m.–6 p.m.
Monday	39	37	46	36	41	34	33	32	22	23
Tuesday	16	31	40	39	42	43	39	37	24	17
Wednesday	24	39	37	45	44	39	38	36	29	27
Thursday	19	33	32	34	42	38	37	39	25	27
Friday	30	37	36	41	48	47	40	43	35	34
Saturday	28	38	46	39	42	48	42	40	31	33
Sunday	3	7	42	14	11	33	36	35	27	26

Use groups 0–5, 6–10, 11–15, etc. to make a frequency table and a cumulative frequency table. Draw the cumulative frequency curve and use it to estimate the median number of cars passing Southwood Post Office per hour.

INTERQUARTILE RANGE

Consider this set of marks obtained by a student in the end-of-year exams.

$$54, \quad 93, \quad 86, \quad 75, \quad 8, \quad 59, \quad 73, \quad 83, \quad 55, \quad 64, \quad 73.$$

These range from 8 to 93. All the marks except one, however, are over 50. Therefore the range of these marks gives a misleading impression of their spread and it would be better to quote the range of a restricted section of the marks.

Now consider again the heights of five-year-olds collected by a local health authority.

The authority may want to compare the range of heights with those from other areas. If there are one or two abnormally short or tall children in any one authority, comparing the full ranges is not as helpful as comparing the ranges of a restricted sections of the heights. Clearly the comparisons are valid only if the *same* sections of each distribution are compared.

The restricted range usually used is the middle half, that is from $\frac{1}{4}$ to $\frac{3}{4}$ of the way through a distribution.

The value that is $\frac{1}{4}$ of the way through is called the lower quartile;

> For n values arranged in order of size,
> the *lower quartile* is the $\dfrac{n+1}{4}$ th value

The value that is $\frac{3}{4}$ of the way through is called the upper quartile;

> For n values arranged in order of size,
> the *upper quartile* is the $\dfrac{3(n+1)}{4}$ th value.

We use Q_1 to denote the lower quartile and Q_3 to denote the upper quartile. (Q_2 is used to denote the median.)
For an ungrouped distribution, these can be found exactly.
For a grouped distribution, we use a cumulative frequency curve to estimate

> Q_1 as the $\dfrac{n}{4}$ th value, and Q_3 as the $\dfrac{3n}{4}$ th value.

The difference between the quartiles is called the *interquartile range*.

The interquartile range is $Q_3 - Q_1$

This is the cumulative frequency curve of the heights of 90 children.

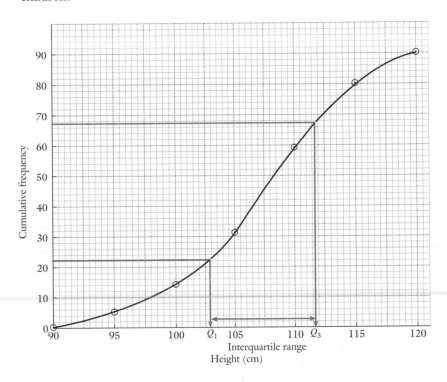

Interquartile range
Height (cm)

Q_1 is the $\dfrac{90}{4}$ th, i.e. the 22.5th value,

and Q_3 is the $\dfrac{(3 \times 90)}{4}$ th, i.e. the 67.5th value.

Reading from the graph, $Q_1 = 103\,\text{cm}$ and $Q_3 = 111.8\,\text{cm}$

Therefore the interquartile range is $Q_3 - Q_1 = 111.8 - 103\,\text{cm}$
$$= 8.8\,\text{cm}$$

(Note that readings from graphs can only give estimates for values.)

EXERCISE 8G   **1** Use the cumulative frequency curves drawn for **Exercise 8E** to find the upper and lower quartiles and the interquartiles ranges.

2 The cumulative frequency curve given below shows the weekly earnings, in pounds, of a group of teenagers.

Use the graph to find

a the median   **b** the upper and lower quartiles.

Hence find the interquartile range.

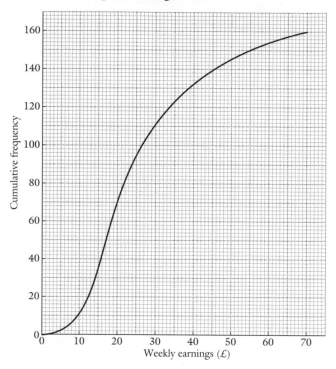

3 The table shows the distribution of the ages of people attending a public concert.

Age range	0–19	20–39	40–59	60–79	80–99
No. of people attending concert	8	26	110	128	56

Copy and complete the following cumulative frequency table and use it to draw a cumulative frequency curve.

Age range	< 20	< 40	< 60	< 80	< 100
No. of people attending concert					

Hence find

a the number of people attending the concert

b the median age

c the upper and lower quartile ages, and the interquartile range.

4 Several darts players were chosen at random and each was asked to throw 50 darts at the bull on a dart board. The number of bulls scored by each person was noted and the following frequency table was constructed.

Number of bulls	Frequency	Number of bulls	Cumulative frequency
0–4	25	$\leqslant 4$	
5–9	20	$\leqslant 9$	
10–14	15	$\leqslant 14$	
15–19	12		
20–24	10		
25–29	3		
30–34	2		
35–39	1		
40–44	1		
45–	0		

Copy this table and complete the third and fourth columns. Draw the cumulative frequency curve using a scale of 2 cm to represent 10 bulls on one axis and 2 cm to represent a cumulative frequency of 10 on the other axis. Use your graph to estimate

a the median   **b** the upper and lower quartiles.

5 In the first round of a golf tournament the following scores were recorded

     70   68   71   67   74   69   69   71   68   70
     71   70   72   69   69   68   71   70   70   72
     72   69   68   70   68   69   67   71   69   70
     68   67   70   70   73   69   71   67   69   68

a Construct a cumulative frequency table for these scores.

b How many rounds of less than 70 were there?

c How many rounds of more than 69 were there?

d Find the median score.

e Explain why you do not need a cumulative frequency curve to find the median.

6 The marks obtained by the candidates sitting a test are given in the following table.

Mark	0–9	10–19	20–29	30–39	40–49	50–59	60–69	70–79
Frequency	3	13	27	43	28	20	12	8

Draw a cumulative frequency curve for these figures. Use 2 cm to represent 10 marks and a cumulative frequency of 20.
Use your graph to estimate

a the median

b the upper and lower quartiles and, hence, the interquartile range

c the pass mark if 75% of the pupils pass.

7 A botanist measured the lengths of 120 leaf specimens from a certain species of tree. The results were as follows:

Length (cm)	< 9	< 9.5	< 10	< 10.5	< 11	< 11.5	< 12	< 12.5	< 13	< 13.5	< 14	< 14.5	< 15
Cumulative frequency	2	6	14	23	35	50	73	92	103	112	116	119	120

Use this data to draw a cumulative frequency graph. From your curve find

a the median length

b the values of the upper and lower quartiles.

MIXED EXERCISE

EXERCISE 8H

1 The following cumulative frequency table gives the percentage marks of 250 pupils in an English examination.

Mark	10	20	30	40	50	60	70	80	90	100
Number of pupils scoring up to and including this mark	5	15	29	52	89	142	197	223	240	250

a How many pupils scored a mark of more than 70?

b How many pupils scored a mark from 41 to 60?

c Plot the values from the table on a graph and draw a smooth curve through your points. (Use a scale of 2 cm to represent 20 marks on the one axis and 2 cm to represent a cumulative frequency of 25 on the other axis.)

d Use your graph to estimate
  i the median   **ii** the upper and lower quartiles.

e State the probability that a pupil chosen at random will have a mark
  i less than or equal to 50   **ii** greater than 60.

2 The bar chart illustrates the distribution of the weekly pocket money of the 240 pupils in Year 10 of a school.

  a How many pupils received from £6 to £10 inclusive?

  b Half the pupils received more than £x per week. Estimate the value of x.

  c The line AB indicates that the value of the lower quartile of the distribution is £1.50. What does this mean?

  d The value of the interquartile range is £2.70. What is the value of the upper quartile? What information does the interquartile range give us about the weekly pocket money of the group?

  e Estimate the total amount of pocket money received by the Year 10 pupils.

  f Hence estimate the mean amount of pocket money received by the Year 10 pupils.

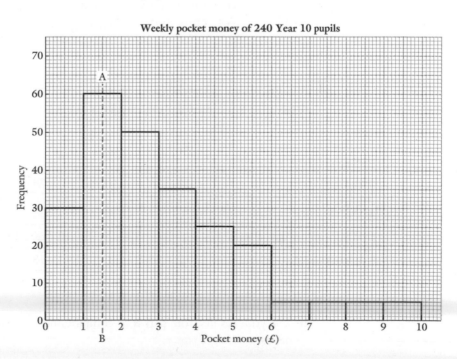

Weekly pocket money of 240 Year 10 pupils

3 The following marks were obtained by the 80 candidates in an English test which was marked out of 60.

54	52	31	47	24	36	27	15	44	26	8	20	46	32
27	31	33	57	39	32	43	32	23	33	31	21	38	28
40	19	52	37	38	39	9	30	47	29	8	13	33	35
48	18	36	39	23	58	34	35	16	21	32	38	34	13
27	32	37	23	37	49	25	38	24	27	48	36	45	18
41	34	43	12	47	24	8	29	37	33				

Use the data to complete the table below.

Interval	Tally	Frequency	Mark	Cumulative frequency
0–9			$\leqslant 9$	
10–19			$\leqslant 19$	
20–29			$\leqslant 29$	
30–39				
40–49				
50–60				

Use the information in your table to draw a cumulative frequency curve and from it estimate

a the median mark

b the upper and lower quartiles

c the number of candidates who passed if the pass mark was 40

d the pass mark if 70% of the candidates passed

e the probability that a pupil selected at random scored less than 30.

f Draw a stem-and-leaf diagram. Use the number of 10s as the stem.

4 An agricultural researcher wanted to compare milk yield from two herds of cows. The following evidence was collected.

Herd A

Yield per cow per day, c litres	$0 \leqslant c < 5$	$5 \leqslant c < 10$	$10 \leqslant c < 15$	$15 \leqslant c < 20$
Frequency	2	6	24	14

Herd B

Yield per cow per day, c litres	$0 \leqslant c < 5$	$5 \leqslant c < 10$	$10 \leqslant c < 15$	$15 \leqslant c < 20$
Frequency	3	14	18	5

a Find the median, the range and the interquartile range for the yield from each herd.

b Use your results from part **a** to compare the yields from these two herds.

5 A group of Year 8 pupils went on a school outing. The diagram shows the distribution of spending money that the pupils had at the beginning of the trip in the form of two bar charts drawn back to back.

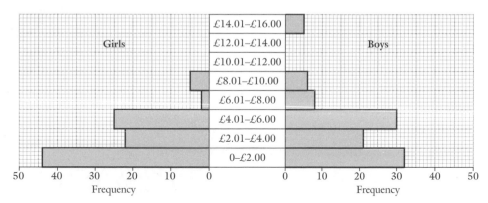

a How many **i** girls   **ii** boys went on this trip?

b What fraction of the **i** boys   **ii** girls had more than £10 to spend?

c Draw up a cumulative frequency table for both sets of data.

d Find the median and interquartile range for each set of data.

e Give two differences between the girls' and the boys' spending money.

STATISTICAL INVESTIGATION

You are asked to investigate the weights of the bags carried to school by pupils in your year.

First decide on the questions that you want to answer. For example,
'Are there any differences between the weights of the bags carried by boys and by girls?'
'What are the differences between the weights of necessary items and clutter?'
'Do pupils who bring packed lunches to school carry heavier loads than those who buy lunch at school?'

When you have decided which question(s) you want to try and answer, you must write down what information you need. Then you must decide how you are going to collect it: will you use a questionnaire or a data collection sheet, will you weigh items yourself or rely on the pupils to provide the answers? Write down your decisions and give some advantages and disadvantages of your choices. Then collect your information.

Now you must decide how you are going to present the information, e.g. will you use bar charts, stem-and-leaf diagrams or other diagrams? You also need to summarise the data – again decide which summaries you want to use and then find them.

Finally draw conclusions from your work and give some answers to your initial questions.

FORMULAS

A newspaper headline reports that a skull has been found in a peat bog. From this single part of the skeleton it is possible to draw several conclusions about the person to whom the skull belonged. For instance the scientists say that it belonged to a man who was between 1.78 m and 1.80 m tall. Have you ever wondered how they do this? They draw on experience which shows that an adult's height is just over three times the distance round their head measured immediately above the eyebrows.

- This is an example of a simple formula.

  The formula might be written: height $= 3 \times$ circumference of head

  or, in letters; $H = 3C$

  where H cm is the height and C cm is the circumference of the head just above the eyebrows.

Experiment can confirm the truth or otherwise of this formula. Perhaps the formula should be $H = 3.2C$ for a particular group of people, or H should be somewhere between $3.1C$ and $3.3C$. The important thing is that there is a relationship which can be used to deduce useful information when only a limited amount of data is available.

- When a definite relationship between two quantities has been established we can find either one when we are given the other. In this case we can calculate the height if we know the circumference of a person's skull.

The relationship between the quantities referred to opposite is not exact but in other cases it may be. For example, if grapefruit cost 33 p each the formula $C = 33n$ enables us to find the exact cost, C pence, when we buy n grapefruit.

In this chapter we look at some formulas that are more complicated than the ones we studied in previous books.

EXERCISE 9A

Discuss with members of a group whether or not each of the following statements is true. If you come to the conclusion that a particular statement is true, list examples to support your conclusion.

1 There are many examples to show that two quantities can be connected by an exact relation.

2 Sometimes two quantities are related, but not in an exact way. (If you can, list some quantities that are very closely related and others that are related but not strongly.)

3 In some cases one quantity can be related to several different quantities at the same time.

4 There are cases where two quantities are not related in any way.

CONSTRUCTING FORMULAS

Electricity bills are presented every quarter. They are made up of a fixed standing charge plus the cost of the number of units used in the quarter.

By using letters for the unknown quantities, we can construct a formula for a quarterly electricity bill.

If $£C$ is the total bill, $£R$ is the standing charge, the cost of one unit is x pence and N units are used, then

the cost of the units is Nx pence or $£\dfrac{Nx}{100}$

therefore $C = R + \dfrac{Nx}{100}$

Notice that pounds are used throughout, that is, the cost of the units was converted from pence to pounds so that we added pounds to pounds, not pounds to pence.

EXERCISE 9B

> A number p is equal to the sum of a number q and twice a number r.
> Write down a formula for p in terms of q and r.
>
> $p = q + 2r$

In questions **1** to **7** write down a formula connecting the given letters.

1 A number a is equal to the sum of two numbers b and c.

2 A number m is equal to twice the sum of two numbers n and p.

3 A number z is equal to the product of two numbers x and y.

4 A number a is equal to twice the product of two numbers b and c.

5 A number d is equal to the difference of two numbers e and f, where
e is greater than f.

6 A number n is equal to the sum of a number p and its square.

7 A number v is equal to the sum of a number u and the product of
the numbers a and t.

8 Cloth is sold at £p per metre. The cost of N metres is £R. Find a
formula for R in terms of N and p.

9 A card has P metres of lace edging wound on it. Nerys buys n
lengths of edging, each x centimetres long. If Q metres of edging is
left on the card, find a formula for Q in terms of P, n and x.

10 A ship moving with constant speed takes x minutes to cover one
nautical mile. It takes X minutes to cover y nautical miles. Find a
formula for X in terms of x and y.

11 A shop sells two brands of baked beans. It has N tins of baked beans
altogether; y of them are one brand and z of them are the other
brand. Find a formula for N in terms of y and z.

12 Fertiliser is applied at the rate of a grams per square metre. It takes
b kilograms to cover a field of area c square metres. Find a formula
for b in terms of a and c.

13 A bag of coins contains x ten-pence coins and y twenty-pence coins.
The total value of the coins is £R. Find a formula for R in terms of
x and y.

**SUBSTITUTING
NUMBERS INTO
FORMULAS**

EXERCISE 9C

Given that $s = ut - \frac{1}{2}gt^2$ find s when $u = 8$, $t = 6$ and $g = -10$

$$s = ut - \frac{1}{2}gt^2$$

When $u = 8$, $t = 6$ and $g = -10$

$$s = (8)(6) - \left(\frac{1}{2}\right)(-10)(6)^2$$
$$= 48 - (-5)(36)$$
$$= 48 - (-180)$$
$$= 48 + 180$$
$$= 228$$

> Notice that we have put each number in brackets; this is
> particularly important in the case of negative numbers.

1 Given that $v = \dfrac{u-t}{3}$, find v when $u = 4$ and $t = -2$

2 Given that $z = \dfrac{1}{x} + \dfrac{1}{y}$, find z when $x = 2$ and $y = 4$

3 If $C = rt$, find C when $r = -3$ and $t = -10$

4 If $x = rt - v$, find x when $r = 2$, $t = 10$ and $v = -4$

5 If $p = x + x^2$, find p when

   a $x = 2$ **b** $x = 3.4$ **c** $x = 0.79$

6 Given that $s = \frac{1}{2}(a + b + c)$, find s when

   a $a = 6$, $b = 9$ and $c = 5$
   b $a = 5.04$, $b = 7.35$ and $c = 4.83$

7 If $p = r(2t - s)$, find p when $r = \frac{1}{2}$, $t = 3$ and $s = -2$

8 If $a = (b + c)^2$, find a when

   a $b = 8$ and $c = -5$
   b $b = 4.1$ and $c = 7.8$

9 If $r = \dfrac{2}{s+t}$, find r when $s = \frac{1}{2}$ and $t = \frac{1}{4}$

10 Given that $a = bc - \frac{1}{2}dc$, find a when $b = 3$, $c = -4$ and $d = 7$

11 Given that $V = \frac{1}{2}(X - Y)^2$, find V when $X = 3$ and $Y = -5$

12 If $P = 2Q + 5RT$, find P when $Q = 8$, $R = -2$ and $T = -\frac{1}{2}$

13 Given that $a = (b - c)(c - d)$, find a when $b = 2$, $c = 4$ and $d = 7$

14 The displacement ($D\,\text{cm}^3$) of an engine is given by the formula

$$D = \pi n \times \left(\frac{b}{2}\right)^2 \times s$$

where n is the number of cylinders, b is the bore of each cylinder in centimetres and s is the length of the stroke in centimetres.
Find the displacement of a 4-cylinder petrol engine that has a bore of 78 mm and a stroke of 84 mm.

15 A person's mean blood pressure (P), in millimetres of mercury, is calculated using the formula

$$P = D + \frac{(S - D)}{3}$$

where D is the diastolic pressure and S is the systolic pressure, both measured in millimetres of mercury (mm/Hg). Use this formula to find Greg's mean blood pressure if his diastolic pressure is 82 mm/Hg and his systolic pressure is 146 mm/Hg.

16 The gross yield of a Government stock, as a percentage, is given by the formula

$$\text{Gross yield} = \frac{I}{P} \times 100$$

where I is the annual rate of interest and P is the price of the stock. Find the gross yield on

a Treasury $11\frac{3}{4}\%$ stock at 121

b Exchequer 4% stock at 94.

Sometimes the letter whose value we need to find is not isolated on one side of the formula. For example, to use the formula $v = u + at$ to find a when $v = 20$, $u = 8$ and $t = 2$, we substitute the values for v, u and t, this gives the equation $20 = 8 + 2a$ which can then be solved to find a.

EXERCISE 9D

Given that $p = q - 5r$ find

a the value of q when $p = 5$ and $r = 2$

b the value of r when $p = 4$ and $q = 24$.

a $\hspace{4cm} p = q - 5r$

Replacing p by 5 and r by 2, gives $5 = q - 10$ ┃ Add 10 to both sides.

$\hspace{6cm} 15 = q$

$\therefore$ if $p = 5$ and $r = 2$, $q = 15$

b $\hspace{4cm} p = q - 5r$

If $p = 4$ and $q = 24$, $\hspace{0.5cm} 4 = 24 - 5r$ ┃ Add 5r to both sides.

$\hspace{4.5cm} 5r + 4 = 24$ ┃ Subtract 4 from each side.

$\hspace{5cm} 5r = 20$ ┃ Divide both sides by 5.

$\hspace{5.5cm} r = 4$

$\therefore$ if $p = 4$ and $q = 24$, $r = 4$

1 If $a = b + 2c$ find

$\hspace{1cm} 10 = b + 2 \times 2 \hspace{0.5cm} 2 \times 2 \hspace{0.5cm} -10 = b$

$\hspace{2.5cm} = b + 2 \times 2 - 10 \hspace{2cm} 6$

a b when $a = 10$ and $c = 2$ $\hspace{0.5cm}$ **b** c when $a = 13$ and $b = 5$

$\hspace{2cm} 6$

2 Given that $x = y - z$ find

a y when $x = 9$ and $z = 4$ $\hspace{0.5cm}$ **b** z when $x = 5$ and $y = 8$.

3 If $p = qr$ find

a q when $p = 36$ and $r = 9$ $\hspace{0.5cm}$ **b** r when $p = 24$ and $q = 8$.

4 Given that $v = u + 8t$ find

a u when $v = 45$ and $t = 3.5$ **b** t when $v = 64$ and $u = 0$.

5 If $s = \frac{1}{2}(a + b + c)$ find

a a when $s = 5.8$, $b = 4.5$ and $c = 3.4$

b c when $s = 4.7$, $a = 2.7$ and $b = 3.8$.

6 At an athletics meeting a points system operates for the long jump event. The number of points scored (P) is calculated using the formula

$$P = a \times (L - b)^2$$

where L is the distance jumped in metres, measured correct to the nearest centimetre, and a and b are non-zero constants.

a The points score for a jump of 6 metres is 0. Which constant does this information enable you to find? What is the value of this constant?

b A jump of 8 metres gives a score of 400 points. Find the value of the other constant.

c What is the shortest jump that will score points?

d How many points are scored for a jump of **i** 6.5 m **ii** 7.34 m?

CHANGING THE SUBJECT OF A FORMULA

In question **16** in **Exercise 9C** the gross yield, G, of a Government stock is given by the formula

$$G = \frac{I}{P} \times 100$$

where I is the annual interest rate and P the price of the stock. This formula can be used to find G directly for different values of I and P.

A financial adviser may want to use this formula to find the interest rate for a gross yield of 10% when the price of the stock is 120.

This can be done by substituting the numbers directly into the formula
to give

$$10 = \frac{I}{120} \times 100$$

and then solving this equation for I. This gives $I = 12$.

The advisor may wish to find values of I for several different
values of G and P. He can repeat the method used above or he
can start by rearranging the formula so that I is alone on the left-
hand side.

We can do this by thinking of $G = \frac{I}{P} \times 100$ as an equation and solving
it for I.

As with any equation, we start by getting rid of any fractions.

Multiplying both sides by P achieves this;

i.e. $\quad P \times G = \not{P} \times \frac{I}{\not{P}} \times 100 \quad \Rightarrow \quad PG = 100I$

(the symbol $\Rightarrow$ means 'gives or implies').

Then dividing both sides by 100 gives $\quad \frac{GP}{100} = I$

i.e. $\quad I = \frac{GP}{100}$

Now I can be found directly for different values of G and P.

When the formula is in the form $G = \frac{I}{P} \times 100$, G is called the *subject*
of the formula.

When the formula is written as $I = \frac{GP}{100}$, I is the subject of the
formula.

Rearranging $G = \frac{I}{P} \times 100$ as $I = \frac{GP}{100}$ is called *changing the subject
of the formula.*

Note that when a formula is entered into a spreadsheet, it has to be in
the form where the required quantity is the subject of the formula. It
is therefore important to be able to change the subject of a formula.
Since changing the subject of a formula is like solving an equation, we
start by solving some equations where the letter is the denominator of
a fraction.

Solve the equation $4 + \dfrac{2}{x} = 7$

$$4 + \dfrac{2}{x} = 7$$

> Multiplying each term by x gets rid of the fraction.

$$4 \times x + \dfrac{2}{x_1} \times x' = 7 \times x$$

$$4x + 2 = 7x$$

$$2 = 3x, \quad \text{i.e.} \quad x = \dfrac{2}{3}$$

> Take $4x$ from both sides.

Check: $\text{LHS} = 4 + 2 \div \dfrac{2}{3} = 4 + 2 \times \dfrac{3}{2} = 7 = \text{RHS}$

Solve

1 $\dfrac{4}{x} = 5$ **4** $\dfrac{2}{3x} = 1$ **7** $1 - \dfrac{7}{x} = 9$

2 $1 + \dfrac{2}{a} = 3$ **5** $\dfrac{5}{y} = 2$ **8** $9 = \dfrac{4}{2t}$

3 $3 - \dfrac{4}{t} = 6$ **6** $\dfrac{3}{p} + 4 = 5$ **9** $2 + \dfrac{5}{3x} = 7$

Make t the subject of the formula $v = u + t$

> We need to 'solve' the formula for t; so we use the same methods as for solving an equation for x.

$$v = u + t$$

Subtracting u from both sides: $v - u = t$

i.e. $t = v - u$

Make the letter in brackets the subject of the formula.

10 $p = s + r$ (s) **15** $X = Y - Z$ (Y)

11 $x = 3 + y$ (y) **16** $r = s + 2t$ (s)

12 $a = b - c$ (b) **17** $k = l + m$ (m)

13 $u = v - 5$ (v) **18** $N = P - Q$ (P)

14 $z = x + y$ (y) **19** $v = u + 10t$ (u)

20 $x = 2y$ (y) **25** $a = 3b$ (b)

21 $v = \frac{1}{2}t$ (t) **26** $X = \frac{1}{10}N$ (N)

22 $a = bc$ (b) **27** $v = ut$ (u)

23 $t = \frac{u}{3}$ (u) **28** $z = \frac{w}{100}$ (w)

24 $l = \frac{m}{k}$ (m) **29** $n = \frac{p}{q}$ (p)

One operation only was needed to change the subject of a formula in the last exercise. In the following exercise, more than one operation is required.

EXERCISE 9F

Make t the subject of the formula $v = u + 2t$

> Start by isolating the term containing t on one side of the formula.

$$v = u + 2t$$

Take u from each side $\quad v - u = 2t$

$\therefore \qquad\qquad\qquad\qquad 2t = v - u$

Divide each side by 2 $\qquad t = \dfrac{(v - u)}{2}$

Make the letter in the bracket the subject of the formula.

1 $p = 2s + r$ (s) **5** $x = 2w - y$ (w)

2 $v = u - 3t$ (t) **6** $l = k + 4t$ (t)

3 $a = b - 4c$ (c) **7** $w = x - 6y$ (y)

4 $V = 2v + 3u$ (v) **8** $N = It - 2s$ (s)

9 $x = \dfrac{3y}{4}$ (y) **13** $V = \dfrac{2R}{I}$ (R)

10 $u = v + 5t$ (t) **14** $p = 2r - w$ (r)

11 $A = P + \frac{1}{10}I$ (I) **15** $a = b + \frac{1}{2}c$ (c)

12 $z = x - \dfrac{y}{3}$ (y) **16** $p = q - \dfrac{r}{5}$ (r)

17 Make u the subject of the formula $v = u + at$.
Find u when $v = 80$, $a = -10$ and $t = 6$

18 Make B the subject of the formula $A = \dfrac{C}{100} + B$
Find B when $A = 20$ and $C = 250$

19 Make C the subject of the formula $P = \dfrac{C}{N}$
Find C when $N = 20$ and $P = \frac{1}{2}$

20 Make x the subject of the formula $z = \frac{1}{2}x - 3t$
Find x when $z = 4$ and $t = -3$

21 A number a is equal to the sum of a number b and twice a number c.

a Find a formula for a in terms of b and c.
b Find a when $b = 8$ and $c = -2$.
c Make b the subject of the formula.

22 A number x is equal to the product of a number z and twice a number y.

a Find a formula for x in terms of z and y.
b Find x when $z = 3$ and $y = 2$.
c Make y the subject of the formula.

23 A number d is equal to the square of a number e plus twice a number f.

a Find a formula for d in terms of e and f.
b Make f the subject of the formula.
c Find f when $d = 10$ and $e = 3$.

24 The heat setting on a gas oven is called its Gas Mark. The formula $F = 25G + 250$ will convert a Gas Mark G into a temperature F measured in degrees Fahrenheit.

a Sally puts a joint into her gas oven which she has previously set at Gas mark 6. What is the temperature inside the oven?
b Make G the subject of this formula.
c Gary wants to bake some bread. The recommended baking temperature is $450\,°\text{F}$.
What Gas Mark should Gary set on the oven?

25 The length of a woman's femur (f cm) and her height (h cm) are approximately related by the formula $h = 3.5f + 40$

a Part of the skeleton of a woman is unearthed on an archaeological dig. The length of the femur is 34 cm. Use the formula to estimate her height.

b Rearrange the formula to make f the subject.

c A woman is 1.58 m tall. Calculate the length of her femur if her measurements fit the formula exactly.

d Sophie was 50 cm long (tall) at birth. Use the formula to find the value for the length of her femur and state why this is impossible.

26 The Body Mass Index (I) for an adult is given by the formula $I = \dfrac{W}{H^2}$ where W is the weight of the person in kilograms and H the height in metres.

Use this formula to express H in terms of I and W hence find the height of a person whose Body Mass Index is 23.4 who weighs 94.2 kg. Give your answer correct to 1 decimal place.

SUBSTITUTING ONE FORMULA INTO ANOTHER

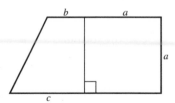

The area of this shape is given by the formula $A = \dfrac{a(b+c)}{2} + a^2$

It is also known that $a = 2b$. We can use this fact to simplify the formula for the area so that A is given in terms of only two letters instead of the three, a, b, and c. We do this by replacing every a in the formula by $2b$,

i.e. $\quad A = \dfrac{(2b)(b+c)}{2} + (2b)^2$

> Notice that we place $2b$ in a bracket; this reduces the likelihood of mistakes.

$\quad\quad = \dfrac{2 \times b \times (b+c)}{2} + 4b^2$

> This can now be simplified by cancelling the common factor, 2, from the fraction.

$\quad\quad = b^2 + bc + 4b^2$

$\therefore \quad A = 5b^2 + bc$

Given $V = u + at$ and $u = 3t$, find V in terms of a and t.

To find V in terms of a and t means that we have to eliminate u from the formula for V. We can do this by replacing u by $3t$, i.e. by substituting $3t$ for u.

$V = u + at$ and $u = 3t$

Substituting $3t$ for u gives $V = (3t) + at$

i.e. $\qquad\qquad\qquad\qquad V = 3t + at$

1 Substitute $2u$ for a in the formula $v^2 = u^2 + 2as$

2 Given that $p = \dfrac{r}{4}$ and $r = 2v$, find p in terms of v.

3 If $p = r - nt$ and $n = 4t$, find a formula for p in terms of r and t.

4 Substitute $2a$ for b in the formula $A = bc + \dfrac{a^2 + b^2}{2}$.

5 If $A = (b - c)(b - a)$ and $b = 4c$, find A in terms of a and b.

6 Use the formula $s = ut + 5t^2$ together with $t = 2s$ to give a relationship between s and u.

7 Given that $P = (V - 2U)^2$ and that $U = \frac{3}{2}V$, find P in terms of V.

8 Find a formula for n in terms of p and q given that $n = \dfrac{r}{p - q}$ and $r = p + q$.

9 If $P = D + \dfrac{(S - D)}{3}$ and $D = 1 - S$, find P in terms of S.

10 Given that $a = (b + c)^2 + (b - c)^2$ and that $b = \frac{1}{2}c$, find a in terms of c.

11 The capacity of this tank is given by
$V = \frac{1}{2}ab(c + d)$.
The surface area of the tank is given by
$A = 2b(c + e) + a(c + d)$.
If $V = 300$, $A = 150$, $c = 5$ and $e = 1$
show that $2b^2 - 25b + 100 = 0$.

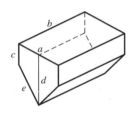

THE nTH TERM OF A SEQUENCE

In a sequence the terms occur in a particular order, that is, there is a first term, a second term and so on. The value of each term depends upon its position in that order. (This is the difference between a sequence of numbers and a set of numbers which can be in any order.)

The letter n is used for a natural number so we can refer to the nth term of a sequence in the same way as we refer to the 4th term or the 8th term.

If we are given a formula such as

$$n\text{th term} = n(n+1)$$

then we can find any term of the sequence by giving n a numerical value.

In this case, the first term is found by substituting 1 for n in $n(n+1)$,

i.e.   1st term $= 1 \times 2 = 2$

Similarly,   2nd term $= 2 \times 3 = 6$ (substituting 2)

                             3rd term $= 3 \times 4 = 12$

                             . . . .

                             10th term $= 10 \times 11 = 110$

and so on.

EXERCISE 9H

The nth term of a sequence is given by the formula

$$n\text{th term} = (n-1)^2$$

Give the first two terms and the eighth term of the sequence.

$n = 1$ 1st term $= (1-1)^2 = 0$

$n = 2$ 2nd term $= (2-1)^2 = 1^2 = 1$

$n = 8$ 8th term $= (8-1)^2 = 7^2 = 49$

Write down the first four terms and the seventh term of the sequence for which the nth term is given.

1 $2n+1$

2 $2n-1$

3 2^n

4 n^2

5 $(n-1)(n+1)$

6 $n+4$

7 $3+2n$

8 $\dfrac{1}{n}$

FINDING AN EXPRESSION FOR THE *n*TH TERM

When the pattern in a sequence is known, an expression for the *n*th term can often be found.

Consider the sequence $\qquad$ 2, 4, 6, 8, …

We need to find the relationship between each term and the number, *n*, that gives its position. It is helpful to start by making a table of values of *n* and the corresponding terms in the sequence.

n	1	2	3	4	…
*n*th term	2	4	6	8	…

The pattern here is that each term is twice its position number, so the 10th term will be 2×10 and the *n*th term is $2n$.

Now we can check that this does give the correct sequence, i.e.

if $n = 1$, 1st term $= 2 \times 1 = 2$, if $n = 2$, 2nd term $= 2 \times 2 = 4$, and so on.

EXERCISE 9I

Find, in terms of *n*, an expression for the *n*th term of each of the following sequences.

1 3, 6, 9, 12, … $\qquad\qquad$ **4** 0, 1, 2, 3, …

2 −1, −2, −3, −4, … $\qquad\quad$ **5** 4, 8, 12, 16, …

3 2, 3, 4, 5, … $\qquad\qquad$ **6** 5, 10, 15, 20, …

Find a formula for the *n*th term of the sequence 1, 4, 7, 10, …

n	1	2	3	4
n th term	1	4	7	10
3*n*	3	6	9	12

The terms increase by 3 each time, so multiples of 3 are involved.

These are each 2 more than the given terms, so take 2 from 3*n*.

nth term $= 3n - 2$

Check that this fits all the given terms.

$n = 1$ $\quad$ 1st term $= 3 \times 1 - 2 = 3 - 2 = 1$

$n = 2$ $\quad$ 2nd term $= 3 \times 2 - 2 = 6 - 2 = 4$

$n = 3$ $\quad$ 3rd term $= 3 \times 3 - 2 = 9 - 2 = 7$

$n = 4$ $\quad$ 4th term $= 3 \times 4 - 2 = 12 - 2 = 10$

Find an expression for the nth term of each of the following sequences.

7 $7, 9, 11, 13, \ldots$

10 $1 \times 3, 2 \times 4, 3 \times 5, \ldots$

8 $0, 3, 6, 9, 12, \ldots$

11 $1, 8, 27, 64, \ldots$

9 $\frac{1}{3}, \frac{1}{4}, \frac{1}{5}, \frac{1}{6}, \ldots$

12 $3, 2, 1, 0, -1, \ldots$

Find an expression for the nth term of the sequence
$4, 7, 12, 19, 28, 39, \ldots$

The differences between consecutive terms are $3, 5, 7, 9, 11, \ldots$
which are not multiples of n so we must try something different.
Consider the square of each position number.

n	1	2	3	4	5	6
nth term	4	7	12	19	28	39
n^2	1	4	9	16	25	36

Comparing the values of n^2 with the terms in the given sequence we see that
every term in the sequence is 3 more than the corresponding value of n^2

i.e. nth term $= n^2 + 3$

In questions **13** to **16** find an expression for the nth term of the sequence.

13 $6, 9, 14, 21, 30, \ldots$

15 $11, 8, 3, -4, \ldots$

14 $3, 9, 19, 33, \ldots$

16 $3, 10, 29, 66, \ldots$

17 Write down the nth term of the sequence $2, 4, 8, 16, 32, \ldots$
Show algebraically that the product of any two terms is also a term
of the sequence.
What term will the product of the nth term and the next term give?

18

One of the races at a school sports day is set out with bean bags
placed at 1 metre intervals along the track.
A competitor starts at S, runs to the first bag, picks it up and returns
it to S. Then she runs to the second bag, picks it up and returns it to
S, and so on.
How far has a competitor run when she has returned

a 1 bean bag   **b** 4 bean bags   **c** n bean bags?

EXERCISE 9J

1 A number z is equal to three times a number x minus a number y. Write down a formula connecting x, y and z.

2 Use the formula $s = \frac{1}{2}(a+b+c)$ to find

  a s when $a = 3.45$, $b = 2.76$ and $c = 4.27$

  b a when $s = 15.1$, $b = 8.2$ and $c = 12.3$

3 Make d the subject of the formula   **a** $C = \pi d$ **b** $a = c + d$

4 Make b the subject of the formula

  a $a = 7b + c$ **b** $a = c - \dfrac{b}{2}$

5 **a** Write down the first four terms and the twelfth term of the sequence for which the nth term is $n(n+3)$

  b Find, in terms of n, an expression for the nth term of the sequence 2, 6, 12, 20, ...

6 Given that $z = \dfrac{x}{y}$ find

  a x when $y = 4$ and $z = 12$

  b y when $x = 20$ and $z = 5$

7 Use the formula $v = u + 10t$ together with $t = 2u$ to find v in terms of u.

EXERCISE 9K

1 A rectangle is 5 cm longer than it is wide. If the perimeter of the rectangle is P cm and the length is x cm find a formula connecting P and x.

2 Given that $P = \dfrac{100I}{RT}$, find P when

  a $I = 3$, $R = 4$ and $T = 2$

  b $I = 63.14$, $R = 5.74$ and $T = 2$

3 Given that $u = v - gt$, find u when

  a $v = 16$, $g = -10$ and $t = 4$

  b $v = 13.2$, $g = 9.8$ and $t = 3.5$

4 Make Q the subject of the formula   **a** $P = \dfrac{Q}{10}$ **b** $P = R + 3Q$

5 **a** Write down the first four terms and the 20th term of the sequence for which the nth term is $5n - 3$.

 b Find, in terms of n, an expression for the nth term of the sequence 0, 3, 8, 15, 24, ...

6 Solve the equation $4 - \dfrac{3}{P} = 6$

7 Substitute $r = h + 2$ into the formula $A = r^2 - 2rh$ to give A in terms of h.

INVESTIGATION

Try this on a friend.
The four suits in a pack of cards are given the following values:
clubs 1, diamonds 2, hearts 3 and spades 4.
Ask your friend to think of one playing card from an ordinary pack of 52 (for example, the 6 of hearts).

a Tell your friend to add on the next highest card (i.e. $6 + 7 = 13$), and multiply the total by 5 ($13 \times 5 = 65$).

b Add on the suit value (in this case add 3 so $65 + 3 = 68$). Finally subtract 5 ($68 - 5 = 63$).

Now ask your friend to tell you the answer (63).
From this you can reveal the number on the original card by stating the number of tens (6) and the suit by assuming that the number of units is the suit number (3 gives hearts).
Your friend therefore chose the 6 of hearts.
Try it several more times. A Jack, Queen and King have values 11, 12 and 13.
Does it always work?

Now start with x as the chosen number and y as the suit number.
Use algebraic methods to prove that it will always work.

PRACTICAL WORK

At the beginning of the chapter we stated that there was a formula connecting a person's height and the circumference of their skull. Investigate the truth of this statement for the students in your group. Is the factor by which the circumference of the skull is multiplied to get the height the same for boys as it is for girls? What is the formula for a group of men or a group of women?

SIMULTANEOUS EQUATIONS

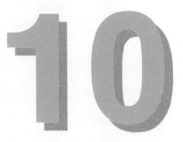

The Bursar has most, but not all, of the information he needs to carry out an audit of the equipment in Highfield School.

There are two types of desk in the school; those which seat two pupils and those which seat three pupils. The Bursar knows that there are 200 desks in total and that 550 pupils can be seated at these desks but not how many there are of each type.

- The Bursar can go round the school and count the numbers of each type of desk but this is likely to take an hour or longer.
- The Bursar can use algebraic methods to find the information he needs. If he knows what he is doing, this will take him about one minute.

EXERCISE 10A

There are several ways in which answers can be obtained to the following problems. Do not attempt to solve these problems but discuss at least two methods for solving them and the efficiency of each method, that is, how long it would take and how accurate an answer it would be likely to give.

1 Glasses suitable for hot drinks are sold either in packs of four with one 'free' glass holder at £4.50 a pack or in packs of six with one 'free' glass holder at £6.20 a pack. Assuming that the glasses are the same price, is the 'free' holder really free?

2 A 20 m long rod is cut into two pieces so that one piece is 2 m longer than half the other piece. How long is the shorter piece?

3 The seats in an aeroplane can be rearranged. The seats are available either in pairs or in groups of four. A total of 160 groups are to be used for 450 passengers. How many two-seat groups are needed?

SOLUTION OF SIMULTANEOUS EQUATIONS

Discussion arising from the last exercise shows that algebraic methods often give the quickest and most accurate answers to problems. In Book 8A we found that it may be possible to solve a problem that involves two unknown quantities by forming two equations and then solving them simultaneously; the method used was to eliminate one of the letters by addition or by subtraction of the given equations. This works when the *coefficient* (number) of one of the letters is the same in both equations. Revision Exercise 1.5 can be used to remind yourself of the method.

In this chapter we look at equations where one or both of the equations need to be adapted before adding or subtracting will eliminate one of the letters.

Consider again the Bursar's problem.

There are two unknown numbers here; the number of two-seat desks and the number of three-seat desks. The extra information given can be used to form two equations, i.e.

if there are x two-seat desks and y three-seat desks,

then $\qquad x + y = 200 \quad [1]$ There are 200 desks in total.

and $\qquad 2x + 3y = 550 \quad [2]$ x two-seat desks seat $2x$ pupils and y three-seat desks seat $3y$ pupils, and there are 550 pupils.

There are not the same number of xs or ys in these two equations so neither adding nor subtracting will make one of the letters disappear. However if we multiply equation [1] by 2 we get $2x + 2y = 400$ and we now have the same number of xs as equation [2]. Therefore this equation and equation [2] can be solved using the same method as in Book 8A.

The solution can be set out as follows.

$$x + y = 200 \qquad\qquad [1]$$

$$2x + 3y = 550 \qquad\qquad [2]$$

[1] × 2 gives $\qquad\qquad 2x + 2y = 400 \qquad\qquad [3]$

[2] − [3] gives $\qquad\qquad\qquad y = 150$

Substituting 150 for y in [1] gives $\quad x + 150 = 200$

so $\qquad\qquad\qquad\qquad x = 50$

Notice that we have numbered the equations and explained what we have done with them – this is essential so that other people can follow the argument.

EXERCISE 10B

Solve the equations $3x - 2y = 1$
$$4x + y = 5$$

$$3x - 2y = 1 \quad [1]$$
$$4x + y = 5 \quad [2]$$

Multiplying [2] by 2 gives $2y$ which is the same number of ys as in [1].

$[2] \times 2$ gives $\qquad 8x + 2y = 10 \quad [3]$
$$3x - 2y = 1 \quad [1]$$

$[1] + [3]$ gives $\qquad 11x = 11$

$\therefore \qquad\qquad\qquad\qquad x = 1$

Write [1] down again to bring the equations together for the next step.

Substitute 1 for x in [2] $\quad 4 + y = 5$

Take 4 from both sides $\qquad y = 1$

Check in [1] $\qquad$ LHS $= 3 - 2 = 1 =$ RHS

Therefore the solution is $x = 1, \ y = 1$

Solve the following pairs of equations.

1 $\quad 2x + y = 7$
$\quad\ \ 3x + 2y = 11$

2 $\ 5x - 4y = -3$
$\quad\ \ 3x + y = 5$

3 $\ 9x + 7y = 10$
$\quad\ \ 3x + y = 2$

4 $5x + 3y = 21$
$\quad\ 2x + y = 3$

5 $6x - 4y = -4$
$\quad\ 5x + 2y = 2$

6 $4x + 3y = 25$
$\quad\ x + 5y = 19$

Solve the equations $\quad 5x + 3y = 7$
$$10x + 4y = 16$$

$$5x + 3y = 7 \quad [1]$$
$$10x + 4y = 16 \quad [2]$$

$[1] \times 2$ gives $\ 10x + 6y = 14 \quad [3]$
$$10x + 4y = 16 \quad [2]$$

$[3] - [2]$ gives $\qquad 2y = -2$
$$y = -1$$

We will subtract [2] from [3] so we write [2] again underneath.

Substitute -1 for y in [1] $\ 5x - 3 = 7$

Add 3 to both sides $\qquad\qquad 5x = 10$
$$x = 2$$

Check in [2] $\qquad$ LHS $= 20 + (-4) = 16 =$ RHS

Therefore the solution is $x = 2, \ y = -1$

Solve the pairs of equations.

7 $5x + 3y = 11$
    $4x + 6y = 16$

10 $9x + 5y = 15$
    $3x - 2y = -6$

8 $2x - 3y = 1$
    $5x + 9y = 19$

11 $4x + 3y = 1$
    $16x - 5y = 21$

9 $2x + 5y = 1$
    $4x + 3y = 9$

12 $7p + 2q = 22$
    $3p + 4q = 11$

EXERCISE 10C

Sometimes we need to multiply both equations before we can add or subtract to eliminate a letter.

Solve the equations $3x + 5y = 6$
            $2x + 3y = 5$

$$3x + 5y = 6 \quad [1]$$
$$2x + 3y = 5 \quad [2]$$

We can get $15y$ in both equations from $[1] \times 3$ and $[2] \times 5$. (Alternatively we could get $6x$ in both equations from $[1] \times 2$ and $[2] \times 3$.)

$[1] \times 3$ gives $9x + 15y = 18 \quad [3]$
$[2] \times 5$ gives $10x + 15y = 25 \quad [4]$
            $9x + 15y = 18 \quad [3]$
$[4] - [3]$ gives $x = 7$

We will subtract $[3]$ from $[4]$ so we write $[3]$ again underneath $[4]$.

Substitute 7 for x in $[2]$ $14 + 3y = 5$
Take 14 from both sides $3y = -9$
Divide both sides by 3 $y = -3$

Check in $[1]$ LHS $= 21 - 15$
            $= 6 =$ RHS

Therefore the solution is $x = 7$, $y = -3$

Solve the following pairs of equations.

1 $2x + 3y = 12$
    $5x + 4y = 23$

5 $14x - 3y = -18$
    $6x + 2y = 12$

2 $3x - 2y = -7$
    $4x + 3y = 19$

6 $6x - 7y = 25$
    $7x + 6y = 15$

3 $2x - 5y = 1$
    $5x + 3y = 18$

7 $5x + 4y = 21$
    $3x + 6y = 27$

4 $6x + 5y = 9$
    $4x + 3y = 6$

8 $9x + 8y = 17$
    $2x - 6y = -4$

9 $9x - 2y = 14$
$7x + 3y = 20$

10 $5x + 4y = 11$
$2x + 3y = 3$

11 $4x + 5y = 26$
$5x + 4y = 28$

12 $2x - 6y = -6$
$5x + 4y = -15$

13 $5x - 6y = 6$
$2x + 9y = 10$

19 $5x + 3y = 8$
$3x + 5y = 8$

20 $7x + 2y = 23$
$3x - 5y = 4$

21 $6x - 5y = 17$
$5x + 4y = 6$

22 $3x + 8y = 56$
$5x - 6y = 16$

23 $7x + 3y = -9$
$2x + 5y = 14$

14 $3p + 4q = 5$
$2p + 10q = 18$

15 $6x + 5y = 8$
$3x + 4y = 1$

16 $7x - 3y = 20$
$2x + 4y = -4$

17 $10x + 3y = 12$
$3x + 5y = 20$

18 $6x - 5y = 4$
$4x + 2y = -8$

24 $7x + 6y = 0$
$5x - 8y = 43$

25 $2x + 6y = 30$
$3x + 10y = 49$

26 $4x - 3y = -7$
$3x + 2y = 16$

27 $17x - 2y = 47$
$5x - 3y = 9$

28 $8x + 3y = -17$
$7x - 4y = 5$

EXERCISE 10D

In this exercise some pairs of equations can be solved without having to multiply either equation, some require one equation to be multiplied by a number and others require both equations to be multiplied by numbers.

Solve the following pairs of equations.

1 $x + 2y = 9$
$2x - y = -2$

2 $x + y = 4$
$x + 2y = 9$

3 $2x + 3y = 0$
$3x + 2y = 5$

4 $3x + 2y = -5$
$3x - 4y = 1$

5 $x + y = 6$
$x - y = 1$

6 $3x - 5y = 13$
$2x + 5y = -8$

7 $3x - y = -10$
$4x - y = -4$

<u>**10**</u> $7x + 3y = 35$
$2x - 5y = 10$

8 $5x + 2y = 16$
$2x - 3y = -5$

<u>**11**</u> $9x + 2y = 8$
$7x + 3y = 12$

9 $2x - 5y = 1$
$3x + 4y = 13$

<u>**12**</u> $3x - 2y = -2$
$5x - y = -15$

Sometimes the equations are arranged in an awkward fashion and need to be rearranged before solving them.

Solve the equations $x = 4 - 3y$
$$2y - x = 6$$

$$x = 4 - 3y \qquad [1]$$
$$2y - x = 6 \qquad [2]$$

We must first arrange the letters in the same order in both equations. By adding $3y$ to both sides, equation [1] can be written $3y + x = 4$.

$[1]$ becomes $\qquad 3y + x = 4 \qquad [3]$
$\qquad\qquad\qquad 2y - x = 6 \qquad [2]$

$[3] + [2]$ gives $\qquad 5y = 10$
$\qquad\qquad\qquad\quad y = 2$

Substitute 2 for y in $[1]$ $\quad x = 4 - 6$
$\qquad\qquad\qquad\qquad\qquad x = -2$

Check in $[2]$ LHS $= 4 - (-2)$
$\qquad\qquad\qquad\quad = 6 = $ RHS

Therefore the solution is $x = -2, \ y = 2$

Solve the following pairs of equations.

13 $y = 6 - x$
$2x + y = 8$

<u>**16**</u> $9 + x = y$
$x + 2y = 12$

14 $x - y = 2$
$2y = x + 1$

<u>**17**</u> $2y = 16 - x$
$x - 2y = -8$

15 $3 = 2x + y$
$4x + 6 = 10y$

<u>**18**</u> $3x + 4y = 7$
$2x = 5 - 3y$

As long as the x and y and number terms are in corresponding positions in the two equations, they need not be in the order we have had so far.

Solve the equations $y = x + 5$
$\qquad\qquad\qquad\qquad y = 7 - x$

$$y = x + 5 \qquad\qquad [1]$$
$$y = 7 - x \qquad\qquad [2]$$

Rewrite [1] as $\quad y = 5 + x \qquad\qquad [3]$

[2] + [3] gives $\qquad 2y = 12$
$\qquad\qquad\qquad\qquad y = 6$

Substitute 6 for y in [1] $6 = x + 5$
$\qquad\qquad\qquad\qquad\qquad x = 1$

Check in [2]   LHS $= 6$
$\qquad\qquad\quad$ RHS $= 1 + 5 = 6 =$ LHS

Therefore the solution is $x = 1, \; y = 6$

Solve the equations.

19 $y = 9 + x$
$\quad\;\; y = 11 - x$

20 $\;\; x = 3 + y$
$\quad\;\; 2x = 4 - y$

21 $y = 4 - x$
$\quad\;\; y = x + 6$

22 $2y = 4 + x$
$\quad\;\;\; y = x + 8$

23 $x + 4 = y$
$\quad\;\;\; y = 10 - 2x$

24 $x + y = 12$
$\quad\;\;\; y = 3 + x$

SPECIAL CASES   Some pairs of equations have no solution and some have an infinite number of solutions.

EXERCISE 10E   Try solving the following pairs of equations. Comment on why the method breaks down.

1 $x + 2y = 6$
$\quad x + 2y = 7$

2 $3x + 4y = 1$
$\quad 6x + 8y = 2$

3 $y = 4 + 2x$
$\quad y - 2x = 6$

4 $9x = 3 - 6y$
$\quad 3x + 2y = 1$

5 Make up other pairs of equations which either have no solution or have an infinite set of solutions.

SOLVING
PROBLEMS

Answers to problems involving two unknown quantities can often be found by forming a pair of simultaneous equations and solving them. In any problem, it is important to

- read the problem carefully and make sure you understand it
- identify the unknown quantities
- allocate letters to the unknown numbers when this is not done for you
- look for information that can be used to give relationships between the unknown numbers
- draw a diagram when the problem is about a geometric figure; this will often help to identify information that is not spelt out in the problem
- check that the answer obtained fits the information given.

EXERCISE 10F

In a right-angled triangle, the two smaller angles are such that one is twice the size of the other. Find the smallest angle.

First we draw a right-angled triangle. The size of the two smaller angles is unknown, so we let one be $x°$ and the other be $y°$. There is no need to write this down, it is sufficient to show the letters clearly in the diagram.

$$y = 2x \quad [1]$$

This information is given in the question.

$$y + x = 90 \quad [2]$$

We know that the sum of the angles in a triangle is 180°.

[1] becomes $y - 2x = 0 \quad [3]$
[2] − [3] gives $3x = 90$
$$x = 30$$

The smallest angle is 30°.

Check: using the information in the problem, if the smallest angle is 30° this gives the other angles as 60° and 90° and they all add up to 180°.

1 The perimeter of a rectangle is 30 cm. The shorter side is x cm long and is 3 cm less than the longer side which is y cm long. Use the information to form two equations in x and y. Find the length of the shorter side.

2 In a pentagon, four of the angles are each $x°$ and the fifth is $y°$ and is half the size of each of the others. Form two equations in x and y. Find the size of the largest angles.

3 The perimeter of an isosceles triangle is 280 mm. The base of the triangle is 70 mm longer than the other two sides. Find the length of one of the equal sides.

A shop makes a profit of 3 pence on each brown roll sold and a profit of 4 pence on each white roll sold. One Saturday the shop sells 180 rolls and makes a profit of £6.20 on these rolls. How many brown rolls were sold?

If x brown rolls and y white rolls were sold

then $3x + 4y = 620$ [1]

> 1 brown roll gives a profit of 3 p, so x brown rolls gives a profit of $3x$ p. Likewise, y white rolls gives a profit of $4y$ p. The profit on the rolls sold is £6.20, i.e. 620 p.

and $x + y = 180$ [2]

> 180 rolls are sold.

[2] × 3 gives $3x + 3y = 540$ [3]

$$3x + 4y = 620 \quad [1]$$
$$3x + 3y = 540 \quad [3]$$

[1] − [3] gives $y = 80$

Substituting 80 for y in [2] gives $x + 80 = 180$
$$x = 100$$

100 brown rolls are sold.

Check: profit on 100 brown rolls is £3, profit on 80 white rolls is £3.20, so total profit is £6.20.

---

4 When three times a number, x, is added to a second number, y, the total is 33. The first number added to three times the second number is 19.

   a Form two equations connecting x and y.

   b Find the two numbers.

5 Find two numbers such that twice the first added to the second is 26 and the first added to three times the second is 28.

6 Find two numbers such that twice the first added to the second gives 27 and twice the second added to the first gives 21.

7 A cutlery manufacturer packs 200 teaspoons into boxes holding either four spoons or six spoons. There are 48 full boxes.

    a If there are a four-spoon boxes and b six-spoon boxes, use the information to find two equations relating a and b.

    b How many boxes hold four spoons?

8 The manager of a bookshop ordered 70 copies of *Teen* magazine; s of these were charged at 95 p each and the remainder, t, were charged at a discounted price of 85 p each.
The bill for all 70 copies was £62.
How many magazines were charged at the discounted price?

9 Sara often goes swimming. She can either pay £2.50 a visit which includes the use of a locker or she can buy a season ticket which costs £50 on top of which she has to pay 50 p for the use of a locker on each visit.
Assume that Sara uses a locker on each visit.
If £C is the cost of n visits

    a find an equation relating C and n if Sara does not buy a season ticket

    b find another equation relating C and n if Sara does buy a season ticket.

    c Solve the two equations simultaneously and hence give the number of visits that Sara needs to make for it to be worth her while buying a season ticket.

10 Airdale Study Centre can accommodate 128 students in 20 rooms, x of which sleep 8 and y of which sleep 4. How many four-bed rooms are there?

11 The charge at a club for renting a tennis court for one hour is £3 for non-members and £1 for members, who also pay an annual fee of £50.

    a Write down an equation for C in terms of n where £C is the cost for a non-member to hire the court for n hours.

    b Write down another equation for C in terms of n where £C is the total cost for a member to hire the court for n hours in one year.

    c Hence find the number of hours a year a court has to be hired for it to be worth paying the membership fee.

12 The equation of a line is $y = mx + c$. The line goes through the points $(1, 3)$ and $(2, 5)$. Find the equation of this line.

Questions **13** to **15** were used for discussion in **Exercise 10A**; now solve them.

13 A 20 m long rod is cut into two pieces so that one piece is 2 m longer than half the other piece. How long is the shorter piece?

14 Glasses suitable for hot drinks are sold either in packs of four with one 'free' glass holder at £4.50 a pack or in packs of six with one 'free' glass holder at £6.20 a pack. Assuming that the glasses are the same price, is the 'free' holder really free?

15 The seats in an aeroplane can be arranged either in pairs or in groups of four. A total of 160 groups are to be used for 450 passengers. How many two-seat groups are needed?

GRAPHICAL SOLUTIONS

In Book 8A we saw that simultaneous equations can be solved graphically. For example, to solve the equations $x + 3y = 6$ [1]
and $3x - y = 6$ [2]
we start by rearranging each equation so that they are in the form $y = \ldots$, that is, we make y the subject of the equation.
[1] becomes $3y = -x + 6$
giving $y = -\frac{1}{3}x + 2$ [3]

[2] becomes $3x = y + 6$
giving $3x - 6 = y$, i.e. $y = 3x - 6$ [4]

Now we can plot the lines whose equations are $y = -\frac{1}{3}x + 2$ and $y = 3x - 6$. This can be done by drawing up a table of values and plotting the points on graph paper, or by using a graphics calculator.

Using the values $0 \leqslant x \leqslant 5$ and $0 \leqslant y \leqslant 5$ gives the following display on a calculator.

Using the TRACE function we move the cursor to a position that is as near as possible to the point where the lines intersect; at this position we see that $x = 2.42$ and $y = 1.19$ (correct to 2 d.p.)

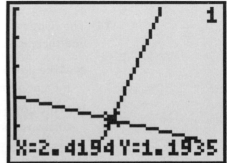

We cannot be certain that the cursor is exactly on the point where the lines cross, so these values are therefore *approximate* solutions.

Better approximations can be obtained by using the ZOOM function.

1 In the text before this exercise we used a graphical method to solve
the equations $x + 3y = 6$
and $\qquad 3x - y = 6$
Solve these equations algebraically.
Comment on the advantages and disadvantages of each method of
solution.

Solve the following equations graphically, either by plotting the graphs
on graph paper or by using a graphics calculator.
In either case use values for x and y in the ranges given.
Check your answers by solving the equations algebraically.

2 $3x + 2y = 9$ $\;0 \leqslant x \leqslant 4,\; -2 \leqslant y \leqslant 5$
$2x - 2y = 3$

3 $2x + 3y = 4$ $\;-2 \leqslant x \leqslant 2,\; 0 \leqslant y \leqslant 4$
$y = x + 2$

4 $x = 2y - 3$ $\;-2 \leqslant x \leqslant 3,\; 0 \leqslant y \leqslant 4$
$y = 2x + 1$

> The examples in this exercise show that, when using
> graphical methods, we cannot always tell whether
> solutions obtained are exact or not.

You will need a graphics calculator.
Investigate graphically the solutions of the following equations.

a $x + y = 9$ $\;0 \leqslant x \leqslant 9$
$x + y = 4$ $\;0 \leqslant y \leqslant 9$

b $y = 2x + 3$ $\qquad 0 \leqslant x \leqslant 4$
$y = 2x - 1$ $\;-1 \leqslant y \leqslant 11$

c $2x + y = 3$ $\qquad 0 \leqslant x \leqslant 3$
$4x + 2y = 7$ $\;-3 \leqslant y \leqslant 4$

d $y = 2x - 4$ $\qquad 0 \leqslant x \leqslant 4$
$2x = y + 4$ $\;-4 \leqslant y \leqslant 4$

Give reasons for what you find.

SUMMARY 3

The number, including the sign, that is multiplied by a letter is called the *coefficient* of that letter. In the expression $3x - 4y$, for example, the coefficient of x is 3 and the coefficient of y is -4.

Product of two brackets

$(a + b)(c + d)$ means $a \times (c + d) + b \times (c + d)$
$$= ac + ad + bc + bd,$$

that is, each term in the second bracket is multiplied by each term in the first bracket. The order in which the terms are multiplied does not matter, but it is sensible to stick to the same order each time,

e.g.

$$(2x - 3)(4x + 5) = (2x)(4x) + (2x)(5) + (-3)(4x) + (-3)(5)$$
$$= 8x^2 + 10x - 12x - 15 = 8x^2 - 2x - 15$$

In particular, when squaring a bracket we can use

$$(x + a)^2 = (x + a)(x + a) = x^2 + 2ax + a^2$$
or $(x - a)^2 = (x - a)(x - a) = x^2 - 2ax + a^2$

The product of two brackets that are the same except for the sign between the two terms is called the **difference between two squares**.

i.e. $(x + a)(x - a) = x^2 - a^2$

Factorising

Factorising is the reverse of the process of expanding (multiplying out) an algebraic expression.

A *common factor* of two or more terms can be seen by inspection and 'taken outside a bracket',

e.g. the terms $2ab + 4bc$ both have $2b$ as factors,

so $2ab + 4bc = 2b(a + 2c)$. (This can be checked by expanding the result.)

To factorise an expression such as $x^2 + 3x - 10$, we look for two brackets whose product is equal to the original expression.
We start by writing $x^2 + 3x - 10 = (x + \ )(x - \ )$.
The sign in each bracket is determined by the signs in the original expression, i.e.

a +ve number term *and* a +ve x-term gives a '+' sign in both brackets
a +ve number term *and* a $-$ve x-term gives a '$-$' sign in both brackets
a $-$ve number term gives '+' in one bracket and '$-$' in the other bracket.

The numbers at the ends of the brackets have to satisfy two conditions:
their product has to be equal to the number at the end of the original
expression,
collecting the product of the outside terms in the brackets and the inside
terms in the brackets must give the x term in the original expression.

In the case of $x^2 + 3x - 10$, the product is 10 so the numbers could be
10 and 1 or 5 and 2.
We try each pair in turn until we find a pair (if there is one) which give
the correct x term when the brackets are expanded;
$$x^2 + 3x - 10 = (x + 5)(x - 2)$$
If we cannot find two numbers that satisfy the conditions, the expression
does not factorise.

Inequalities in two unknowns

An inequality in two unknowns
can be represented by a
region of the x–y plane.
For example, the
unshaded area in the diagram
represents the inequality
$y \leqslant -1.5x + 6$
The boundary line is
$y = -1.5x + 6$ and is
included in the region; this
is indicated in the diagram
by a solid line.

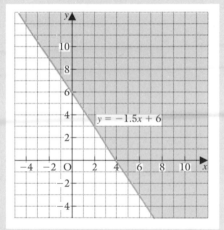

In this diagram, the
unshaded region contai:
the points whose
coordinates satisfy the
three inequalities
$y \leqslant -1.5x + 6$
$y \geqslant 1$
$x > -4$
The broken line shows
that points on the line
$x = -4$ are not included
in the region.

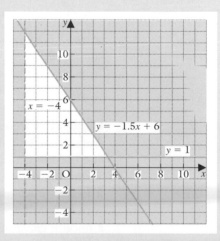

Formulas

The formula $v = u + st$ gives v in terms of u, s and t; v is called the *subject of the formula*.

When the formula is rearranged to give $s = \dfrac{v - u}{t}$, s is the subject.

The process of rearranging $v = u + st$ as $s = \dfrac{v - u}{t}$ is called *changing the subject of the formula*. It is achieved by thinking of $v = u + st$ as an equation which has to be 'solved' to find s.

Start by isolating the term containing s on one side of the formula:

take u from both sides: $v - u = st$

divide both sides by t: $\dfrac{v - u}{t} = s$, i.e. $s = \dfrac{v - u}{t}$

The nth term of a sequence is sometimes expressed in terms of n, the position number of the term. Any term of the sequence can then be found by giving n a numerical value.

For example, when the nth term $= 3n - 2$,

the 10th term is given by substituting 10 for n, i.e. by $3(10) - 2 = 28$

Simultaneous equations

A pair of simultaneous equations in two unknowns can be solved algebraically by eliminating one of the letters. It may be necessary to multiply one or both equations by numbers to make the coefficients of one of the letters the same in each equation.

For example, to solve $2y - x = 7$ [1]

and $3y + 4x = 5$ [2]

$[1] \times 4$ gives $\qquad 8y - 4x = 28$ [3]

then $[2] + [3]$ eliminates x to give $11y = 33$

so $y = 3$ and, from [1], $x = -1$

STATISTICS

Mean value

The mean value of a frequency distribution is given by $\dfrac{\Sigma fx}{\Sigma f}$

where x is the value of an item and f is its frequency, and Σ means 'the sum of all such items'.

In the case of grouped data, x is the midclass value and the mean obtained is an estimate.

Cumulative frequency

Cumulative frequency is the sum of the frequencies of all values up to and including a particular value.

A *cumulative frequency polygon* is drawn by plotting the cumulative frequencies against the upper ends of the groups and joining the points with straight lines. A *cumulative frequency curve* is obtained by drawing a smooth curve through the points.

For example, the grouped frequency distribution given in the table

Heights of tomato plants

Height, h cm	$20 \leqslant h < 30$	$30 \leqslant h < 40$	$40 \leqslant h < 50$	$50 \leqslant h < 60$
Cumulative frequency	5	20	45	55

gives this cumulative frequency curve.

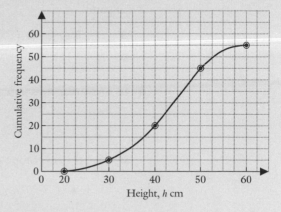

The *median* of a grouped distribution of n values is the $\frac{n}{2}$th value in order of size and is denoted by Q_2.

The median can be estimated from a cumulative frequency curve.

The *lower quartile* is the value that is $\frac{1}{4}$ of the way through a set of values arranged in order of size and is denoted by Q_1.

The *upper quartile* is the value that is $\frac{3}{4}$ of the way through a set of values arranged in order of size and is denoted by Q_3.

For a grouped distribution, Q_1 is the $\frac{n}{4}$th value, and Q_3 is the $\frac{3n}{4}$th value.

Q_1 and Q_3 can be estimated from a cumulative frequency curve.

Interquartile range

The interquartile range is the difference between the upper and lower quartiles,

i.e. $Q_3 - Q_1$

The diagram shows the median, the upper and lower quartiles and the interquartile range of the distribution described above.

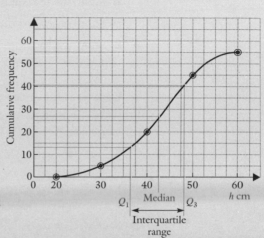

1 Expand

    a $3(1-x)$ **b** $(a+b)(a+2b)$ **c** $(x-3)(x-6)$

2 Expand

    a $(x+7)(x-3)$ **b** $(x-4)(3x+5)$ **c** $(3x-2)(5x+1)$

3 Expand

    a $(x+5)^2$ **b** $(2x+3y)(2x-3y)$
    c $(x+1)(x+2)+x(x+3)$

4 a Draw diagrams to represent the regions described by the following inequalities. Use axes for values of x and y from -5 to 5.

    i $2 \leqslant x \leqslant 4,\ 1 \leqslant y \leqslant 3$ **ii** $1 \leqslant x < 3,\ 2 < y < 3$

  b Give the sets of inequalities that describe the unshaded regions.

    i

    ii

5 Draw three sets of axes for values of x and y from -6 to 6 and use one set for each part.

  a Show shaded the region defined by the inequality $x+y > 2$.

  b Show unshaded the region defined by the inequality $y \leqslant 3x-2$

  c Show unshaded the region defined by the inequalities
    $y < 3x,\ y \geqslant 4-x$

6 Draw axes for values of x and y from -6 to 6. Draw suitable straight lines and leave unshaded the region defined by the set of inequalities

    $x+y < 5,\ x \geqslant 1,\ y > 2x$

7 Factorise **a** $4a+12$ **b** x^2-5x **c** b^3-3b^2

8 Factorise **a** $x^2+11x+18$ **b** x^2-6x+8 **c** $x^2-10x+25$

9 Factorise

a $x^2 - 2x - 15$

c $18 + 9x + x^2$

b $x^2 + 2x - 8$

d $14 + 5x - x^2$

10 a Factorise **i** $a^2 - b^2$ **ii** $p^2 - 4q^2$ **iii** $16x^2 - y^2$

b Find, without using a calculator **i** $67^2 - 33^2$ **ii** $7.31^2 - 2.69^2$

REVISION EXERCISE 3.2 (Chapters 8 to 10)

1 Forty boxes of oranges were examined and the number of bad oranges in each box was noted.

Number of bad oranges	0–5	6–11	12–17	18–23
Frequency	25	8	5	2

a What is the modal group for this distribution?

b Estimate the mean number of bad oranges per box.

2 Use the frequency table from question **1** to make a cumulative frequency table for this data.

Draw the cumulative frequency curve and use it to find the median number of bad oranges per box, the upper and lower quartiles and the interquartile range.

3 The masses of a group of boys were measured and the results recorded in the following table.

Mass, m kg	60–64	64–69	69–74	74–79	79–84	84–89
Frequency	3	7	13	15	11	6

a How many boys were there in the group?

b How many boys had a mass of more than 79 kg?

c Use the data to draw a cumulative frequency curve and use your curve to find

i the median mass   **ii** the upper and lower quartiles.

4 Write down a formula connecting the given letters.

a A number x is equal to three times the sum of two numbers p and q.

b A number z is twice the product of two numbers x and y.

c A number z is the sum of x and y minus their product.

5 Given that $a = b^2 + c$ find a when

   a $b = 2$ and $c = 3$

   b $b = -1$ and $c = 4$

   c $b = -3$ and $c = -2$

6 a Write down the first four terms and the tenth term of the sequence for which the nth term is $3n - 2$.

   b Find, in terms of n, an expression for the nth term of the sequence $4, 8, 12, 16, \ldots$

7 a Solve the equation $4 = \dfrac{12}{x}$

   b Given that $R = 3st$ and that $s = 5t$ find R in terms of t.

8 Solve the simultaneous equations.

   a $7x + y = 25$ **b** $7x + 5y = 30$

      $2x + y = 5$ $4x - y = 21$

9 Solve the simultaneous equations.

   a $4x - 3y = 11$ **b** $3x + 7y = 27$

      $6x + 4y = 8$ $5x + 2y = 16$

10 a Two numbers, x and y, are such that three times the first number added to twice the second number is 24 and the first added to twice the second is 23.

     i Form two equations relating x and y.

     ii Solve the equations to find the two numbers.

   b Solve the following simultaneous equations graphically, either by plotting the graphs on graph paper or by using a graphics calculator. In either case use values for x and y in the ranges given.

$$3x + 2y = 11 \qquad 2 \leqslant x \leqslant 4, \ -1 \leqslant y \leqslant 5$$
$$5x - 2y = 10$$

REVISION EXERCISE 3.3
(Chapters 5 to 10)

1 Expand

   a $5(2-3a)$ **c** $(p-3)(p-8)$

   b $(3x+2y)(1-2z)$ **d** $(3x+4)(3-x)$

2 Expand and simplify

   a $(3x-1)^2$ **b** $(x+2)(x+3)+4(x+1)$

3 Give the set of inequalities that defines the unshaded region.

a
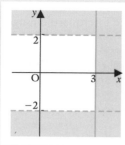

b
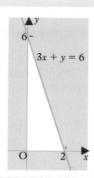

c
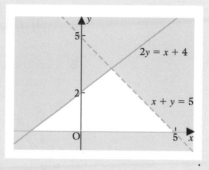

4 Copy the diagram onto squared paper and shade the region defined by the inequalities $y<x+4$, $y\leqslant 4-3x$ and $2y+x>0$

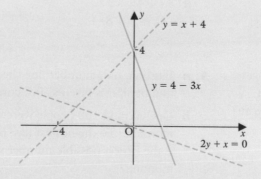

5 Factorise

   a $15-5x$ **c** $x^2-10x+21$

   b $x^2-2x-15$ **d** $2x^2+4x+8$

6 Factorise

   a $12-7x+x^2$ **b** $12+x-x^2$

   c $9-a^2$ **d** $25-x^2$

7 The lengths of 35 leaves taken from the same tree are given in the table.

Length, h cm	$8 \leqslant h < 8.5$	$8.5 \leqslant h < 9$	$9 \leqslant h < 9.5$	$9.5 \leqslant h < 10$
Frequency	3	10	18	4

a Estimate the mean length of the leaves.

b Construct a cumulative frequency table and draw a cumulative frequency curve.

c Estimate
  i the median length of these leaves
  ii the upper and lower quartiles, and the interquartile range.

8 Grapefruit are sold at x pence each. The cost of n grapefruit is C pence. Find a formula for C in terms of x and n.

9 a If $R = (p + 2q)^2$ find R when
  i $p = 10$ and $q = -3$ **ii** $p = 2.1$ and $q = 3.7$

b Given that $V = 2xy^2 + \dfrac{y^4}{x}$ and that $y = 2x$, find V in terms of x.

10 Solve the simultaneous equations.

a $3x + 4y = 17$
  $2x + y = 8$

b $x = 5 + 3y$
  $2y = 3x - 8$

REVISION EXERCISE 3.4 (Chapters 1 to 10)

1 a Illustrate on a number line the range of possible values of a number if it is given as 0.7 correct to 1 decimal place.

b One evening 100 people were asked if they had drunk coffee that morning. To the nearest ten, 60 people said that they had. What is the largest possible number who had not drunk coffee that morning?

c Find $\left(2 - \dfrac{1}{1 + \frac{1}{3}}\right) \div \left(4 + \dfrac{1}{2 + \frac{3}{4}}\right)$

2 Two dice are rolled one after the other. Find the probability of getting

a a 3 on the first dice

b a double 3

c a double 3 or a double 4.

3 After a pay rise of 3% Sheila's weekly wage is £226.60. How much did she earn before the rise?

4 a A page of text in a book is 42 lines with an average of 16 words per line. It is rewritten with an average of 14 words per line. How many lines will be needed?

b In a mixed choir of 156 there are 72 men. What is the ratio of the number of men to the number of women?

c The cost of a 25 kg bag of potatoes is £5.75. At the same price per kilogram how much will a bag containing 12 kg cost?

5 Expand   **a** $5a(2b+3c)$ **c** $(7x+3)(x+1)$

                  b $(1-a)(a+3b)$ **d** $(x-4y)^2$

6 Draw x- and y-axes on squared paper for $-3 \leqslant x \leqslant 5$ and $0 \leqslant y \leqslant 5$. Shade the region defined by the inequalities

$$x > -2, \qquad y > 1, \qquad x+y \leqslant 3, \qquad y < x+4$$

7 Factorise   **a** $x^2 + 6x$ **c** $8 + x^2 - 6x$

                     b $x^2 - 10x + 24$ **d** $p^2 - 36$

8 The bar chart shows the number of flowers on a new variety of plant grown from the seeds in one packet.

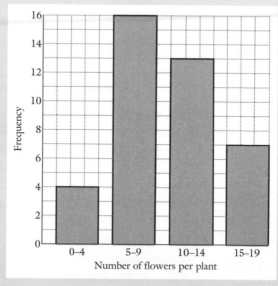

a Make a frequency table and estimate the mean number of flowers per plant.

b Make a cumulative frequency table for this data and draw the cumulative frequency curve. Hence find estimates for the median, the upper and lower quartiles and the interquartile range.

9 a Make T the subject of the formula $I = \dfrac{PRT}{100}$. Find T when
$I = 28$, $P = 350$ and $R = 4$.

b Make s the subject of the formula $v^2 = u^2 + 2as$.

10 Solve the simultaneous equations.

a $6x - 7y = 31$
$7x + 6y = 22$

b $y = x + 6$
$2y = 15 + x$

**REVISION
EXERCISE 3.5
(Chapters 1 to 10)**

1 a Write down the reciprocal of **i** 20 **ii** $\frac{1}{8}$ **iii** 1.8

b Use dot notation to write the following fractions as decimals
i $\frac{9}{11}$ **ii** $\frac{11}{12}$ **iii** $\frac{3}{13}$

c Express as a fraction
i $0.0\dot{5}$ **ii** $0.00\dot{5}$ **iii** $0.0\dot{8}$ **iv** $0.0\dot{6}$

d Simplify $\left(\frac{3}{7} \times \frac{5}{9}\right) \div \left(5\frac{1}{4} - 4\frac{4}{15}\right)$

2 The probability that the postman calls before I leave for school in the morning is $\frac{5}{6}$. He never calls at the moment I am leaving. Find the probability that

a the postman calls after I have left for school

b next Monday and Tuesday the postman will call before I leave.

3 The selling price of a patio set is £177. This includes VAT at 18%. Find the price excluding VAT.

4 a Express $3 : 7$ in the form $1 : n$.

b A scale model of a car is such that the model car is 6.5 cm long and the actual car is 4.68 m long. The height of the car is 1.44 m. How high is the model?

5 Expand **a** $3x(5y - 2z)$ **c** $(5x + 3y)(z + 4)$

b $(a + 4)(a + 6)$ **d** $(a + 3b)^2$

6 Use inequalities to describe the unshaded region in this diagram.

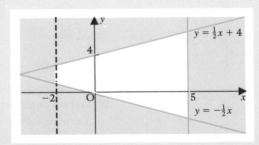

7 Factorise
    a $3t - 6t^2$
    c $x^2 - 4x - 12$

    b $3x^2 + 9x - 12$
    d $x^2 + 8x + 16$

8 The duration of the telephone calls made from an office on one particular day were recorded and the frequency polygon summarises the results.

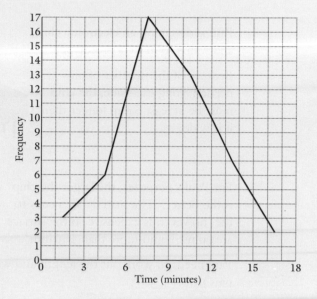

    a How many telephone calls were made?

    b Estimate the range of the times taken.

    c Copy and complete the table.

Middle value, t minutes	1.5					
Frequency						

    d Estimate the mean duration of a telephone call that day.

9 Make the letter in brackets the subject of the formula.

    a $z = x - 2y$ (x)
    b $p = q + 2r$ (r)
    c $a = 3b - c$ (b)

10 The perimeter of a triangle is 172 mm. The shortest side is 50 mm shorter than each of the other two sides, which are equal. If the length of the shortest side is x mm and the length of each of the longer sides is y mm, form two equations in x and y and solve them simultaneously to find the lengths of the three sides.

QUADRATIC EQUATIONS

11

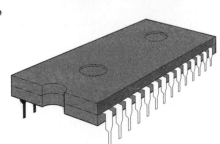

Paul is designing an electronic chip and wants to make a prototype.

Because of the number of pins needed along each side, the chip must be at least 5 mm longer than it is wide.

For Paul's purposes, the larger the chip is the better. The material he needs to use, however, is expensive so the surface area must be no more than $150\,\text{mm}^2$.

Paul needs to work out the length and width of the chip (excluding the pins).

- He can try various values for the width until he finds one that fits the constraints. As we saw in Book 8A, this is a 'hit and miss' affair; it can take ages to find a suitable value. A more organised search can be made by forming an equation and using trial and improvement methods.
- Trial and improvement methods give solutions as accurately as needed. The disadvantage is that they involve a lot of arithmetic which is tedious without the help of, for example, a spreadsheet.

There are, however, quicker methods for solving some equations and we look at one of these in this chapter.

EXERCISE 11A

1 Amy needs to find a value for x that satisfies the equation

$$x^2 - 6x = 17.01$$

Discuss the advantages and disadvantages of these methods.

a Trial and improvement

b Drawing a graph

c Using a graphics calculator.

2 There are two values of x that satisfy the equation
$$x^2 - 6x = 17.01$$

  a Discuss which of the methods in question **1** will find both values of x.

  b Discuss whether any of the methods in question **1** can tell you whether the solutions are exact values or whether they can only be given correct to a number of significant figures.

QUADRATIC
EQUATIONS

Discussion from the last exercise shows that there are disadvantages in using numerical and graphical methods for solving equations. One disadvantage is the time it can take to find a solution so, when it is possible, using algebraic methods is more efficient.

We can always solve equations such as $2x - 3.5 = 5x + 8.5$. Equations like this, that do not involve powers of the unknown letter, are called *linear equations*.

Equations such as $x^2 - 6x = 17.01$, that have a term involving the square of the unknown letter, are called *quadratic equations*.

In this chapter we will solve some straightforward quadratic equations. The method we use depends on the product of two numbers being zero.

EXERCISE 11B

Find the value of $(x + 3)(x - 7)$ if

  a $x = 8$ **b** $x = 7$ **c** $x = -3$

  a If $x = 8$, $(x + 3)(x - 7) = (8 + 3)(8 - 7)$
  $$= (11)(1)$$
  $$= 11$$

  b If $x = 7$, $(x + 3)(x - 7) = (10)(0)$
  $$= 0$$

  c If $x = -3$, $(x + 3)(x - 7) = (-3 + 3)(-10)$
  $$= (0)(-10) = 0$$

1 Find the value of $(x - 4)(x - 2)$ if

  a $x = 6$ **b** $x = 4$ **c** $x = 2$

2 Find the value of $(x - 5)(x - 9)$ if

  a $x = 5$ **b** $x = 10$ **c** $x = 9$

3 Find the value of $(x-7)(x-1)$ if

   a $x=1$ **b** $x=8$ **c** $x=7$

4 Find the value of $(x-3)(x+5)$ if

   a $x=6$ **b** $x=3$ **c** $x=-5$

5 Find the value of $(x-4)(x+6)$ if

   a $x=0$ **b** $x=-6$ **c** $x=4$

6 Find the value of $(x+4)(x+5)$ if

   a $x=4$ **b** $x=-5$ **c** $x=0$

When we consider the product of two factors, the questions above show that if one or both of them are zero so is the product.
Further, if neither factor is zero, the product is *not* zero.

In general we can say

> if $A \times B = 0$
>
> then either $A = 0$ or/and $B = 0$

In questions **7** to **12** find, if possible, the value or values of A.
Note that if $A \times 0 = 0$ then A can have any value.

7 $A \times 6 = 0$ **10** $0 \times A = 0$

8 $A \times 7 = 0$ **11** $3 \times A = 12$

9 $A \times 4 = 0$ **12** $8 \times A = 8$

<u>13</u> If $AB = 0$ find   **a** A if $B = 2$ **b** B if $A = 10$

<u>14</u> If $AB = 0$ find   **a** A if $B = 5$ **b** B if $A = 5$

Find a and b if $a(b-3) = 0$

Either $a = 0$ or/and $b - 3 = 0$
i.e. either $a = 0$ or/and $b = 3$

Find a and b if

15 $a(b-1) = 0$ **18** $a(b-4) = 0$

16 $a(b-5) = 0$ **19** $(a-3)b = 0$

17 $a(b-2) = 0$ **20** $(a-9)b = 0$

SOLVING
QUADRATIC
EQUATIONS

Consider the equation $(x-1)(x-2)=0$

This is true either when $x-1=0$ or when $x-2=0$,

i.e.   either when $x=1$ or when $x=2$

There are, therefore, two values of x that satisfy the equation

$$(x-1)(x-2)=0$$

Expanding the left-hand side gives $x^2-3x+2=0$

This is a quadratic equation and we have solved it.

Any quadratic equation written in the form

$$(x-a)(x-b)=0$$

can be solved because we know that

either $x-a=0$ and/or $x-b=0$,

i.e. $x=a$ and/or $x=b$

EXERCISE 11C

What values of x satisfy the equation $x(x-9)=0$?

$$x(x-9)=0$$

Therefore either $x=0$ or $x-9=0$

i.e.   either $x=0$ or $x=9$

What values of x satisfy the following equations?

1 $x(x-3)=0$ **6** $x(x-6)=0$

2 $x(x-5)=0$ **7** $x(x-10)=0$

3 $(x-3)x=0$ **8** $(x-7)x=0$

4 $x(x+4)=0$ **9** $x(x+7)=0$

5 $(x+5)x=0$ **10** $(x+9)x=0$

What values of x satisfy the equation $(x-3)(x+5)=0$?

$$(x-3)(x+5)=0$$

Either $\qquad x-3=0 \qquad$ or $\qquad x+5=0$

i.e. either $\qquad x=3 \qquad$ or $\qquad x=-5$

What values of x satisfy the following equations?

11 $(x-1)(x-2)=0$ **16** $(x-8)(x+11)=0$

12 $(x-5)(x-9)=0$ **17** $(x-3)(x+5)=0$

13 $(x-10)(x-7)=0$ **18** $(x+7)(x-2)=0$

14 $(x-4)(x-7)=0$ **19** $(x+2)(x+3)=0$

15 $(x-6)(x-1)=0$ **20** $(x+4)(x+9)=0$

21 $(x+1)(x+8)=0$ **26** $(x+6)(x+7)=0$

22 $(x-p)(x-q)=0$ **27** $(x+10)(x+11)=0$

23 $(x+a)(x+b)=0$ **28** $(x-a)(x-b)=0$

24 $(x-4)(x+1)=0$ **29** $(x+a)(x-b)=0$

25 $(x+9)(x-8)=0$ **30** $(x-c)(x+d)=0$

31 $(2x-3)(2x+1)=0$ **32** $(4x+7)(3x-1)=0$

SOLUTION BY FACTORISATION

Quadratic equations do not often come in the form

$$(x-4)(x-1)=0.$$

They are more likely to arise in the form $x^2-5x+4=0$.

The previous exercise shows that if the left-hand side of such a quadratic equation can be expressed as two linear factors, we can use these factors to solve the equation.

e.g. $\qquad x^2-5x+4=0$

gives $\quad (x-4)(x-1)=0$

from which we see that either $x-4=0$ or $x-1=0$

i.e. $\qquad x=4 \qquad$ or $\qquad x=1$

EXERCISE 11D

> Solve the equation $x^2 - 10x + 9 = 0$
>
> If $\qquad\qquad\qquad\qquad x^2 - 10x + 9 = 0$
>
> then $\qquad\qquad\qquad\quad (x-1)(x-9) = 0$
>
> $\therefore$ either $\qquad x - 1 = 0 \qquad$ or $\qquad x - 9 = 0$
>
> i.e. $\;x = 1 \quad$ or $\quad 9$
>
> *Check*: when $x = 1$, LHS $= 1 - 10 + 9 = 0 =$ RHS
>
> $\qquad\qquad$ when $x = 9$, LHS $= 81 - 90 + 9 = 0 =$ RHS

Solve the equations.

1 $\;x^2 - 3x + 2 = 0$ $\qquad\qquad$ **6** $\;x^2 - 6x + 5 = 0$

2 $\;x^2 - 8x + 7 = 0$ $\qquad\qquad$ **7** $\;x^2 - 12x + 11 = 0$

3 $\;x^2 - 5x + 6 = 0$ $\qquad\qquad$ **8** $\;x^2 - 6x + 8 = 0$

4 $\;x^2 - 7x + 10 = 0$ $\qquad\qquad$ **9** $\;x^2 - 8x + 12 = 0$

5 $\;x^2 - 7x + 12 = 0$ $\qquad\qquad$ **10** $\;x^2 - 13x + 12 = 0$

> Solve the equation $x^2 + 2x - 8 = 0$
>
> $$x^2 + 2x - 8 = 0$$
> $$(x+4)(x-2) = 0$$
>
> $\therefore$ either $\qquad x + 4 = 0 \qquad$ or $\qquad x - 2 = 0$
>
> i.e. $\;x = -4 \quad$ or $\quad 2$
>
> *Check*: when $x = -4$, LHS $= 16 - 8 - 8 = 0 =$ RHS
>
> $\qquad\qquad$ when $x = 2$, LHS $= 4 + 4 - 8 = 0 =$ RHS

Solve the equations.

11 $\;x^2 + 6x - 7 = 0$ $\qquad\qquad$ **16** $\;x^2 - 12x - 13 = 0$

12 $\;x^2 - 2x - 8 = 0$ $\qquad\qquad$ **17** $\;x^2 + x - 6 = 0$

13 $\;x^2 - x - 12 = 0$ $\qquad\qquad$ **18** $\;x^2 - 4x - 12 = 0$

14 $\;x^2 - 2x - 15 = 0$ $\qquad\qquad$ **19** $\;x^2 + x - 20 = 0$

15 $\;x^2 + 7x - 18 = 0$ $\qquad\qquad$ **20** $\;x^2 - 5x - 24 = 0$

Solve the equation $x^2 + 9x + 8 = 0$

$$x^2 + 9x + 8 = 0$$
$$(x+1)(x+8) = 0$$
Either $\quad x+1 = 0 \quad$ or $\quad x+8 = 0$
i.e. $\qquad x = -1 \quad$ or $\quad -8$

Solve the equations.

21 $x^2 + 3x + 2 = 0$ $\qquad$ **26** $x^2 + 7x + 6 = 0$

22 $x^2 + 8x + 7 = 0$ $\qquad$ **27** $x^2 + 7x + 10 = 0$

23 $x^2 + 8x + 15 = 0$ $\qquad$ **28** $x^2 + 14x + 13 = 0$

24 $x^2 + 8x + 12 = 0$ $\qquad$ **29** $x^2 + 16x + 15 = 0$

25 $x^2 + 11x + 18 = 0$ $\qquad$ **30** $x^2 + 9x + 18 = 0$

Solve the equation $x^2 - 49 = 0$

$$x^2 - 49 = 0$$
$$(x+7)(x-7) = 0$$
Either $\quad x+7 = 0 \quad$ or $\quad x-7 = 0$
i.e. $\qquad x = -7 \quad$ or $\quad 7$

Solve the equations.

31 $x^2 - 1 = 0$ $\qquad$ **36** $x^2 - 4 = 0$

32 $x^2 - 9 = 0$ $\qquad$ **37** $x^2 - 25 = 0$

33 $x^2 - 16 = 0$ $\qquad$ **38** $x^2 - 100 = 0$

34 $x^2 - 81 = 0$ $\qquad$ **39** $x^2 - 144 = 0$

35 $x^2 - 169 = 0$ $\qquad$ **40** $x^2 - 36 = 0$

The equations we have solved by factorising have all been examples of the equation $ax^2 + bx + c = 0$ when $a = 1$. We consider next the case when $c = 0$,

e.g. the equation $3x^2 + 2x = 0$

Since x is common to both terms on the left-hand side we can rewrite this equation as

$$x(3x + 2) = 0$$

Then, either $x = 0$ or $3x + 2 = 0$

i.e. $x = 0$ or $3x = -2$

i.e. $x = 0$ or $-\frac{2}{3}$

EXERCISE 11E

Solve the equations.

1 $x^2 - 2x = 0$ **6** $x^2 - 5x = 0$

2 $x^2 - 10x = 0$ **7** $x^2 + 3x = 0$

3 $x^2 + 8x = 0$ **8** $x^2 + x = 0$

4 $2x^2 - x = 0$ **9** $3x^2 - 5x = 0$

5 $4x^2 - 5x = 0$ **10** $5x^2 - 7x = 0$

11 $2x^2 + 3x = 0$ **16** $6x^2 + 7x = 0$

12 $8x^2 + 5x = 0$ **17** $12x^2 + 7x = 0$

13 $x^2 - 7x = 0$ **18** $x^2 + 4x = 0$

14 $3x^2 + 5x = 0$ **19** $7x^2 - 2x = 0$

15 $7x^2 - 12x = 0$ **20** $14x^2 + 3x = 0$

Sometimes a quadratic equation has two answers, or *roots*, that are exactly the same.

Consider $x^2 - 4x + 4 = 0$

then $(x - 2)(x - 2) = 0$

i.e. either $x - 2 = 0$ or $x - 2 = 0$

i.e. $x = 2$ or $x = 2$

i.e. $x = 2$ (twice)

Such an equation involves a *perfect square*. As with any quadratic equation, it has two answers, or roots, but they are equal. We say that such an equation has a repeated root.

EXERCISE 11F

> Solve the equation $x^2 + 14x + 49 = 0$
>
> $$x^2 + 14x + 49 = 0$$
> $$(x + 7)(x + 7) = 0$$
> Either $\qquad x + 7 = 0 \qquad$ or $\qquad x + 7 = 0$
>
> i.e. $\qquad\qquad x = -7 \qquad$ (twice)

Solve the equations.

1 $\;x^2 - 2x + 1 = 0$ $\qquad\qquad$ **5** $\;x^2 - 6x + 9 = 0$

2 $\;x^2 - 10x + 25 = 0$ $\qquad\quad$ **6** $\;x^2 - 8x + 16 = 0$

3 $\;x^2 + 8x + 16 = 0$ $\qquad\quad$ **7** $\;x^2 + 2x + 1 = 0$

4 $\;x^2 + 6x + 9 = 0$ $\qquad\qquad$ **8** $\;x^2 + 20x + 100 = 0$

9 $\;x^2 + 18x + 81 = 0$ $\qquad\quad$ **13** $\;x^2 - 12x + 36 = 0$

10 $\;x^2 - 22x + 121 = 0$ $\qquad$ **14** $\;x^2 - 40x + 400 = 0$

11 $\;x^2 - x + \frac{1}{4} = 0$ $\qquad\qquad$ **15** $\;x^2 - 16x + 64 = 0$

12 $\;x^2 + 10x + 25 = 0$ $\qquad\;$ **16** $\;x^2 + \frac{4}{3}x + \frac{4}{9} = 0$

Quadratic equations do not always present themselves already arranged in the form $ax^2 + bx + c = 0$.

Returning to Paul's problem, on page 226, where he needs to find the dimensions of a rectangle whose area is 150 mm^2 and whose length is 5 mm more than its width, we start by drawing a diagram with the known information on it.

Using x mm for the width, we see from the diagram that he needs to solve the equation

$$x(x + 5) = 150$$

150 mm² x mm

$(x + 5)$ mm

The equation $x(x + 5) = 150$ needs to be rearranged so that it is in the form

$$(x^2 \text{ term}) \quad \text{then} \quad (x \text{ term}) \quad \text{then} \quad (\text{number}) = 0$$

We start by multiplying out the bracket, $\quad x^2 + 5x = 150$

Then we take 150 from each side, $\quad x^2 + 5x - 150 = 0$

Now we are ready to factorise, giving $\;(x + 15)(x - 10) = 0$

EXERCISE 11G   Solve the equations.

1 $x^2 - x = 30$ **7** $x^2 - x = 6$

2 $x^2 - 6x = 16$ **8** $x^2 + 6x = 7$

3 $x^2 + 9x = 36$ **9** $x^2 - x = 12$

4 $x^2 = 2x + 8$ **10** $x^2 = 3x + 10$

5 $x^2 = 2x + 24$ **11** $x^2 = 6x - 8$

6 $x^2 = 12x - 35$ **12** $x^2 = 5x + 50$

Solve the equation $\ 21 = 10x - x^2$

$$21 = 10x - x^2$$
$$21 - 10x + x^2 = 0$$
$$x^2 - 10x + 21 = 0$$
$$(x - 7)(x - 3) = 0$$

Collect the terms on the LHS; this makes the x^2 term positive. Now rearrange so that the x^2 term is first, followed by the x term, then the number term.

Either $x - 7 = 0$,   so $x = 7$

or $x - 3 = 0$,   so $x = 3$

Solve the equations.

13 $10 = 7x - x^2$ **17** $12 = 8x - x^2$

14 $7 = 8x - x^2$ **18** $20 = 9x - x^2$

15 $8 = 6x - x^2$ **19** $35 = 12x - x^2$

16 $21 = 10x - x^2$ **20** $15 = 8x - x^2$

Solve the equation $\ 2x^2 + 10x - 12 = 0$

$$2x^2 + 10x - 12 = 0$$
$$2(x^2 + 5x - 6) = 0$$
$$x^2 + 5x - 6 = 0$$
$$(x + 6)(x - 1) = 0$$

The LHS has a common factor, 2; we start by taking it out.

$x^2 + 5x - 6$ must be zero since 2 is not.

Either $x + 6 = 0$ giving $x = -6$

or $x - 1 = 0$ giving $x = 1$

Solve the equations.

21 $2x^2 - 8x = 0$ **22** $5x^2 - 15x + 10 = 0$

23 $3x^2 + 9x + 6 = 0$

24 $3x^2 - 24x + 36 = 0$

25 $2x^2 - 4x = 0$

26 $3x^2 - 9x = 0$

27 $6x^2 + 18x + 12 = 0$

28 $8x^2 - 4x = 0$

Solve the equation $x(x-2) = 15$

$$x(x-2) = 15$$
$$x^2 - 2x = 15$$
$$x^2 - 2x - 15 = 0$$
$$(x-5)(x+3) = 0$$

Either $\quad x - 5 = 0 \quad \Rightarrow \quad x = 5$

or $\quad\quad x + 3 = 0 \quad \Rightarrow \quad x = -3$

Check: when $x = 5$, LHS $= 5(3) = 15 =$ RHS

when $x = -3$, LHS $= (-3)(-5) = 15 =$ RHS

Solve the equations.

29 $x(x+1) = 12$

30 $x(x-1) = x+3$

31 $x(x-5) = 24$

32 $x(x+3) = 5(3x-7)$

Solve the equation $(x-3)(x+2) = 6$

$$(x-3)(x+2) = 6$$
$$x^2 - x - 6 = 6$$
$$x^2 - x - 12 = 0$$
$$(x-4)(x+3) = 0$$

Either $\quad x - 4 = 0 \quad$ or $\quad x + 3 = 0$

i.e. $\quad\quad x = 4 \quad$ or $\quad -3$

Solve the equations.

33 $(x+2)(x+3) = 56$

34 $(x+9)(x-6) = 34$

35 $(x-2)(x+6) = 33$

36 $(x+3)(x-8) + 10 = 0$

37 $(x-5)(x+2) = 18$

38 $(x+8)(x-2) = 39$

39 $(x+1)(x+8) + 12 = 0$

40 $(x-1)(x+10) + 30 = 0$

In general, to solve a quadratic equation by factorising

- multiply out brackets
- collect all the terms on one side of the equal sign (choose the side where x^2 is positive)
- arrange the terms in the order (x^2 term), (x term), (number) and then look for factors.

Solve the equations.

1 $x^2 - x - 20 = 0$ **11** $12x^2 + 16x = 0$

2 $x^2 = 4x - 4$ **12** $3x^2 + 5x = 0$

3 $x^2 - 36 = 0$ **13** $x^2 - 2x - 15 = 0$

4 $x^2 = 7 - 6x$ **14** $(2x - 3)(x + 1) = 0$

5 $2x^2 + 12x + 18 = 0$ **15** $x^2 = 2x + 35$

6 $2 - x = x^2$ **16** $x^2 + 13x + 12 = 0$

7 $5x(x - 1) = 4x^2 - 4$ **17** $12x^2 - 48 = 0$

8 $15 - x^2 + 2x = 0$ **18** $x(x + 6) = 3x + 10$

9 $x^2 + 12x + 32 = 0$ **19** $2x(x + 9) = 0$

10 $x(x + 8) = x + 30$ **20** $(x - 4)^2 = 25$

When trying to form an equation from information given in a problem, start by identifying unknown numbers.
Use a letter to represent an unknown number. State what this letter stands for, either by writing a short sentence or by marking it clearly on a diagram. Use the information given in the problem to find a relationship between the letter and other numbers.

When the solutions of an equation have been found, remember to check that they fit the information given; remember also that not all solutions of an equation give solutions to a problem. On page 234 for example, Paul's problem gave the equation $(x + 15)(x - 10) = 0$. The solutions of this equation are $x = -15$ and $x = 10$. A length, however, cannot be negative so $x = -15$ is *not* a solution to Paul's problem.

EXERCISE 11I

I think of a positive number x, square it and then add three times the number I first thought of. The answer is **54**. Form an equation in x and solve it to find the number I first thought of.

> The square of x is x^2 and three times x is $3x$.
> Adding them together gives $x^2 + 3x$ and this must be 54.

$$x^2 + 3x = 54$$

i.e. $\qquad x^2 + 3x - 54 = 0$

$$(x - 6)(x + 9) = 0$$

Either $\qquad x - 6 = 0 \qquad$ or $\qquad x + 9 = 0$

i.e. $\qquad\qquad\quad x = 6 \qquad$ or $\qquad -9$

Reject -9. > The number must be positive.

The number I first thought of was therefore 6.

Check: > Use the number found with the original instructions. $\quad 6^2 + 3 \times 6 = 54$

1 The square of a number x is 16 more than six times the number.

   a Write 'the square of a number x' in symbols.

   b Write '16 more than six times x' in terms of x.

   c Hence form an equation in x and solve it.

2 When five times a number x is subtracted from the square of the same number, the answer is 14.

   a Write 'five times a number x is subtracted from the square of x' in mathematical notation.

   b Hence form an equation in x and solve it.

3 Peter had x marbles.

   a The number of marbles Fred had was six less than the square of the number Peter had.
      Write down the number of marbles Fred had in terms of x.

   b Write down, in terms of x, the total number of marbles they both had.

   c Together they had 66 marbles. Form an equation in x and solve it.

   d How many marbles did Fred have?

4 I think of a positive number x. If I square it and add it to the number I first thought of the total is 42. Find the number I first thought of.

5 Ahmed is x years old and his father is x^2 years old. The sum of their ages is 56 years. Form an equation in x and solve it to find the age of each.

6 Peter is y years old and his sister is 5 years older. If the product of their ages is 84, form an equation in y and solve it to find Peter's age.

7 Kathryn is x years old. If her mother's age is two years more than the square of Kathryn's age, and the sum of their ages is 44 years, form an equation in x and solve it to find the ages of Kathryn and her mother.

A rectangle is 4 cm longer than it is wide. If it is x cm wide and has an area of 77 cm^2, form an equation in x and solve it to find the dimensions of the rectangle.

77 cm^2 x cm

The area of a rectangle is given by length × breadth.

$(x + 4)$ cm

Area $= (x + 4) \times x$ cm^2 But the area is 77 cm^2.

i.e. $(x + 4)x = 77$

$x^2 + 4x = 77$

$x^2 + 4x - 77 = 0$

$(x - 7)(x + 11) = 0$

Either $x - 7 = 0$ or $x + 11 = 0$

i.e. $x = 7$ or -11

The breadth of a rectangle cannot be negative, so reject -11,

$\therefore$ $x = 7$.

The rectangle measures $(7 + 4)$ cm by 7 cm,

i.e. 11 cm by 7 cm.

---

8 A rectangle is x cm wide and is 3 cm longer than it is wide.

  a Draw a diagram showing this information, clearly marking the length of the rectangle in terms of x.

  b Write down the area of the rectangle in terms of x.

  c If the area is 28 cm^2, form an equation in x and solve it to find the dimensions of the rectangle.

9 The base of a triangle is x cm long and its perpendicular height is 5 cm less than the length of its base.

   a Draw a diagram showing this information, clearly marking the perpendicular height of the triangle in terms of x.

   b Write down an expression for the area of the triangle in terms of x.

   c The area of the triangle is 25 cm^2. Form an equation in x and solve it.

   d What is the height of the triangle?

10 A rectangle is 5 cm longer than it is wide and its area is 66 cm^2. Find the dimensions of the rectangle.

11 A skier is sliding down a gentle slope. The distance she covers in t seconds is given by $(t^2 + t)$ metres. Find how long it takes the skier to cover 110 metres.

12 A rectangular lawn is bordered on two adjacent sides by a path as shown in the diagram.

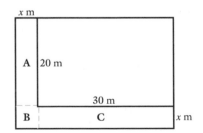

   a Express in terms of x each of the areas marked by the letters **A**, **B** and **C**.

   b The area of the path is 104 m^2. Form an equation in x and solve it to find the width of the path.

13 The formula $P = n(n + 2)$ is used to work out the cost, £P, for storing n tonnes of butter for a year. Find the number of tonnes of butter that can be stored for a cost of £2600.

14 **a** The square of an unknown positive integer, x, and the square of the consecutive (the next larger) integer are added. Write an expression for this sum in terms of x.

   b The sum of these two squares is 181. Find the smaller integer.

SOLVING
EQUATIONS BY
TRIAL AND
IMPROVEMENT

In this chapter we have solved quadratic equations by factorisation. Not all quadratic equations can be solved this way and we will return to this topic in Book 10A.

There are, however, many other types of equation for which there is no method of finding an exact solution. In such cases we start by giving x a value that looks reasonable and *try* it in the equation. Then we adjust that value to give an *improvement*.

Solution by trial and improvement was introduced in Book 8A. The next worked example is a reminder of the process.

EXERCISE 11J

Find a solution to the equation $x^2 - \dfrac{2}{x} = 6$ correct to 1 decimal place.

1 is clearly too small so we try 2 as the first trial value for x.

If this gives the value of $x^2 - \dfrac{2}{x}$ as less than 6 (i.e. too small), we try next a value greater than 2, and vice-versa.

We can keep track of results by placing them in a table.

Try $x =$	Value of $x^2 - \dfrac{2}{x}$	Compared with 6	
2	$2^2 - \dfrac{2}{2} = 3$	too small	This shows that $2 < x < 3$
3	$3^2 - \dfrac{2}{3} = 8.33\ldots$	too big	
2.5	$2.5^2 - \dfrac{2}{2.5} = 5.45\ldots$	too small	
		But not by much, so try $x = 2.6$ next.	
2.6	$2.6^2 - \dfrac{2}{2.6} = 5.99\ldots$	too small	
		Only just.	Now we see that $2.6 < x < 2.61$
2.61	$2.61^2 - \dfrac{2}{2.61} = 6.04\ldots$	too big	

$x = 2.6$ correct to 1 decimal place.

For each equation, use trial and improvement to find a positive solution correct to 1 decimal place.

1 $x^3 + 7x = 12$ (Start by trying $x = 1$.)

2 $x^3 - 2x - 5 = 0$ (Start by trying $x = 2.$)

3 $x^2 + x + \dfrac{1}{x} = 11$ (Start by trying $x = 2.$)

4 $x^3 + 3x = 7$ **5** $x + \dfrac{8}{x} = 12$

Sometimes when a trial number gives a result that is too small, trying a larger trial number gives a result that is even smaller. When this happens, try a number smaller than the first one.

The equation $2x - x^3 + 1 = 0$ has a solution between $x = 1$ and $x = 2.$
Find the solution correct to 1 decimal place.

Try $x =$	$2x - x^3 + 1$	Compared with 0	
1.5	0.625	> 0	
1.3	1.403	> 0	This is even bigger than the value given by 1.5, so we try a value larger than 1.5.
1.6	0.104	> 0	This is nearer to 0 than the value given by $x = 1.5$, so we will try a number a little larger than 1.6.
1.7	-0.513	< 0	Now we see that $1.6 < x < 1.7$. We will try 1.65 next.
1.65	$-0.192\ldots$	< 0	

Now we can see that $1.6 < x < 1.65,$

i.e. $x = 1.6$ correct to 1 decimal place.

For each equation, use trial and improvement to find a positive solution correct to 1 decimal place.

6 $\dfrac{10}{x} - x^2 = 6$ (Start with $x = 2.$)

7 $24 - x^3 + 2x = 0$

8 $x(7 - x)(2x + 1) = 8$ (Start with $x = 6.$)

Use a spreadsheet to find a positive solution to the equation
$\frac{12}{x} + x = 9$, giving the answer correct to 2 decimal places.

	A	B	C
1	12/x + x = 9		
2		Try x =	12/x + x
3		2	8
4		3	7
5		1.5	9.5
6		1.6	9.1
7		1.7	8.7588235
8		1.65	8.9227273
9		1.63	8.9919632
10		1.62	9.0274074
11		1.625	9.0096154
12			ERR
13			ERR

Enter the formula +12/B3 + B3; then use the fill command down the column.

$1.5 < x < 2$

$1.6 < x < 1.7$

$1.62 < x < 1.63$

$1.625 < x < 1.63$, so $x = 1.63$ (correct to 2 d.p.)

$x = 1.63$ (correct to 2 d.p.)

For each equation, use a spreadsheet to find a positive solution correct to 2 decimal places.

9 $x^3 + x^2 + x = 45$

10 $8x^2 - x^3 = 10$

11 $x^2 - \frac{6}{x} = 16$

12 $x^4 - x = 10$

INVESTIGATION

The sequence of positive integers is $1, 2, 3, 4, \ldots$
The sequence of square numbers is $1, 4, 9, 16, \ldots$
The sequence of triangular numbers is $1, 3, 6, 10, \ldots$

a Find a relationship between the nth term of the triangular sequence and the corresponding terms of the first two sequences.

b Find a relationship between the nth term of the triangular sequence and the first n terms of the sequence of positive integers.

c The sum of the first n terms of the first sequence is 630. How many terms have been added?

GRAPHS

Freda hopes to find a relationship between the distance slid by a puck down a slope and the time for which it has been sliding.

She collected the following data.

Time, in seconds, after leaving the top of the slope	1	2.7	3.5	4.5	4.9
Distance, in metres, from the top of the slope	0.05	0.4	0.6	1	1.2

To find a relationship between the time and distance, Freda could

- guess and see whether the figures in her table fit the guess. This is not likely to give an answer quickly, if ever.
- hope to spot a relationship from the figures in her table. This is not usually possible unless the relationship is a simple one.
- plot the points on graph paper. Freda may then recognise the points as lying on a curve whose equation is of a form she knows.

The ability to recognise the kind of equation that gives a particular shape of curve requires a wide knowledge of different forms of graph.
This knowledge is also useful when a graph has to be drawn from its equation because, if we know what shape to expect, mistakes are less likely to be made.

EXERCISE 12A

1 Gail needs to solve all these equations: $x^2 - 2x - 4 = 0$,
$$x^2 - 2x - 4 = 6,$$
$$2x^2 - 4x + 1 = 0,$$
$$1 + 2x - x^2 = 0.$$

Discuss what these equations have in common.
Discuss also the advantages and disadvantages of methods for solving them that you are familiar with.

2 Cheryl knows that, for a given voltage, the current flowing in a circuit is inversely proportional to the resistance.

She takes the following readings during an experiment.

Resistance (ohms)	0.5	0.7	1	1.2	1.5	2	2.4
Current (amps)	5.6	4	2.8	2.3	1.9	1.4	1.5

Cheryl then has to plot the values on a graph to confirm the relationship between them. Discuss what knowledge would help her to decide whether the relationship is confirmed and if she has made a mistake in any of her readings.

3 Peter is designing a swimming pool for a CDT project. He wants to include a sketch showing how the depth of water in the pool increases when it is filled at a constant rate.

This is the cross-section of the pool. Discuss how he could do this and what he needs to know.

Discussion from the last exercise shows that graphs have many uses. In this chapter we look at some of those and extend our knowledge of curves.

STRAIGHT LINES AND PARABOLAS

From Book 8A we know that an equation of the form $y = mx + c$ gives a straight line where m is the gradient of the line and c is the intercept on the y-axis.

We also know that an equation of the form $y = ax^2 + bx + c$ (a quadratic expression in x) gives a curve whose shape is called a parabola and looks like this.

When the x^2 term is negative, the curve is 'upside down'.

USING GRAPHS TO SOLVE QUADRATIC EQUATIONS

One of the problems discussed in the last exercise concerned solving the equations $x^2 - 2x - 4 = 0$,

$$x^2 - 2x - 4 = 6,$$
$$2x^2 - 4x + 1 = 0,$$
$$1 + 2x - x^2 = 0$$

You may have noticed that, in one form or another, the terms '$x^2 - 2x$' appear in all these equations. This suggests that we can use one graph to solve all these equations. The worked example in the next exercise shows how this can be done.

EXERCISE 12B

Use the graph of $y = x^2 - 2x - 4$ to solve the equations

a $x^2 - 2x - 4 = 0$ **b** $x^2 - 2x - 10 = 0$ **c** $2x^2 - 4x + 1 = 0$ **d** $1 + 2x - x^2 = 0$

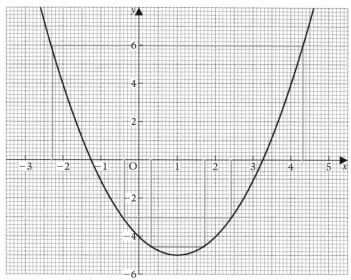

a $x^2 - 2x - 4 = 0$

> When this graph crosses the x-axis, the value of y is 0, i.e. $x^2 - 2x - 4 = 0$.
> The graph crosses the x-axis where $x = -1.25$ and $x = 3.25$.

$x = -1.25$
or 3.25

> This agrees with what we discovered in Chapter 11: a quadratic equation has two solutions.

b $x^2 - 2x - 10 = 0$

> To use the graph of $y = x^2 - 2x - 4$ to solve the equation $x^2 - 2x - 10 = 0$,
> we must convert the LHS to $x^2 - 2x - 4$; we can do this by adding 6 to both sides.

$x^2 - 2x - 10 + 6 = 6$,   i.e. $x^2 - 2x - 4 = 6$

> From the equation of the graph of $y = x^2 - 2x - 4$, we see that if $x^2 - 2x - 4 = 6$ then $y = 6$.
> Therefore the values of x when $y = 6$ give the solutions to the equation.

$x = -2.3$ or 4.3

c $2x^2 - 4x + 1 = 0$

> To use the graph we must convert the LHS to $x^2 - 2x - 4$.
> We can do this by dividing both sides by 2, then subtracting $4\frac{1}{2}$ from both sides.

$x^2 - 2x + \frac{1}{2} = 0$
$x^2 - 2x - 4 = -4\frac{1}{2}$

From the graph, when $y = -4\frac{1}{2}$, $x = 0.3$ or 1.7.

d $1 + 2x - x^2 = 0$

> To convert $1 + 2x - x^2 = 0$ to the form $x^2 - 2x - 4 = ?$,
> first add $x^2 - 2x$ to both sides, then subtract 4 from both sides.

$1 = x^2 - 2x$
$-3 = x^2 - 2x - 4$

i.e. $x^2 - 2x - 4 = -3$ so $y = -3$
From the graph, when $y = -3$, $x = -0.4$ or 2.4

1 Use the graph of $y = x^2 - 3x - 3$, which is given below, to solve the equations

a $x^2 - 3x - 3 = 0$

b $x^2 - 3x - 3 = 5$

c $x^2 - 3x - 7 = 0$

d $x^2 - 3x + 1 = 0$

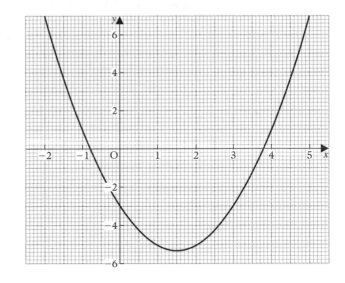

2 The graph of $y = 4 + 4x - x^2$ is given below. Use the graph to solve the equations

a $4 + 4x - x^2 = 0$

b $4 + 4x - x^2 = 5$

c $1 + 4x - x^2 = 0$

d $x^2 - 4x + 2 = 0$

e $x^2 = 4x - 3$

f $2x^2 - 8x - 1 = 0$

g Use the graph to solve the equation $4 + 4x - x^2 = 8$. What do you notice?

h Is it possible to use this graph to solve the equation $x^2 - 4x - 14 = 0$? If it is possible give the solutions. If it is not possible, explain why.

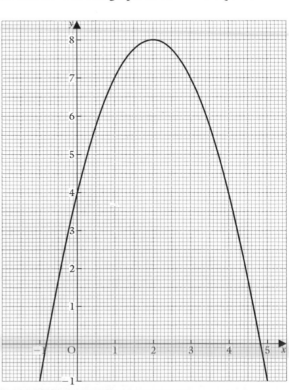

3 Copy and complete the following table which gives values of $(2-x)(x+1)$ for values of x from -3 to 4.

x	-3	$-2\frac{1}{2}$	-2	-1	$-\frac{1}{2}$	0	$\frac{1}{2}$	1	$1\frac{1}{2}$	2	3	$3\frac{1}{2}$	4
$(2-x)$	5	$4\frac{1}{2}$		3	$2\frac{1}{2}$		$1\frac{1}{2}$	1		0	-1	$-1\frac{1}{2}$	-2
$(x+1)$	-2	$-1\frac{1}{2}$		0	$\frac{1}{2}$		$1\frac{1}{2}$	2		3	4	$4\frac{1}{2}$	5
$(2-x)(x+1)$	-10	$-6\frac{3}{4}$		0	$1\frac{1}{4}$		$2\frac{1}{4}$	2		0	-4	$-6\frac{3}{4}$	-10

> A spreadsheet can be used to find the values of $(2-x)(x+1)$ for the different values of x.

Hence draw the graph of $y=(2-x)(x+1)$ for values of x from -3 to 4. Take 2 cm as 1 unit for x and 1 cm as 1 unit for y. Use your graph to solve these equations.

a $x^2-x-2=0$ **b** $2x^2-2x+1=0$ **c** $x^2=x+1$

4 The table gives values of y for certain values of x on the curve given by the equation $y=2x^2-7x+8$. Complete this table.

> Remember that $2x^2$ means square x first and then double the result.

x	0	$\frac{1}{2}$	1	$1\frac{1}{2}$	2	$2\frac{1}{2}$	3	$3\frac{1}{2}$	4
y	8	5		2	2	3		8	12

Use the table to draw the graph of $y=2x^2-7x+8$ for values of x from 0 to 4. Take 4 cm as 1 unit for x and 2 cm as 1 unit for y. Use the graph to solve the equation $2x^2-7x+8=4$.

5 What graph would you draw to solve the equation

a $x^2+2x-4=0$ **b** $3-5x-x^2=0$?

6 Copy and complete the following table.

x	-4	$-3\frac{1}{2}$	-3	$-2\frac{1}{2}$	-2	$-1\frac{1}{2}$	-1	$-\frac{1}{2}$	0	$\frac{1}{2}$	1
7	7	7		7	7	7		7	7	7	
$-6x$	24	21		15	12	9		3	0	-3	
$-2x^2$	-32	$-24\frac{1}{2}$		$-12\frac{1}{2}$	-8	$-4\frac{1}{2}$		$-\frac{1}{2}$	0	$-\frac{1}{2}$	
$7-6x-2x^2$	-1	$3\frac{1}{2}$		$9\frac{1}{2}$	11	$11\frac{1}{2}$		$9\frac{1}{2}$	7	$3\frac{1}{2}$	

Hence draw the graph of $y=7-6x-2x^2$ for values of x from -4 to 1. Take 4 cm as 1 unit on the x-axis and 1 cm as 1 unit on the y-axis.

a Use your graph to solve the equation $7-6x-2x^2=0$.

b Draw the line $y=5$. Write down the x values of the points where the line $y=5$ meets the curve $y=7-6x-2x^2$. Find, in as simple a form as possible, the equation for which these x values are the roots.

7 Using squared paper and values of x from -6 to 0, draw a graph to show that the equation $x^2+6x+10=0$ cannot be solved.

CUBIC GRAPHS   When the equation of a curve contains x^3 (and possibly terms involving x^2, x or a number), the curve is called a *cubic* curve.

These equations all give cubic curves:

$$y = x^3 + x, \quad y = 2x^3 - 5, \quad y = x^3 - 2x^2 + 6$$

We start by plotting the simplest cubic graph, whose equation is $y = x^3$.

The table gives values of x^3 for some values of x from -3 to 3.

x	-3	-2	-1.5	-1	-0.5	0	0.5	1	1.5	2	3
x^3	-27	-8	-3.4	-1	-0.1	0	0.1	1	3.4	8	27

Plotting these points and joining them with a smooth curve gives this curve.

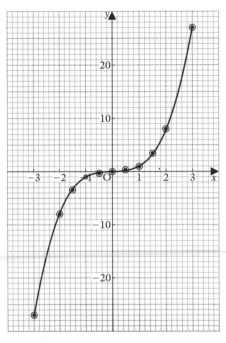

EXERCISE 12C   **1** Copy and complete the following table which gives values of $\frac{1}{5}x^3$, correct to one decimal place, for values of x from -3 to $+3$.

x	-3	-2.5	-2	-1.5	-1	-0.5	0	0.5	1	1.5	2	2.5	3
x^3	-27	-15.6	-8	-3.4		-0.13		0.13	1				27
$\frac{1}{5}x^3$	-5.4	-3.1	-1.6	-0.7		-0.03		0.03	0.2				5.4

Hence draw the graph of $y = \frac{1}{5}x^3$ for values of x from -3 to $+3$. Take 2 cm as unit on each axis.

Use your graph to solve the equations   **a** $\frac{1}{5}x^3 = 4$ **b** $x^3 = -15$

2 Make your own copy of the graph of $y = x^3$.

a On the same axes draw the line $y = x + 6$.

b Give the values of x where the curve and straight line intersect.

The graph of $y = (x-1)(x+1)(x-3)$ is given below.
Use the graph to find

a the values of x when   **i** $y = 1$ **ii** $y = -8$

b the solutions to the equations   **i** $(x-1)(x+1)(x-3) = 0$
                                   ii $(x-1)(x+1)(x-3) = 4$

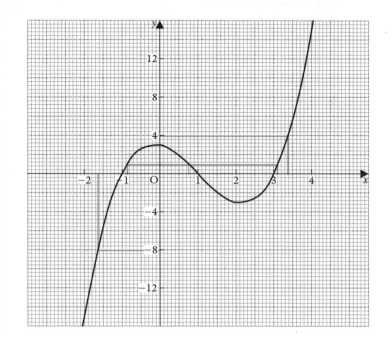

a   **i**   Drawing the line $y = 1$, we see that there are three values of x where $y = 1$.

$y = 1$ where $x = -0.85,\ 0.75,\ 3.1$

    ii   The line $y = -8$ cuts the curve once, so there is one value of x where $y = -8$

$y = -8$ where $x = -1.6$

b   **i** $(x-1)(x+1)(x-3) = 0$ where $y = 0$,
       i.e. where the curve crosses the x-axis, which is when
       $x = -1,\ 1$ and 3

    ii $(x-1)(x+1)(x-3) = 4$ where the curve cuts the
       line $y = 4$,
       i.e. when $x = 3.4$

3 Copy and complete the table which gives the values of y when $y = x(x-2)(x-4)$.

x	0	0.5	1	1.5	2	2.5	3	3.5	4
$x-2$		-1.5						1.5	
$x-4$		-3.5						-0.5	
y		2.625						-2.625	

Hence draw the graph of $y = x(x-2)(x-4)$, using 4 cm for 1 unit on each axis.

a Use your graph to find **i** the lowest value **ii** the highest value of $x(x-2)(x-4)$ within the given range of values for x.

b Find the solutions of the equation $x(x-2)(x-4) = 2$ within the given range.

c If the values of x were extended, do you think there may be another solution to the equation in part **b**?

4 Copy and complete the table which gives the value, correct to 2 decimal places, of $\frac{1}{3}x^3 - 2x + 3$ for values of x from -2 to 2.

x	-2	-1.5	-1	-0.5	0	0.5	1	1.5	2
$\frac{1}{3}x^3$	-2.67	-1.13	-0.33		0		0.33	1.13	
$-2x$	4	3	2		0		-2	-3	
$+3$	3	3	3		3		3	3	
$\frac{1}{3}x^3 - 2x + 3$	4.33	4.87	4.67		3		1.33	1.13	

Hence draw the graph of $y = \frac{1}{3}x^3 - 2x + 3$ using 4 cm for 1 unit on each axis. Estimate the value(s) of x where the graph crosses the x-axis.

5 The graph of $y = 1 - x + 2x^2 - x^3$ is given below.

a Write down the values of x where the curve crosses the x-axis. What can you deduce about the number of solutions to the equation $1 - x + 2x^2 - x^3 = 0$?

b How many solutions are there to the equation $1 - x + 2x^2 - x^3 = -1$? Explain your answer and give the solutions.

c Give a value of c if the equation $1 - x + 2x^2 - x^3 = c$
i has only one solution **ii** has three solutions.

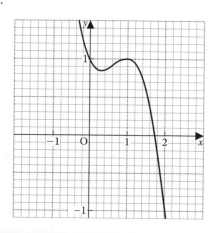

6 Without drawing the curve, or working out a table of values, explain how you can state the values of x where the curve $y = (x-2)(x-3)(x-4)$ crosses the x-axis. *Sketch* the curve.

THE SHAPE OF A CUBIC CURVE

From the last exercise we see that a cubic curve looks like

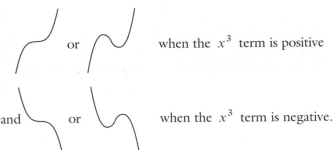

or   when the x^3 term is positive

and   or   when the x^3 term is negative.

RECIPROCAL GRAPHS

The equation $y = \dfrac{a}{x}$ where a is a number, is called a *reciprocal equation*. The simplest reciprocal equation is $y = \dfrac{1}{x}$.

Making a table showing values of y for some values of x from -4 to $-\frac{1}{4}$ and from $\frac{1}{4}$ to 4 gives

x	-4	-3	-2	-1	$-\frac{1}{2}$	$-\frac{1}{4}$	$\frac{1}{4}$	$\frac{1}{2}$	1	2	3	4
y	-0.25	-0.33	-0.5	-1	-2	-4	4	2	1	0.5	0.33	0.25

Plotting these points on a graph and joining them with a smooth curve gives this distinctive two-part shape.

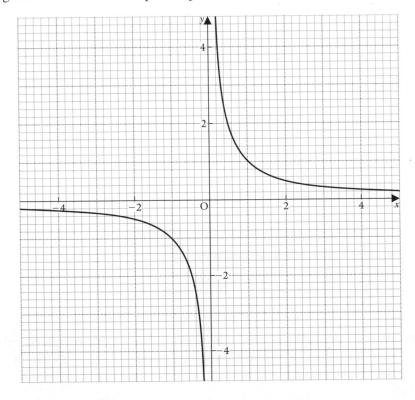

1 Discuss why the value $x = 0$ was not used in the table made to draw the graph of $y = \dfrac{1}{x}$.

2 Use the graph of $y = \dfrac{1}{x}$, drawn opposite, to answer these questions.

   a Give the value of y when $x = 2.5$

   b What is the value of x when $y = 2.5$?

   c What is the value of y when $x = 0.2$?

   d What happens to the value of y when x gets smaller than 0.2?

   e Why is there no point on the curve shown when $x = 0$?

   f How many forms of symmetry does the graph have?

3 Draw the graph of $y = \dfrac{2}{x}$ for values of x from -4 to $-\frac{1}{2}$ and from $\frac{1}{2}$ to 4. Use 2 cm for 1 unit on both axes.

   a Why is there no point on the graph where $x = 0$?

   b Give the value of y when   **i** $x = 2.6$ **ii** $x = -1.8$

4 Draw the graph of $y = \dfrac{12}{x}$ for values of x from 1 to 12. Use 1 cm for 1 unit on both axes.

   a Give the lowest value of y in the given range, and the value of x at which it occurs.

   b If the graph was drawn for values of x from 1 to 100, what would the lowest value of y be?

   c If the graph could be continued for values of x as large as you choose, what would the lowest value of y be then?

   d Is there a value of x for which $y = 0$?

5 *Sketch* the graph of $y = \dfrac{1}{x}$ for values of x from -10 to $-\frac{1}{10}$ and from $\frac{1}{10}$ to 10.

   a What happens to the value of y as the value of x increases beyond 10?

   b Is there a value of x for which $y = 0$? Explain your answer.

   c Is there a value of y for which $x = 0$? Explain your answer.

RECIPROCAL CURVES

An equation of the form $y = \dfrac{a}{x}$, where a is a constant (that is, a number), gives a distinctive two-part curve called a *reciprocal curve*.

The shape is also called a *hyperbola*.

Notice that there is a break in the graph where $x = 0$.
This is because there is no value for y when $x = 0$;

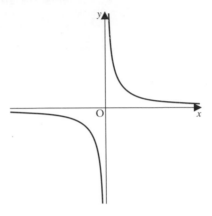

> we cannot divide by zero.

RECOGNISING CURVES

We are now in a position to look at a curve and recognise that its equation could be

- $y = mx + c$
- $y = ax^2 + bx + c$
- $y = ax^3 + bx^2 + cx + d$
- $y = \dfrac{a}{x}$
- none of these.

EXERCISE 12E

For questions **1** to **4**, write down the letter that corresponds to the correct answer.

1 The equation of this curve could be

A $y = x^2$

B $y = \dfrac{1}{x}$

C $y = x^3$

D $y = 4x - x^2$

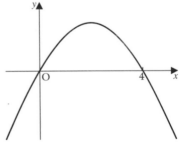

2 The equation of this curve could be

A $y = x^2 + x - 9$

B $y = (x - 3)(x + 3)(x + 1)$

C $y = \dfrac{9}{x}$

D $y = x^3$

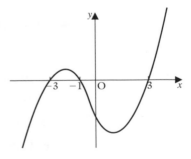

3 The equation of this curve could be

A $y = \dfrac{12}{x}$

B $y = x^2 - 9$

C $y = 9 - x^2$

D $y = x^3$

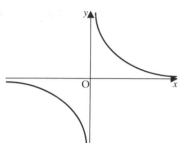

4 The equation of this curve could be

A $y = x^2$

B $y = 4 - x^2$

C $y = x^2 - 2x + 6$

D $y = x^3 - 4x^2 + 3$

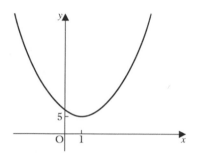

For question **5** to **13**, use squared paper. Draw x- and y-axes but do not scale them. *Sketch* the graph for each of the following equations.

5 $y = 3$ **8** $y = x$ **11** $y = x + 1$

6 $y = x^2$ **9** $y = x^2 + 1$ **12** $y = 1 - x^3$

7 $y = \dfrac{1}{x}$ ($x \neq 0$)   **10** $y = x^3$ **13** $x = 3$

For the remaining questions, write down the letter that corresponds to the correct answer.

14 The graph representing $y = x^2 - 1$ could be

A   **B**   **C**   **D**

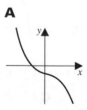 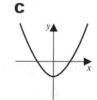 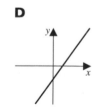

15 The graph representing $y = \dfrac{3}{x}$ could be

A   **B**   **C**   **D**

16 The graph representing $y = 10x^2$ could be

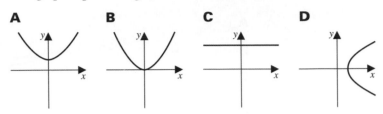

A   B   C   D

17 The graph representing $y = x + 2$ could be

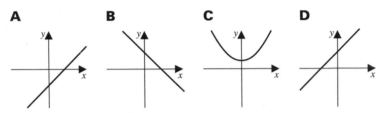

A   B   C   D

18 The graph representing $y = -x^2$ could be

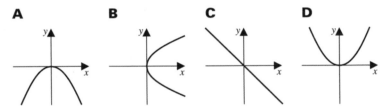

A   B   C   D

19 The graph representing $y = x^3 + 1$ could be

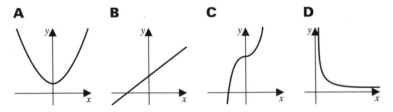

A   B   C   D

20 The graph representing $y = x(x - 1)(x - 2)$ could be

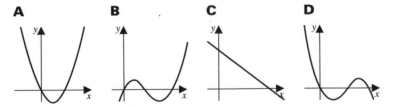

A   B   C   D

21 In questions **1** to **4**, explain why you cannot be certain that the equation chosen is the equation of the graph.

USING GRAPHS

For question **2** in **Exercise 12A**, Cheryl needed to use the following readings to confirm that the current flowing in a circuit is inversely proportional to the resistance.

Resistance (ohms)	0.5	0.7	1	1.2	1.5	2	2.4
Current (amps)	5.6	4	2.8	2.3	1.9	1.4	1.5

If the resistance is R ohms and the current is I amps, Cheryl wants to confirm that I is inversely proportional to R, that is that the product of corresponding values of R and I is constant.

This means that the relationship between I and R should be of the form $IR = k$.

Rearranging to make I the subject gives $I = \dfrac{k}{R}$ where k is a constant.

Cheryl also knows that a relationship of the form $y = \dfrac{k}{x}$ gives a curve whose shape is

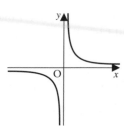

when x and y can have positive and negative values.

If x and y can only take positive values, we get just the right-hand side of this curve, i.e.

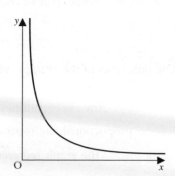

The resistance and the current have positive values only, so this is the shape that Cheryl expects to find when she plots the values of I against the values of R in the table.

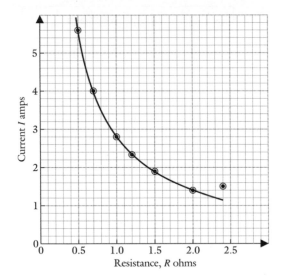

Allowing for slight errors in readings, these points do lie on a curve of the shape expected, except for one. Either a mistake has been made with this point (Cheryl may have made an error with her readings), or it could be that the relationship between R and I no longer applies when $R = 2.4$.

Assuming that $I = \dfrac{k}{R}$, we can substitute a pair of values *from the graph* to find the value of k,

i.e.   when $R = 2$, $I = 1.4$ so $1.4 = \dfrac{k}{2}$

$$\Rightarrow \qquad k = 2 \times 1.4 = 2.8$$

$\therefore$ $I = \dfrac{2.8}{R}$

(We can use another pair of coordinates to check the value found for k.)

INVERSE PROPORTION

The quantities in the example above are inversely proportional.

> Any two quantities, x and y, that are inversely proportional, are related by the equation $y = \dfrac{k}{x}$ and the graph representing them is a hyperbola.

EXERCISE 12F

1 This is a graph relating pressure and volume. The equation of the graph is $p = \dfrac{k}{v}$.

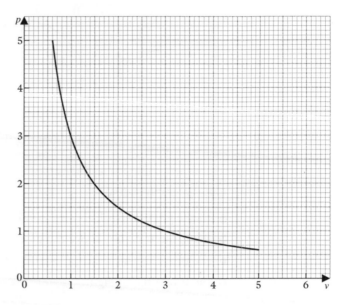

a Use the graph to find

   i the value of p when $v = 1.22$
   ii the value of v when $p = 3.6$

b Use the answer to part **a i** to find the value of k. Check your answer using the values found in part **a ii**.

2 Water in a plastic bottle cools down in the freezer. The graph shows how its temperature is changing.

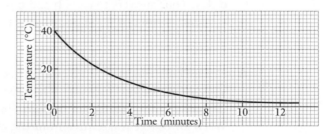

a What is the temperature of the water after

   i 4 minutes   **ii** 8 minutes?

b How long does it take to cool down to 14 °C?

c Explain why this graph shows that the temperature is not inversely proportional to the time.

3 This graph shows the height of a cricket ball above ground at different times during its flight.

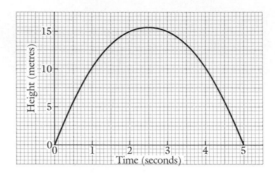

Use the graph to find

a the height of the ball above ground 1.5 seconds after it has been hit

b the greatest height above the ground reached by the ball

c the times at which the ball is 10 m above the ground

d the total time of flight.

4 In question **3**, the relationship between the height, h metres, and the time, t seconds, is $h = at^2 + bt + c$

a Use the point on the curve where $t = 0$ to show that $c = 0$.

b Use the point on the curve where $t = 5$ to find a relationship between a and b.

c Use the point on the curve where $t = 1$ to find another relationship between a and b.

d Use your answers to parts **b** and **c** to find the values of a and b. Hence write down the equation of the curve.

5 Gravity pulls objects towards the surface of the Earth making the speed of the objects increase. The effect of this pull decreases as the distance from the surface of the Earth increases.
When an object is d metres from the centre of the Earth, the rate at which its speed increases is g m/s per second, where g is given by

$$g = \frac{4.1 \times 10^{14}}{d^2}$$

(Note that this relationship applies only when d is greater than the radius of the Earth, i.e. outside the Earth's surface.)

This graph illustrates the relationship.

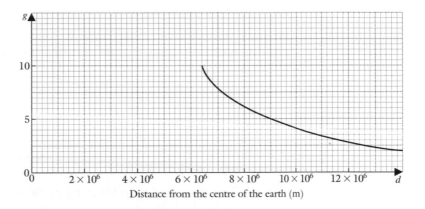

Distance from the centre of the earth (m)

a What is the radius of the Earth?

b What is the value of g when an object is 4.1×10^6 metres above the *surface* of the Earth?

c When $g = 2.5$, what is the value of d? How far is the object above the Earth's surface when $g = 2.5$?

Note that, in question **5**, the relationship between g and d is of the form

$$g = \frac{\text{number}}{d^2}.$$

This form of the relationship is called the *inverse square law*.

For example, if $y = \dfrac{4}{x^2}$, x and y are related by the inverse square law.

6 The relationship between x and y for the graph shown is
$y = ax^3 + b$.
Use the information on the graph to find the values of a and b.

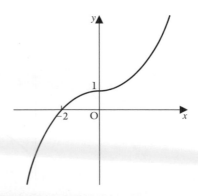

GRADIENTS OF CURVES

The gradient of a straight line is the same at any point on the line, that is, the gradient is constant. The gradient, or slope, of a curve however, changes from point to point.

We can give an estimate of the gradient of a curve but it is difficult to calculate because it changes as we move along the curve.

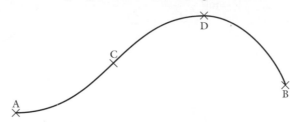

Moving along this curve from left to right, the gradient is zero at A, increases gradually to reach its maximum value at C and then decreases to zero again at D. The gradient then becomes negative and gets more negative (that is, the downhill slope increases) as we approach B.

EXERCISE 12G

Describe the way in which the gradient changes as we move along each curve from A to B.

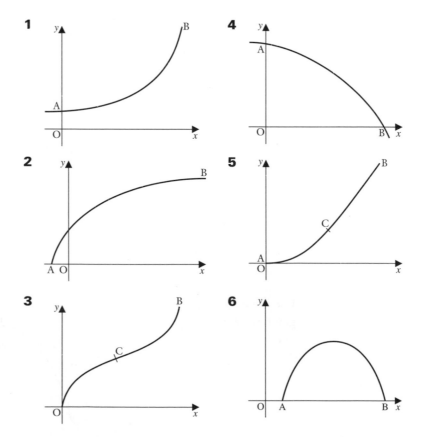

**INTERPRETATION
OF GRADIENT**

Gradient gives the rate at which the quantity on the vertical axis is changing as the quantity on the horizontal axis increases.

For example, this graph shows the temperature of a saucepan of water.

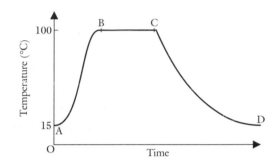

Although there are only two numbers on the axes, we can describe what is happening in general terms.

The water temperature starts at 15 °C at A, increases slowly at first but then more rapidly until it levels off at its maximum value at B. The gradient is positive for the whole of this section of the graph.

From B to C the temperature does not change (the water is boiling and its temperature cannot go above 100 °C). The gradient for this section is zero.

From C to D the temperature is falling, fairly fast to begin with and then more slowly. For this section of the curve the gradient is negative.

EXERCISE 12H

1 The graph shows the temperature of a bowl of soup, from the moment it is served.

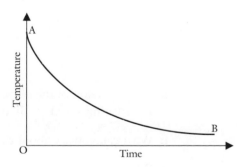

   a Describe how the temperature of the soup changes with time.

   b Roughly, how is the temperature changing near to B ?

   c What roughly, do you think is the temperature of the soup at B ?

2 The graph shows, over several years, the number of people travelling by air.

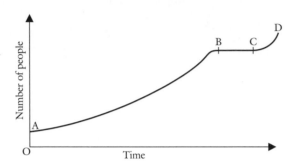

a Does the section of the curve from A to B represent an increase or a decrease in the number of people travelling by air? Is this rate of change constant?

b Describe how the number of people travelling by air changes for the sections from B to C and from C to D.

3 The graph shows the speed of a car between two sets of traffic lights.

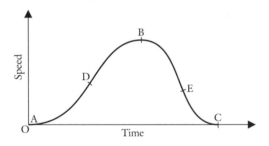

a Which section of the curve indicates
   i increasing speed   **ii** decreasing speed?

b What can you say about the speed of the car at the point B?

c What colour is the second set of lights? Justify your answer.

d When is the increase in speed greatest?

e When is the car losing speed at the greatest rate?

4 The graph shows how the surface area changes as a balloon is blown up.

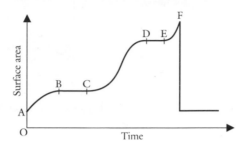

   a What do you think happens at F?

   b Describe how the surface area changes between
      i A and B   **ii** B and C.

   c What do you think accounts for the gradient of the curve for the sections B to C and D to E?

5 A stone is thrown into the air. The graph shows the distance of the stone from the ground.

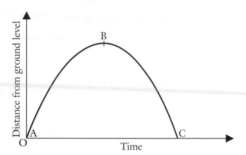

   a Describe how the gradient of this curve varies throughout the flight.

   b Where on the curve is the speed of the stone
      i greatest   **ii** least?

6

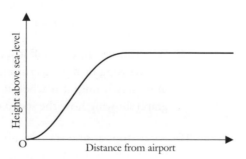

The graph shows the height of an aircraft above sea-level as the aircraft flies away from the airport. Describe how the height of the aircraft changes for the whole curve.

7

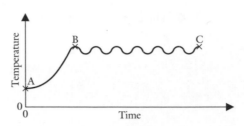

The graph shows the temperature of an oven controlled by a thermostat.

a Why does the curve not start at zero on the vertical axis?

b Describe what is happening for the part of the curve between A and B.

c What do you think can account for the shape of the curve between B and C?

8 The graph shows the speed of a motor cycle on a section of a cross-country race.

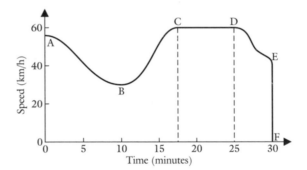

a Which sections of the curve represent the rider
  i going uphill   **ii** going downhill   **iii** falling off?

b For roughly how long is the speed constant?

c What, roughly, is the rate at which the speed is changing for the section from B to C?

9 When an aircraft takes off it accelerates rapidly from zero speed until it reaches roughly half its cruising speed. It then accelerates more and more slowly until it reaches its cruising speed of 400 km/h. Sketch a graph showing how the speed of the aircraft changes during this time.

10 A car stops at one set of traffic lights. When these lights go green it moves off but has to stop again at the next set of traffic lights which are 500 yards from the first set. Sketch a graph showing the distance of the car from the *first* set of lights during the time that the car takes to move between the two sets.

11 Derek and Carlos both ran in a marathon (26 miles). Derek started off running fairly fast, covering the first four miles in 30 minutes, but then gradually slowed down to finish the race in 6 hours. Carlos ran at a steady speed throughout the race and finished in $5\frac{1}{2}$ hours. Sketch, on the same axes, graphs illustrating the two runs. Hence estimate how far Carlos had run when he overtook Derek.

12 The graph illustrates a 100 metre race between Phyllis and Gita.

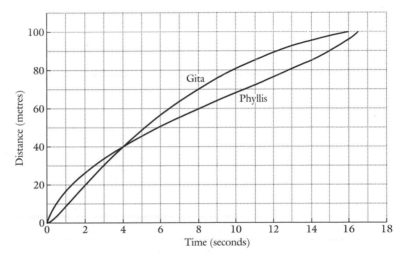

a Who won the race and in what time ?

b What was the difference in the times taken to finish the race ?

c Who ran faster at the start ?

d How far from the start did one competitor overtake the other ?

13 Liquid is poured at a constant rate into each of the containers whose cross-sections are given below. Sketch a graph showing how the depth of liquid in the container increases as the liquid is poured in.

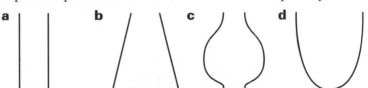

14 In **Exercise 12A**, you were asked to discuss how Peter could sketch a graph showing how the depth of water in a swimming pool varied as the pool was filled at a constant rate.

This is the cross-section of the pool.

Draw the graph for Peter.

15 The diagram shows some grain storage tanks with different shapes.
Grain is drawn off at the same constant rate from each tank.

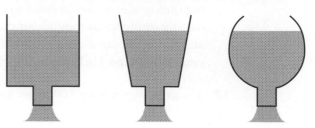

Make three copies of this diagram:

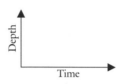

Sketch a graph, one on each diagram, showing how the depth of
grain in the tank varies with time.

Curve stitching
Interesting curves can be produced by joining points that are equally
spaced on straight lines.

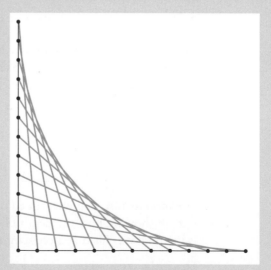

This example uses two straight lines drawn at right angles.
The coloured lines form what is called the *envelope* of the curve.

a Draw two lines, approximately 10 cm long that bisect each other, as shown in the diagram. Mark equally spaced points on the lines about 5 mm apart. Use a ruler or compasses, which should give a more accurate result if used carefully, to mark the points. Use a ruler and a coloured line to join the points as shown in the diagram, then continue the pattern to complete the curve. Repeat the pattern on the other half of the diagram to give two curves.

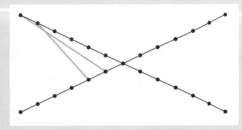

b Experiment with two lines drawn at different angles.

c Now try experimenting with more than two lines.

d This envelope is produced by joining points on circles.

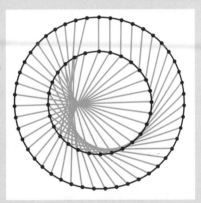

See whether you can produce any interesting curves by experimenting with circles and with a mixture of straight lines and circles.

AREAS AND VOLUMES

One of the components in the design for a new computer desk is a conduit (a channel for cables).

These are the drawings for the conduit which has a uniform cross-section.

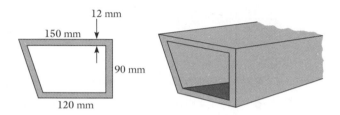

The conduit is to be manufactured in lengths of **2.5** metres. Angela is asked to organise shelving for storing **5000** lengths of this conduit.

To do this Angela needs to know how much space they require. She can start by finding the volume occupied by one length of conduit.

- To find the volume occupied by one length, just the external measurements are needed, i.e.

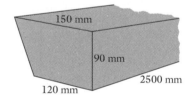

The solid has a constant cross-section. Therefore the volume is found by multiplying the area of the cross-section by the length.

Now the shape of the cross-section is this trapezium.

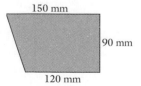

We can find its area by dividing it into a rectangle and a triangle, but trapeziums occur often enough in real objects to justify finding a formula for the area of any trapezium.

**AREA OF A
TRAPEZIUM**

Any trapezium can be divided into two triangles.

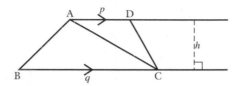

Triangles ABC and ADC both have the same height, as the height of each is the distance, h, between the parallel sides.

Now,   area $\triangle ADC = \frac{1}{2}$ base $\times$ height $= \frac{1}{2}p \times h$

area $\triangle ABC = \frac{1}{2}$ base $\times$ height $= \frac{1}{2}q \times h$

Therefore area ABCD $= \frac{1}{2}ph + \frac{1}{2}qh$

$= \frac{1}{2}(p+q) \times h$

This formula is easier to remember in words; $(p+q)$ is the sum of the parallel lengths, so

> the area of a trapezium is equal to
> $\frac{1}{2}$(sum of the parallel sides) $\times$ (distance between them).

EXERCISE 13A

Find the area of this trapezium

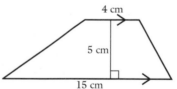

Area $= \frac{1}{2}$(sum of parallel sides) $\times$ (distance between them)

$= \frac{1}{2}(15 + 4) \times 5 \, \text{cm}^2$

$= \frac{1}{2} \times 19 \times 5 \, \text{cm}^2 = 47.5 \, \text{cm}^2$

Find the area of each of the following trapeziums.

1

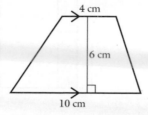

2

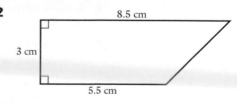

3

4

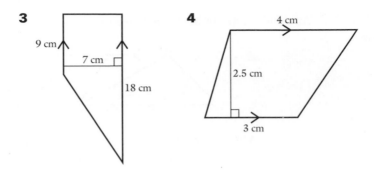

For questions **5** to **7** use squared paper and draw axes for x and y using ranges $-6 \leqslant x \leqslant 6$ and $-6 \leqslant y \leqslant 6$ and a scale of one square to 1 unit. Plot the points and join them up in alphabetical order to form a closed shape. Find, in square units, the area of the resulting figure.

5 A(6, 1), B(4, −3), C(−2, −3), D(−3, 1)

6 A(4, 4), B(−2, 2), C(−2, −2), D(4, −3)

7 A(3, 5), B(−4, 4), C(−4, −2), D(3, −5)

8 This is the cross-section of the conduit described on page 270. Find its area.

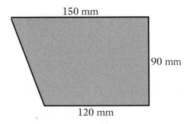

For the remaining questions you will also need to find areas of squares, rectangles, triangles and circles. If you need to remind yourself of the formulas for these, they are on page 6.

9 The diagram shows the end wall of a lean-to conservatory. The shaded area is covered with cedar wood, and the hatched area is glazed.

Find

a the total area of the end wall
b the area of the door
c the area of the glazed section
d the area which is covered with cedar wood.

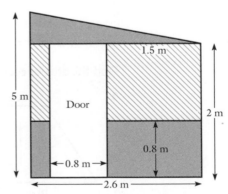

10 Find the areas of these cork gaskets.

a

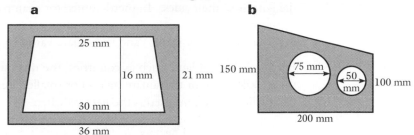

25 mm

16 mm 21 mm 150 mm

30 mm

36 mm

b

75 mm

50 mm 100 mm

200 mm

11 This is a classic patchwork quilt pattern called the 'captain's wheel'. It is based on a 4 unit by 4 unit square.

a Find the area of the patterned grey part of the design if the length of one side of the large square is 12 cm.

b A quilt is made from 800 of these squares. What area of the quilt is green?

12 This company logo is made from a semicircle on the shorter parallel side of a trapezium. Find its area.

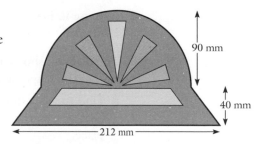

90 mm

40 mm

212 mm

13 Wood is used to make the top of this child's posting box. It is drawn on a 1 cm grid and then scaled down.

The wood weighs 0.35 grams per square centimetre. Find the mass of the top of this posting box.

UNITS OF AREA

Area is measured in standard sized squares. These squares are defined by the length of their sides. In metric units, for example, we may use the square centimetre (cm^2). In Imperial units we use the square inch (sq in), the square foot (sq ft) and the square yard (sq yd).

For large areas of land, such as countries, the units used are usually the square kilometre or the square mile. For smaller areas, such as those of fields, the metric unit used is the *hectare* where

$$1 \text{ hectare } = 10\,000 \text{ square metres}$$

The Imperial unit used for the areas of fields and so on, is the *acre*, where

$$1 \text{ acre } = 4840 \text{ square yards}$$

Acres and hectares can be interchanged using the approximation

$$1 \text{ hectare } \approx 2.5 \text{ acres}$$

(This relationship is correct to 2 significant figures; more accurately 1 hectare = 2.471 acres, which is correct to 3 decimal places.)

EXERCISE 13B

1 Given that 1 yard = 3 feet and that 1 foot = 12 inches, express

   a 2.5 sq yd in square feet   **b** 1.5 sq ft in square inches.

2 A rectangular field measures 250 yards by 130 yards. How many acres does the field cover?

3 A house in France is advertised for sale with 2.6 hectares of ground. How many acres is this?

4 The diagram shows a plot of land in the shape of a trapezium.

   a Find the area of the plot in hectares, correct to 1 decimal place.

   b How many acres does this plot cover?

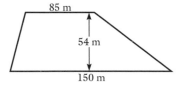

5 The area of the United Kingdom is 94 214 square miles. Express this area in **a** acres **b** hectares **c** square kilometres.

6 The area of Scotland is 30 405 square miles. Use the relationship

$$1 \text{ mile } = 1.6093 \text{ km correct to 4 decimal places}$$

to give the area of Scotland in square kilometres to the nearest square kilometre. State with reasons whether your answer is accurate to the nearest whole number.

ARCS AND SECTORS OF CIRCLES

Another shape that occurs frequently is a slice of a circle. This wood beading, for example, has a cross-section of this shape.

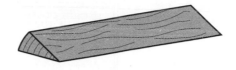

Part of the circumference of a circle is called an *arc*.

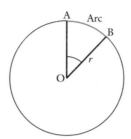

The length of an arc depends upon the radius of the circle and the angle enclosed at the centre of the circle by the radii from the two ends of the arc. This angle is $A\widehat{O}B$ in the diagram; it is called the angle *subtended* by the arc AB at the centre, O, of the circle.

The length of the arc AB as a fraction of the circumference is $\dfrac{A\widehat{O}B}{360°}$,

i.e.   length of arc AB $= \dfrac{A\widehat{O}B}{360°}$ of the circumference.

The circumference of a circle is $2\pi r$ so

$$\text{length of arc AB} = \frac{A\widehat{O}B}{360°} \times 2\pi r$$

The slice of the circle enclosed by the arc and the radii is called a *sector*.

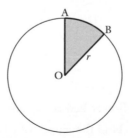

The area of the sector AOB as a fraction of the area of the circle is also equal to $\dfrac{A\widehat{O}B}{360°}$

i.e. $$\text{area of sector AOB} = \frac{A\widehat{O}B}{360°} \times \pi r^2$$

EXERCISE 13C

Find **a** the length of the arc
b the area of the sector.

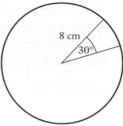

a Length of arc $= \dfrac{30°}{360°} \times$ circumference

$= \dfrac{30}{360} \times 2 \times \pi \times 8$ cm $= 4.188\ldots$ cm

The length of the arc is 4.19 cm correct to 3 s.f.

b Area of sector $= \dfrac{30°}{360°} \times$ area of circle

$= \dfrac{30}{360} \times \pi \times 8^2 = 16.75\ldots$ cm^2

The area of the sector is 16.8 cm^2 correct to 3 s.f.

In each question from **1** to **6** find

a the length of the arc   **b** the area of the sector.

1
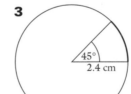

3

5
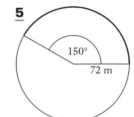

2

4
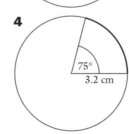

6
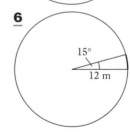

7 A flower bed is a quadrant of a circle.

  a Find the length of edging needed for the curved edge of the bed.

  b Find the area of the flower bed.

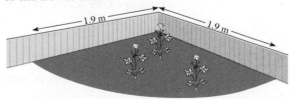

8 The diagram shows the plan of a herb garden. The beds are sectors of a circle, each one of which contains an angle of 40° at the centre. The radius of the circle is 0.9 m.

   a What length of edging is required to surround all these beds?

   b One handful of fertilizer covers one quarter of a square metre of soil.
   How many handfuls are needed to cover all the beds?

9 A silver earring pendant is part of a sector of a circle. Find the area of silver.

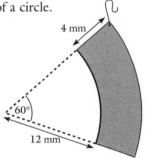

10 In this pattern, the green sections are congruent sectors of a circle of radius 9 cm and the grey sections are congruent sectors of a circle of radius 7 cm.
Which of the grey or green sections of the pattern covers the greater area? Justify your answer.

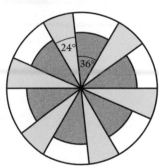

11 Harry is making a marquetry pattern.
The diagram shows one of the pieces he has cut.
The arcs are quadrants of circles.
Find **a** the perimeter of the piece
      b its area.

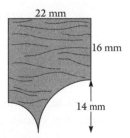

12 This is part of a map showing a wood.
The scale on the map is 2 cm to 1 km.
Estimate the area of the wood in **a** hectares **b** acres.

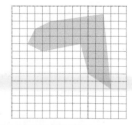

VOLUME OF A PRISM

A prism is a solid whose cross-section is the same all the way through. These are prisms.

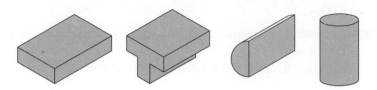

The volume of a prism is equal to

(area of cross-section) × (length)

Returning to Angela's task of organising shelving for lengths of conduit, introduced on page 270, we will now calculate the volume occupied by one length.

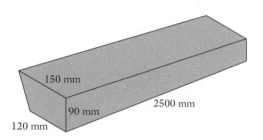

The volume of one length is

$$\left[\tfrac{1}{2}(\, 150 + 120 \,) \times 90\right] \times 2500 \text{ mm}^3$$

$$= 30\,375\,000 \text{ mm}^3$$

$$= 30\,375\,000 \div 1000^3 \text{ m}^3$$

$$= 0.030\,375 \text{ m}^3$$

> This is a large number of small units; we will convert them into cubic metres as this is a more sensible unit to use.

Therefore, assuming that the measurements are very accurate, the minimum volume that 5000 lengths of conduit will occupy is

$$5000 \times 0.030\,375 \text{ m}^3 = 151.875 \text{ m}^3$$

IMPERIAL UNITS OF VOLUME

Volume is measured in standard sized cubes. In Imperial units these are the cubic inch (cu in), the cubic foot (cu ft) and the cubic yard (cu yd).

EXERCISE 13D

1 a Discuss why 151.875 m^3 is the *minimum* volume that is needed to store 5000 lengths of conduit.

b Discuss how much space should be allowed for storage.

Find the volume of the solid in the diagram.

We draw the cross-section, not the whole solid.
We can find its area by dividing the shape into 2 rectangles.

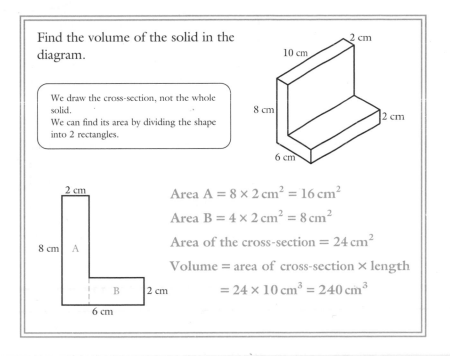

Area A $= 8 \times 2 \text{ cm}^2 = 16 \text{ cm}^2$

Area B $= 4 \times 2 \text{ cm}^2 = 8 \text{ cm}^2$

Area of the cross-section $= 24 \text{ cm}^2$

Volume $=$ area of cross-section $\times$ length

$= 24 \times 10 \text{ cm}^3 = 240 \text{ cm}^3$

Find the volumes of the solids illustrated in questions **2** to **7**. In each case draw a diagram of the cross-section but do not draw a picture of the solid.

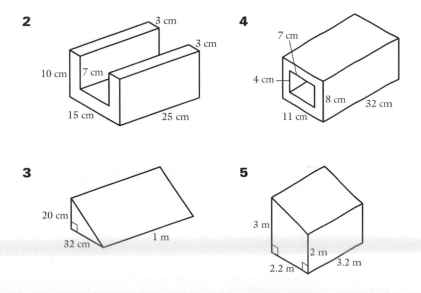

6

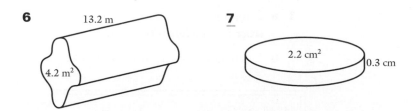

7

In each question from **8** to **11**, find the volume of the solid whose cross-section and length are given. Give the answer in the unit indicated in brackets.

8

Length = 1 m
(cm³)

10

Length = 4 m
(m³)

9

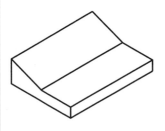

Length = 24 cm
(cm³)

11

Length = 16 cm
(cm³)

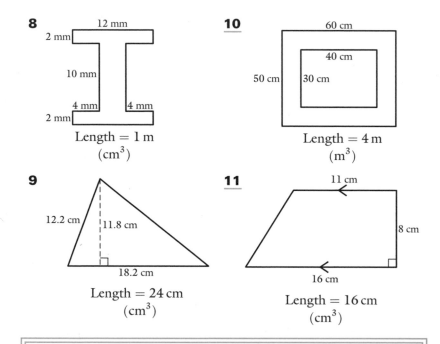

The volume of the solid shown in the diagram is 144 cm³ and the area of its cross-section is 14 cm². Find its length.

Let its length be *l* cm

Volume = area of cross-section × length

$$144 = 14 \times l$$
$$l = \frac{144}{14} = 10.28\ldots$$

i.e.   the length is 10.3 cm correct to 3 s.f.

12 The volume of a solid of uniform cross-section is $72\,\text{cm}^3$. The area of its cross-section is $8\,\text{cm}^2$. Find the length of the solid.

13 The volume of a solid of uniform cross-section is $32\,\text{m}^3$. Its length is $10\,\text{m}$. Find the area of its cross-section.

14

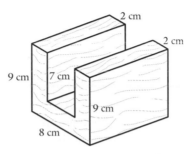

The volume of this mitre block is $396\,\text{cm}^3$.

Find   **a** the area of the cross-section

       b the length of the block.

15

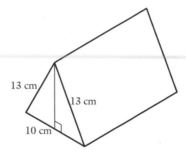

The cross-section of the solid is an isosceles triangle. The volume of the solid is $1200\,\text{cm}^3$.

Find   **a** the area of the triangle

       b the length of the solid.

16

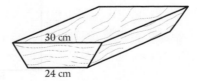

The cross-section of the block of wood is a trapezium. The height of the trapezium is $10\,\text{cm}$ and the volume of the block is $7800\,\text{cm}^3$. Find the length of the block.

Angela has to organise shelving to store 5000 lengths of conduit so she needs to consider the weight that the shelves will have to support. This means that she must find the mass of the conduits. To do this she has to know the mass of one unit of volume, that is, *the density* of the material from which they are made.

The volume of material in one length of conduit can be worked out from the information in the drawings so we can find the volume of material in 5000 lengths. The density of the material can then be used to calculate the total mass of the conduits to be stored on the shelves. You are invited to work this out in question **8** in the next exercise.

The questions in this exercise are a mixture of problems involving volumes.

1 A drop of oil of volume $2.5\,\text{cm}^3$ is dropped on to a flat surface and spreads out to form a circular pool of even thickness and area $50\,\text{cm}^2$.

How thick is the oil **a** in centimetres **b** in millimetres?

2

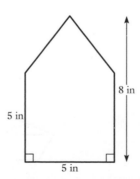

8 in

5 in

5 in

A cuboid of metal measuring 6 in by 8.2 in by 9.5 in is recast into the shape of a prism. The cross-section of the prism is shown in the diagram. How long is the prism?

Water comes out of a pipe of cross-section 3.2 cm^2 at a speed of 0.5 m/s.

a What volume of water is delivered by the pipe in one second?

b How many litres per minute is this?

a 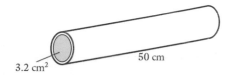 Imagine 0.5 m, i.e. 50 cm, of pipe being emptied in 1 second.

50 cm

3.2 cm^2

Volume = area of cross-section × length
$$= 3.2 \times 50 \text{ cm}^3$$
$$= 160 \text{ cm}^3$$
∴ 160 cm^3 of water is delivered in 1 second.

b Volume of water delivered in one minute is 160 × 60 cm^3
$$= \frac{16\emptyset \times 6\emptyset}{10\emptyset\emptyset} \text{ litres}$$
$$= 9.6 \text{ litres}$$

3 The cross-section of a pipe is 4.8 cm^2. Water comes out of the pipe at 30 cm/s. How much water is delivered in 1 second? Give your answer in litres.

4 Water comes out of a pipe at 60 cm/s. The cross-section of the pipe is a circle of radius 0.5 cm.
How many litres of water are delivered

a in 1 second **b** in 1 minute?

5 The diagram shows the net for a prism with an isosceles triangular cross-section. Find

a the lengths of the sides of the triangular cross-section

b the area of the triangular ends

c the volume of the prism

d the surface area of the prism.

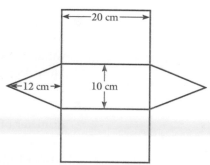

6 It is claimed that 1 litre of paint will cover $10\,\mathrm{m}^2$ of wall. What is the thickness of the paint applied to the wall?

7

The diagram shows the side view of a swimming bath of width 25 m.

a Find the volume of water in the bath when it is full. Give your answer in cubic metres.

b The bath is emptied through a pipe whose cross-sectional area is $200\,\mathrm{cm}^2$; the water runs out at $1.5\,\mathrm{m/s}$. What volume of water is removed in 1 second?

c Find how long it would take to empty the bath if four similar pipes are used each removing water at the same steady rate as in part **b**.

8 This is the drawing of the cross-section of the conduit we considered at the start of this chapter, showing the internal and external measurements.

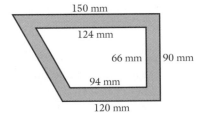

a Find the volume of material needed to make 5000 lengths of conduit, each **2.5** metres long. Decide how accurately you should give the answer and what unit to use.

b The material used to make these conduits has a density $0.75\,\mathrm{g/cm}^3$. Find the mass, in kilograms, of 5000 lengths of conduit.

9 Three identical cylindrical tumblers of height 9 cm and base radius 3 cm are completely filled with water. Their contents are poured, without any spillage, into a cylindrical jug with a diameter of 10 cm. How deep is the water in the jug?

10 A one litre cylindrical tin of paint has a height of 11 cm and the diameter of its base is 11.5 cm. If the can is opened how far below the lid would you expect the level of the paint to be? Give your answer to the nearest millimetre.

11

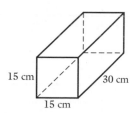 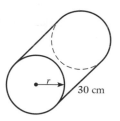

The diagrams show two parcels, each of length 30 cm. The cross-sections of the two parcels are equal in area. If the first is a square of side 15 cm find, correct to three significant figures, the radius of the second.

Compare the volumes of the two parcels.

String is tied around the parcels. The string passes around each parcel in the middle of the length and once from end to end. Which parcel requires the greater length of string, and by how much?

12

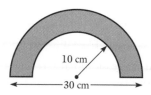

The diagram shows the cross-section through a coping-stone which is used on the top of a wall. Each stone is 50 cm long. Find the mass of the stone if the density of the material it is made from is 3.5 g/cm^3.

13

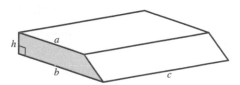

The diagram shows a prism whose cross-section is a trapezium. Find, in terms of a, b, c and h, a formula for

a the volume of the prism

b the surface area of the prism.

14 A cylindrical tin is 15 inches high and holds 5 gallons of paint when it is full. Find the diameter of the tin.
(1 gallon = 0.1605 cubic feet.)

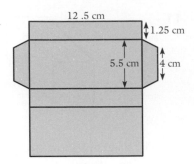

The diagram (not drawn to scale) shows the part of the wrapper round a bar of chocolate which is in contact with the bar. The density of the chocolate is $1.5 \, \text{g/cm}^3$. What is the mass of the bar?

THE DIMENSIONS OF A FORMULA

On page 271 we found a formula for the area of a trapezium. It is possible to check whether such a formula does give an area.

Formulas for finding lengths, areas and volumes all contain letters, each of which represents a number of units of length or area or volume.

An expression that has only one 'length unit' letter (or the sum of two or more such letters) is *one-dimensional* and gives a length as a number of miles, centimetres, kilometres, etc.

When an expression contains the product of two length symbols it is *two-dimensional* and represents an area as a number of (length unit)^2 such as cm^2, m^2, etc.

An expression with three length symbols multiplied together (or an area symbol multiplied by a length symbol) is *three-dimensional*. This gives a volume as a number of (length unit)^3, e.g. m^3, mm^3, etc.

Sometimes in a formula there is also a number, or a symbol that stands for a number such as π. These do not represent a number of units of length, area or volume and so do not affect the dimensions of an expression.

Suppose, for example, that d is a number of length units, then

$$2d \quad \text{is one-dimensional}$$
$$4\pi d^2 \quad \text{is two-dimensional}$$
$$\pi d^3 \quad \text{is three-dimensional.}$$

Checking units and dimensions can help us to decide whether a given quantity represents length, area or volume.

For example, if a sentence refers to '$z \, \text{cm}^3$', then z must be a number of volume units.

Similarly, if a sentence contains 'a cm', 'b cm' and '$X = ab$' then ab must be (a number of cm) $\times$ (a number of cm), that is, a number of cm^2. Therefore X represents a number of area units.

On the other hand, suppose we are told that the formula for the volume, V cubic units, of a container is $V = 3\pi ab$ where a and b are numbers of length units. V is three-dimensional but $3\pi ab$ is two-dimensional. So the formula must be incorrect.

EXERCISE 13F

1 State whether each of the following quantities is a length, an area or volume.

   a 10 cm   **c** 85 cm^2 **e** 630 mm

   b 21 cm^3 **d** 4 m^3 **f** 93 km^2

2 State whether each of the following quantities should be measured in length or area or volume units.

   a Diameter of a circle   **d** Perimeter

   b Amount of air in a room   **e** Region inside a square

   c Space inside a sphere   **f** Surface of a sphere

3 a, b, and c represent numbers of centimetres. Give a suitable unit (e.g. cm^2) for X.

   a $X = a + b$ **c** $X = 4ab$ **e** $X = 4\pi a^2$

   b $X = abc$ **d** $X = \pi c$ **f** $X = \frac{4}{3}\pi a^3$

In questions **4** and **5**, a, b and c represent numbers of length units, A, B and C represent numbers of area units and V represents a number of volume units.

4 State whether each of the letters P to Y used in the following formulas represents numbers of length or area or volume units.

   a $P = \pi bc$ **d** $Q = A + B$ **g** $R = \dfrac{ab}{c}$

   b $S = \pi a^2 b$ **e** $T = 4bA$ **h** $U = \dfrac{2A}{c}$

   c $W = 2a + 3b$ **f** $X = a^2 + b^2$ **i** $Y = \dfrac{V}{a}$

5 Some of the following formulas are wrongly constructed. State which formulas are incorrect and justify your statement.

   a $B = ac$ **c** $C = a^2 + b^3$ **e** $V = ab + c$

   b $C = \pi a^2$ **d** $V = 6a^2 b$ **f** $A = a(b + c)$

6 Peter had to find the area of a circle whose diameter was 18 cm. He wrote down

$$\text{Area} = 2\pi r \,\text{cm}^2$$
$$= 2 \times \pi \times 9 \,\text{cm}^2$$
$$= 56.5 \,\text{cm}^2$$

Louise knew nothing about circle formulas but was able to tell Peter that he was wrong. How did she know?

7 Jane copied a formula that she was supposed to use to find the volume of a solid but she found that she could not read the index number. All she had was $V = \pi x^? y$. She knew that x and y were numbers of length units. What is the index number?

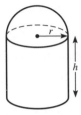

This bollard is a cylinder with a hemisphere (half a sphere) on top. From the list given below, pick out the formula for

a the overall height of the bollard

b the surface area of the bollard

c the volume of the bollard.

$$P = 3\pi r^2 + 2\pi rh \qquad Q = \pi r^2 + h \qquad R = r + h$$
$$S = \pi r^2 h + \tfrac{2}{3}\pi r^3 \qquad T = \pi rh + \tfrac{4}{3}\pi r^3$$

a ⎛ Height is one-dimensional so the formula must have single letter terms. ⎞

The overall height is given by $R = r + h$.

b ⎛ Area is two-dimensional so each term in the formula must have a product of two letter terms. ⎞

The surface area is given by $P = 3\pi r^2 + 2\pi rh$.

c ⎛ Volume is three-dimensional so each term in the formula must have a product of three letter terms. ⎞

The volume is given by $S = \pi r^2 h + \tfrac{2}{3}\pi r^3$.

8

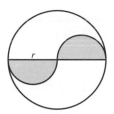

From the list of expressions given below, choose the correct one for
a the perimeter of the shaded region

b the area of the unshaded region.

$$2r + \pi r^2 \qquad 2r + \pi r \qquad \pi r^2 - \tfrac{1}{4}\pi r^2 \qquad \pi r^2 - 2\pi r$$

9

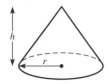

Which of the following formulas could give the volume of this
cone?

$$V = \pi rh \qquad V = \tfrac{1}{3}\pi r^2 h \qquad V = \pi r^2 + \tfrac{1}{3}h \qquad V = \pi r\sqrt{r^2 + h^2}$$

Give reasons for your answer.

10

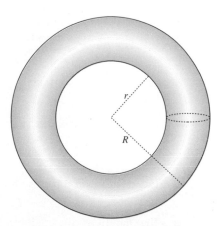

The mathematical name for this ring doughnut is a torus. Which of
the following formulas could give

a its surface area **b** its volume?

$$\tfrac{1}{4}\pi^2(R+r)(R-r)^2, \qquad \pi^2(R^3 - r^2),$$

$$\tfrac{3}{4}(\pi^2 R^2 - 3\pi Rr + R^3), \qquad \pi^2(R^2 - r^2)$$

11 Alice wasn't sure of the formula for the surface area of a closed cylinder of radius r and height h. Her elder brother thought that the formula was $A = 2\pi r^2(r+h)$ while her younger brother thought it was $A = \pi(2r+h)$. Her father suggested $A = 2\pi r(r+h)$. One of these formula is correct. Which one is it? Justify your choice.

12 This solid is a cylinder of radius a cm on a square base of side $2a$ cm.

Find a formula for

a the volume

b the surface area.

Check that the dimensions of your answers are correct.

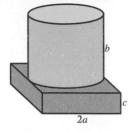

PUZZLE

It is possible to cut this 'cake' into 8 identical pieces in several ways. What is the least number of cuts with a knife needed to do this?

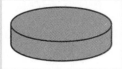

INVESTIGATION

A cone can be made from a sector of a circle, i.e.

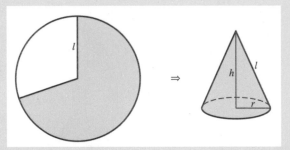

a Find an expression for the circumference of the base of the cone in terms of r.

b What length in the left-hand diagram is given by the expression found in part **a**?

c Find an expression for the area of the curved surface of the cone in terms of r and ℓ.

d Find an expression for the area of the curved surface of the cone in terms of r and h.

TRANSFORMATIONS

14

Helen designs cushion covers. She spends a great deal of time on the original design of, say, a petal but when she is satisfied she needs to be able to manipulate it into different sizes and positions. If she can do lots of different things with the basic design she has much more scope to produce an interesting cushion cover. Normally she works the design using her computer. To investigate as many different designs as possible, she would like to

- reflect the basic design about a line
- rotate it about different points
- make versions of it that are both larger and smaller
- move the original design around so that it looks exactly the same except for the fact that it is in a different position
- combine some of these transformations.

To be able to do all these things she must use her previous knowledge of transformations plus some of the additional work studied in this chapter.

1 Discuss how you would paint the word AMBULANCE on the front of an ambulance so that it can be read in the mirror of a car travelling in front of it.

2 Give examples of common everyday objects that have been designed using reflections and/or rotations. State clearly how each transformation has been used.

3

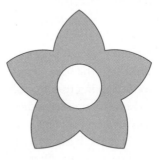

Discuss how to draw, on a computer if you wish, this accurate machine part that is to be made from a circular metal disc and has five axes of symmetry. How would you get over the design problem if the finished part is to be cut from a circle of diameter
a 1 cm **b** 4 m?

NEGATIVE SCALE FACTORS

In Book 8A we considered enlarging various shapes using positive and fractional scale factors. This work is revised in Revision Exercise 1.6 at the beginning of this book. Now we extend that work to include negative scale factors.

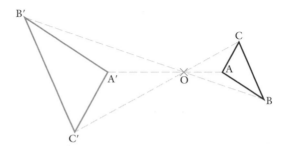

As you can see in the diagram above it is possible to produce an image twice the size of the object by drawing the guidelines backwards rather than forwards from the centre O so that OA$'$ = 2OA, and so on.
To show that we are going the opposite way we say that the scale factor is −2.

The image is the same shape but has been rotated through a half turn compared with the image produced by a scale factor of +2.
The following diagrams show enlargements which have scale factors of −3 and +3.

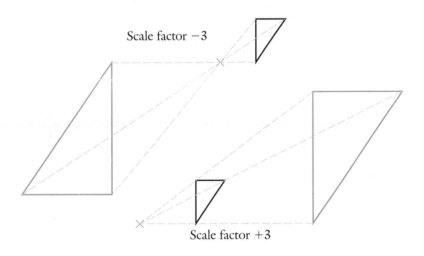

Scale factor −3

Scale factor +3

In questions **1** and **2** give the centre of enlargement and the scale factor.

1

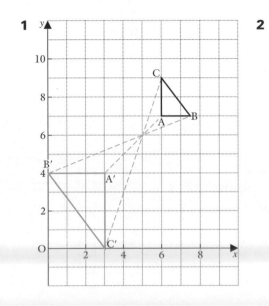

2

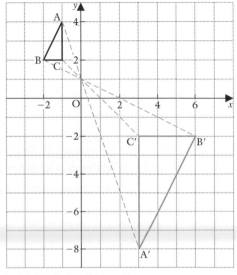

Copy the diagrams in questions **3** to **6** using 1 cm to 1 unit. Find the centre of enlargement and the scale factor. Remember: to find the centre of enlargement draw lines similar to those given in questions **1** and **2**.

3

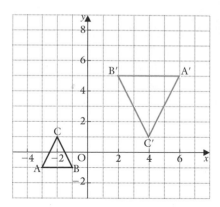

5

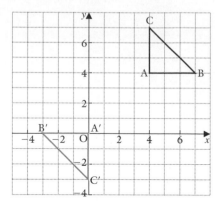

4

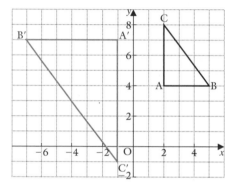

6

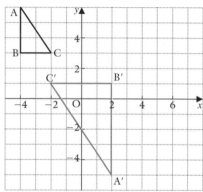

In questions **7** and **8** draw axes for x and y from -6 to 6.
Draw the object and image and find the centre of enlargement and the scale factor.

7 Object: $\triangle ABC$ with $A(6, -1)$, $B(4, -3)$, $C(4, -1)$
Image: $\triangle A'B'C'$ with $A'(-3, 2)$, $B'(1, 6)$, $C'(1, 2)$

8 Object: $\triangle ABC$ with $A(2, 3)$, $B(4, 3)$, $C(2, 6)$
Image: $\triangle A'B'C'$ with $A'(2, 3)$, $B'(-4, 3)$, $C'(2, -6)$

9 **a** If $A'B'C'D'$ is the image of ABCD under enlargement, give the centre and the scale factor.

  b What other transformation would map ABCD to $A'B'C'D'$?

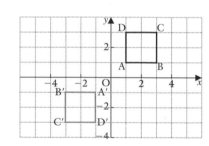

10 On plain paper, draw an object such as a pin man in the top left-hand corner. Mark the centre of enlargement somewhere between the object and the centre of the page. By drawing guidelines, draw the image with a scale factor of −2.

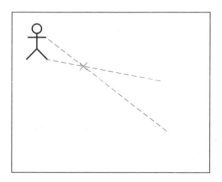

In questions **11** and **12** copy the diagrams and find the images of the triangles using P as the centre of enlargement and a scale factor of −2.

11   **12**

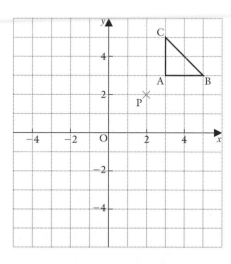
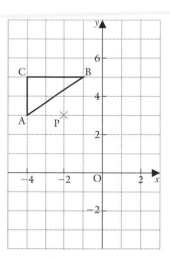

13 Draw axes for x from −10 to 4 and for y from −2 to 2.
Draw △ABC with A(2, 1), B(4, 1) and C(2, 2).
If the centre of enlargement is (1, 1) and the scale factor is −3, find the image of △ABC.

INVARIANT POINTS

A point which is its own image, that is, such that the object point and its image are in the same place, is called an *invariant point*. With reflection, the invariant points lie on the mirror line. The mirror line is an *invariant line*.

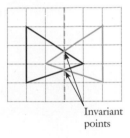

Invariant points

REFLECTIONS: FINDING THE MIRROR LINE

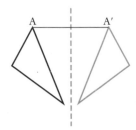

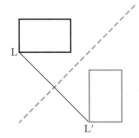

We saw in the work on reflection in Book 8A, that the object points and the corresponding image points are at equal distances from the mirror line, and the lines joining them (for example, AA′ and LL′) are perpendicular (at right angles) to the mirror line.

EXERCISE 14C

Find the mirror line if $\triangle A'B'C'$ is the image of $\triangle ABC$.

The mirror line is halfway between A and A′ and perpendicular to AA′.

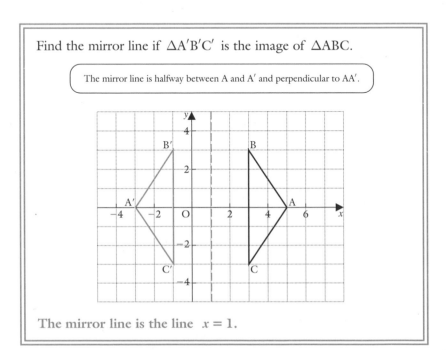

The mirror line is the line $x = 1$.

In each question from **1** to **4** copy the diagram and draw the mirror line.
Hence write down its equation.

1

3

2

4

Draw axes for x and y from -5 to 5 for each of the questions **5** to **8**.

5 Draw square PQRS: P($1, 1$), Q($4, 1$), R($4, 4$), S($1, 4$). Draw
square P$'$Q$'$R$'$S$'$: P$'$($-2, 1$), Q$'$($-5, 1$), R$'$($-5, 4$), S$'$($-2, 4$).
Draw the mirror line so that P$'$Q$'$R$'$S$'$ is the reflection of PQRS and
write down its equation.

6 Draw $\triangle$XYZ: X($2, 1$), Y($4, 4$), Z($-2, 4$), and $\triangle$X$'$Y$'$Z$'$:
X$'$($2, 1$), Y$'$($4, -2$), Z$'$($-2, -2$). Draw the mirror line so that
$\triangle$X$'$Y$'$Z$'$ is the reflection of $\triangle$XYZ and write down its equation.
Are there any invariant points? If there are, name them.

7 Draw $\triangle$ABC: A($-2, 0$), B($0, 2$), C($-3, 3$), and $\triangle$PQR:
P($3, -1$), Q($4, -4$), R($1, -3$). Draw the mirror line so that
$\triangle$PQR is the reflection of $\triangle$ABC. Which point is the image of A?
Are there any invariant points? If there are, name them.

8 Draw lines AB and PQ: A($2, -1$), B($4, 4$), P($-2, -1$), Q($-5, 4$).
Is PQ a reflection of AB? If it is, draw the mirror line. If not, give a
reason.

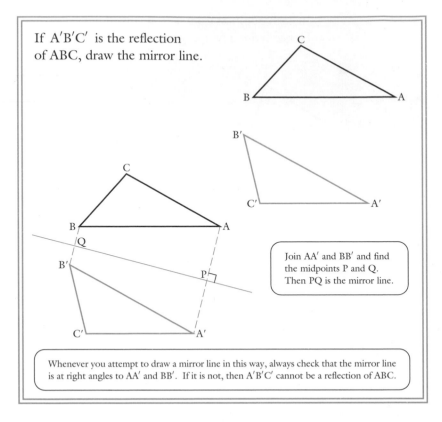

If $A'B'C'$ is the reflection of ABC, draw the mirror line.

Join AA' and BB' and find the midpoints P and Q. Then PQ is the mirror line.

Whenever you attempt to draw a mirror line in this way, always check that the mirror line is at right angles to AA' and BB'. If it is not, then $A'B'C'$ cannot be a reflection of ABC.

9 Trace the diagrams and draw the mirror lines.

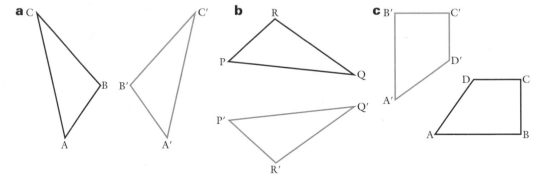

a **b** **c**

10 On squared paper draw $\triangle ABC$: A(3, 1), B(4, 5), C(1, 4), and $\triangle A'B'C'$: A'(0, -2), B'(-4, -3), C'(-3, 0). Draw the mirror line so that $\triangle A'B'C'$ is the image of $\triangle ABC$.

11 Draw axes for x and y from -2 to 8, using 1 cm to 1 unit. B is the point (-2, 0) and B' is the point (6, 2). Draw the mirror line so that B' is the reflection of B.

12 Find the gradient and y-intercept of the mirror line in question **11**. Hence find the equation of the mirror line.

ROTATIONS

In Book 8A, we saw that it is often easy to spot both the centre of rotation and the angle of rotation. In the next exercise we revise this work.

EXERCISE 14D

For each diagram state the centre of rotation and the angle of rotation. △ABC is the object in each case.

1

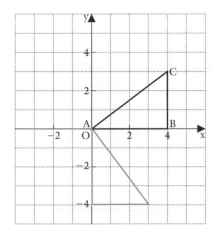

4

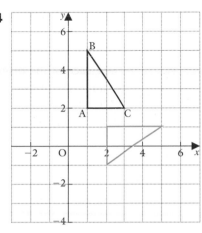

2

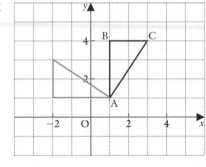

5

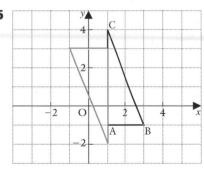

3

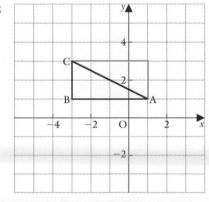

6

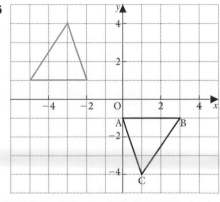

FINDING THE ANGLE OF ROTATION

When the centre of rotation is known, the angle of rotation can be found by joining both an object point and its image to the centre.

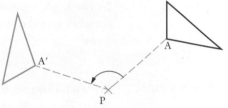

In the diagram above, A′ is the image of A and P is the centre of rotation. Join both A and A′ to P. $\widehat{APA'}$ is the angle of rotation.

In this case the angle of rotation is 120° anticlockwise.

EXERCISE 14E

Trace each of the diagrams and, by drawing in the necessary lines, find the angle of rotation when △ABC is rotated about the centre P to give △A′B′C′.

1

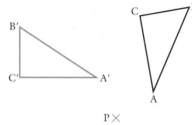

2

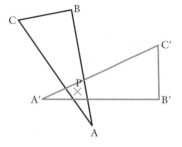

COMPOUND TRANSFORMATIONS

If we reflect an object in the x-axis and then reflect the resulting image in the y-axis we are carrying out a compound transformation.

EXERCISE 14F

In questions **1** to **4** copy the diagram and carry out the given compound transformation. Label the final image P.

1

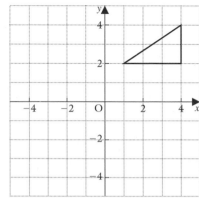

2

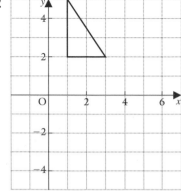

A reflection in the x-axis followed by a reflection the y-axis.

A rotation of 90° clockwise about the point (3, 2) followed by a reflection in the x-axis.

3

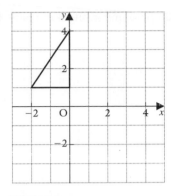

A rotation of $180°$ about the point $(0, 2)$ followed by a reflection in the line $x = 2$.

4

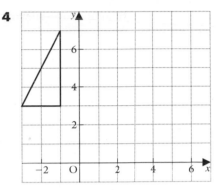

A reflection in the y-axis followed by a translation which is 2 units parallel to the x-axis to the right and 3 units parallel to the y-axis down.

In questions **5** to **8** describe a compound transformation that maps $\triangle ABC$ onto $\triangle A'B'C'$.

5

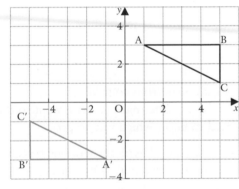

7

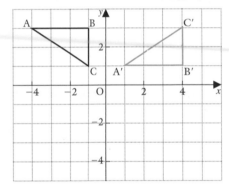

6

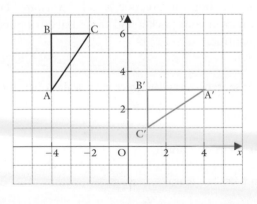

8

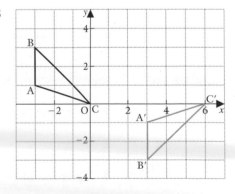

9 The shape labelled **A** can be drawn using a drawing program on a computer.

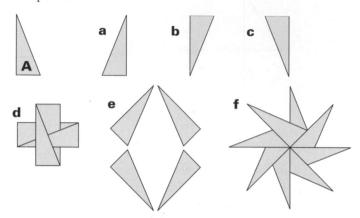

The program has the following commands to transform a shape:

Flip <u>T</u>op-bottom, Flip <u>L</u>eft-right, <u>R</u>otate 45 ° clockwise

Which of these commands, either singly or in combination, have been used to make the shapes **a**, **b**, and **c**? Use the underlined letters to describe the commands.

A shape can be dragged anywhere on the screen so it is possible to build up transformations of the basic shape **A** to make interesting patterns. Describe the transformations of **A** that will give the patterns **d**, **e** and **f**.

VECTORS

If you arranged to meet your friend 3 km from your home, this information would not be enough to ensure that you both went to the same place. You would also need to know which way to go.

Two pieces of information are required to describe where one place is in relation to another: the distance and the direction. Quantities which have both *magnitude* (size) and *direction* are called *vectors*.

A quantity which has magnitude but not direction is called a *scalar*. For example, the amount of money in your pocket and the number of pupils in your school are scalar quantities.

EXERCISE 14G

State whether the following sentences refer to vector or scalar quantities.

1 There are 24 pupils in my class.

2 To get to school I walk 2 km due north.

3 There are 11 players in a cricket team.

4 John walked at 6 km per hour.

5 The vertical cliff face is 50 m high.

6 Flight BA 101 is flying due west at 350 mph.

7 Castle Millington is 50 miles due east of Junction 48 on the M7.

8 Give other examples of **a** vector quantities **b** scalar quantities.

REPRESENTING VECTORS

We can represent a vector by a line segment (i.e. a straight line whose length represents the size of the vector) and indicate its direction with an arrow.

For example

We use **a**, **b**, **c**, ... to name the vectors.
When writing by hand it is difficult to write **a**, which is in heavy type, so we use a̲.

VECTORS IN THE FORM $\begin{pmatrix} a \\ b \end{pmatrix}$

When vectors are drawn on squared paper, we can describe them in terms of the number of squares we need to go across and up. For example, in the diagram below, the vector **a** corresponds to a movement of 4 across and 2 up and we write $\mathbf{a} = \begin{pmatrix} 4 \\ 2 \end{pmatrix}$

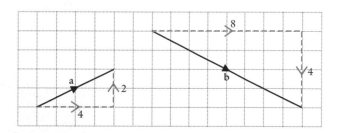

The vector **b** can be described as 8 across and 4 down. As with coordinates, we use negative numbers to indicate movement down or movement to the left.

Therefore $\mathbf{b} = \begin{pmatrix} 8 \\ -4 \end{pmatrix}$

Notice that the top number represents movement across and that the bottom number represents movement up or down.

EXERCISE 14H

Write the following vectors in the form $\begin{pmatrix} p \\ q \end{pmatrix}$.

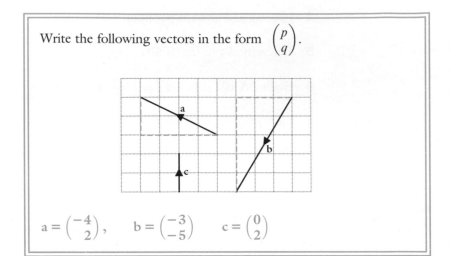

$$a = \begin{pmatrix} -4 \\ 2 \end{pmatrix}, \qquad b = \begin{pmatrix} -3 \\ -5 \end{pmatrix} \qquad c = \begin{pmatrix} 0 \\ 2 \end{pmatrix}$$

Write the following vectors in the form $\begin{pmatrix} p \\ q \end{pmatrix}$.

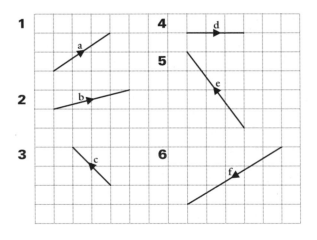

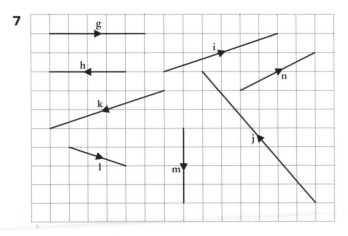

On squared paper draw the following vectors. Label each vector with its letter and an arrow.

8 $a = \begin{pmatrix} 3 \\ 5 \end{pmatrix}$ **11** $d = \begin{pmatrix} 6 \\ -12 \end{pmatrix}$ <u>**14**</u> $g = \begin{pmatrix} -4 \\ 3 \end{pmatrix}$

9 $b = \begin{pmatrix} -4 \\ -3 \end{pmatrix}$ **12** $e = \begin{pmatrix} 6 \\ 10 \end{pmatrix}$ <u>**15**</u> $h = \begin{pmatrix} -1 \\ -5 \end{pmatrix}$

10 $c = \begin{pmatrix} 2 \\ -4 \end{pmatrix}$ **13** $f = \begin{pmatrix} -2 \\ 5 \end{pmatrix}$ <u>**16**</u> $i = \begin{pmatrix} -6 \\ 2 \end{pmatrix}$

17 What do you notice about the vectors in questions **8** and **12**, and in questions **10** and **11**?

The coordinates of the starting point of the vector $\begin{pmatrix} 5 \\ 2 \end{pmatrix}$ are (2, 1). Find the coordinates of its other end.

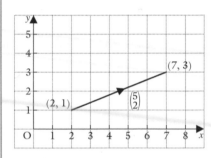

Starting at (2, 1) we go 5 units to the right followed by 2 units up.

The coordinates of its other end are (7, 3).

In each of the following questions you are given a vector followed by the coordinates of its starting point. Find the coordinates of its other end.

18 $\begin{pmatrix} 3 \\ 3 \end{pmatrix}, (4, 1)$ <u>**24**</u> $\begin{pmatrix} -3 \\ 4 \end{pmatrix}, (2, -4)$

19 $\begin{pmatrix} 3 \\ 1 \end{pmatrix}, (-2, -3)$ <u>**25**</u> $\begin{pmatrix} -6 \\ -6 \end{pmatrix}, (-3, -2)$

20 $\begin{pmatrix} -6 \\ 2 \end{pmatrix}, (3, 5)$ <u>**26**</u> $\begin{pmatrix} 4 \\ 3 \end{pmatrix}, (-2, -3)$

21 $\begin{pmatrix} -4 \\ -3 \end{pmatrix}, (5, -2)$ <u>**27**</u> $\begin{pmatrix} 5 \\ -3 \end{pmatrix}, (2, -1)$

22 $\begin{pmatrix} 5 \\ 2 \end{pmatrix}, (3, -1)$ <u>**28**</u> $\begin{pmatrix} -5 \\ 2 \end{pmatrix}, (-4, -3)$

23 $\begin{pmatrix} 4 \\ -2 \end{pmatrix}, (4, 2)$ <u>**29**</u> $\begin{pmatrix} -4 \\ -2 \end{pmatrix}, (-3, -1)$

The coordinates of the other end of the vector $\begin{pmatrix} 6 \\ 4 \end{pmatrix}$ are (8, 6). Find the coordinates of its starting point.

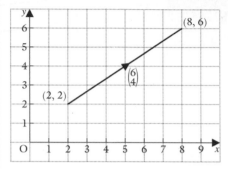

Starting at (8, 6) we have to go 6 units to the left and 4 units down.

The coordinates of the starting point of the vector are (2, 2).

In each of the following questions a vector is given followed by the coordinates of its other end. Find the coordinates of its starting point.

30 $\begin{pmatrix} 10 \\ 2 \end{pmatrix}$, (4, 1)   **33** $\begin{pmatrix} 8 \\ 6 \end{pmatrix}$, (6, 3)   **36** $\begin{pmatrix} 4 \\ -2 \end{pmatrix}$, (-3, 2)

31 $\begin{pmatrix} 5 \\ -1 \end{pmatrix}$, (3, -4)   **34** $\begin{pmatrix} -3 \\ 4 \end{pmatrix}$, (-2, 1)   **37** $\begin{pmatrix} -2 \\ 6 \end{pmatrix}$, (-3, -4)

32 $\begin{pmatrix} -5 \\ -2 \end{pmatrix}$, (-2, -4) **35** $\begin{pmatrix} -6 \\ -3 \end{pmatrix}$, (-5, 2)   **38** $\begin{pmatrix} 1 \\ 4 \end{pmatrix}$, (-5, -2)

TRANSLATIONS

We can describe a translation only by stating what the movement or *displacement* is. The easiest way to do this is to give the vector that describes the displacement.

EXERCISE 14I

Describe the transformation that maps $\triangle ABC$ onto $\triangle A'B'C'$.

Consider the movement or displacement from A to A'.

The transformation is given by the vector $\begin{pmatrix} 6 \\ -2 \end{pmatrix}$.

In questions from **1** to **4** use a vector to describe the transformation that maps the object onto its green image.

1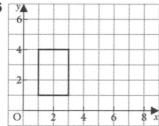

3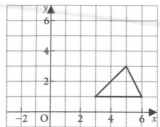

2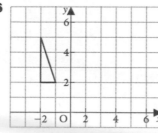

4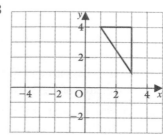

In each question from **5** to **8**, copy the diagram and draw the image of the given object under a translation defined by the given vector.

5

The vector is $\begin{pmatrix} 5 \\ 1 \end{pmatrix}$.

7

The vector is $\begin{pmatrix} -4 \\ 2 \end{pmatrix}$.

6

The vector is $\begin{pmatrix} 5 \\ -3 \end{pmatrix}$.

8

The vector is $\begin{pmatrix} -4 \\ -3 \end{pmatrix}$.

9 In each question from **5** to **8** use a vector to describe the transformation that maps the image onto the given object.

MIXED EXERCISE

EXERCISE 14J

1 a Copy the diagram and find the centre of enlargement and the scale factor when $\triangle ABC$ is enlarged to give $\triangle A'B'C'$.

b Draw the reflection of $\triangle A'B'C'$ in the line $x = 3$ followed by a translation defined by the vector $\begin{pmatrix} 4 \\ -4 \end{pmatrix}$. Label it Q.

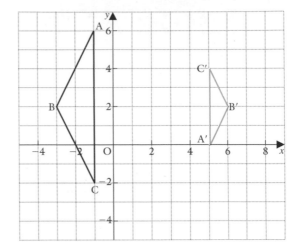

2 a On x- and y-axes scaled from 0 to 12, plot the points $A(6, 3)$, $B(9, 9)$, $C(12, 3)$.

b An enlargement, scale factor $\frac{1}{3}$ and centre $(0, 0)$, transforms $\triangle ABC$ onto $\triangle A_1 B_1 C_1$. Draw $\triangle A_1 B_1 C_1$.

c A translation of $\triangle A_1 B_1 C_1$ onto $\triangle A_2 B_2 C_2$ is defined by the vector $\begin{pmatrix} 8 \\ 2 \end{pmatrix}$. Draw $\triangle A_2 B_2 C_2$. What are the coordinates of B_2?

d $\triangle A_2 B_2 C_2$ can be transformed onto $\triangle ABC$ by an enlargement. Give the scale factor and the centre of enlargement.

3 a Copy the diagram onto squared paper. Draw the mirror line so that $\triangle A_1 B_1 C_1$ is the image of $\triangle ABC$. Write down the coordinates of the points where this line crosses the axes.

b $\triangle ABC$ is mapped onto $\triangle A_2 B_2 C_2$ by a rotation. Give the angle of rotation.

c $\triangle A_2 B_2 C_2$ is reflected in the line $y = -3$ to give $\triangle A_3 B_3 C_3$. Mark $\triangle A_3 B_3 C_3$ on your diagram and write down the coordinates of C_3.

d Describe the transformation that maps $\triangle A_1 B_1 C_1$ onto $\triangle A_3 B_3 C_3$.

e Write down the vector that defines the translation that maps **i** C_3 onto C_1 **ii** C_1 onto C **iii** C onto C_2.

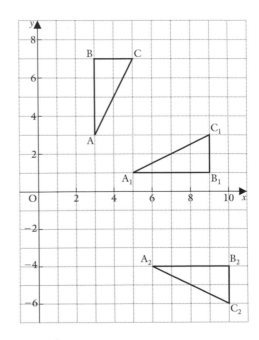

INVESTIGATIONS

1 The sketch shows a plan of Paul's office. He wants to move a filing cabinet from position P to position Q. This can only be done by rotating it through 90° about any one of its corners, but he can do this as many times as he likes. Is it possible for him to do this if the filing cabinet must be facing the same way after the move as it was at the start? Mark

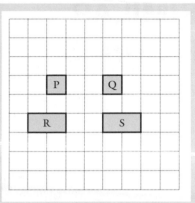

all the squares on the grid to which it is possible to move the cabinet.

Repeat the investigation for a desk that occupies two squares on the grid, and needs to be moved from R to S.

2

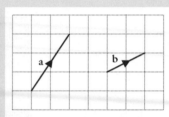

This diagram shows the vectors $\mathbf{a} = \begin{pmatrix} 2 \\ 3 \end{pmatrix}$ and $\mathbf{b} = \begin{pmatrix} 2 \\ 1 \end{pmatrix}$.
The point $(6, 7)$ can be reached from the origin by adding $\mathbf{a}$ and $\mathbf{b}$, and then adding $\mathbf{a}$ to the result. This is shown in the diagram.

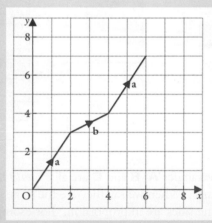

a In how many different ways can these vectors be combined to get from the origin to the point $(6, 7)$? Show them on a diagram.

b The vector in the direction opposite to $\mathbf{a}$ is called $-\mathbf{a}$ so as

$$\mathbf{a} = \begin{pmatrix} 2 \\ 3 \end{pmatrix}, -\mathbf{a} = \begin{pmatrix} -2 \\ -3 \end{pmatrix}.$$

Similarly $-\mathbf{b} = \begin{pmatrix} -2 \\ -1 \end{pmatrix}$.

How many more ways are possible if you must always remain on the given grid but you are allowed to combine the vectors $\mathbf{a}, \mathbf{b}, -\mathbf{a}$ and $-\mathbf{b}$?

c How many of the points on the above grid is it possible to get to from the origin if combinations of $\mathbf{a}, \mathbf{b}, -\mathbf{a}$ and $-\mathbf{b}$ are allowed?

SIMILAR FIGURES

In the last chapter we studied different types of transformation. One type of transformation was an enlargement. In the diagram below, rectangle **A** is enlarged by a factor of 2 with P as the centre of enlargement. This transformation gives rectangle **B**. Rectangle **A** is also enlarged by a scale factor of $\frac{1}{3}$ with Q as the centre of enlargement. This gives rectangle **C**, which is smaller than **A** since the scale factor is a fraction less than 1.

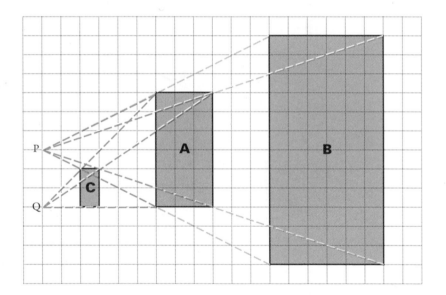

The three rectangles **A**, **B** and **C** are all *exactly the same shape* but are *different in size*. When we say they are the same shape we mean that the corresponding sides are in proportion. For example, the long and short sides in **B** are twice the length of the long and short sides in **A**. Likewise the long and short sides in **C** are $\frac{1}{3}$ of the length of the long and short sides in **A**. When one figure is an enlargement of another (it can be larger or smaller) we say that the two shapes are *similar*.

It doesn't matter whether one shape is then rotated or reflected or translated – the two shapes are still similar. In the diagram given opposite any one shape is similar to each of the other four.

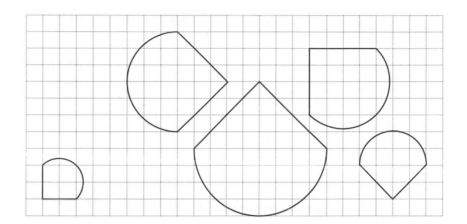

EXERCISE 15A   State whether or not the pairs of figures in questions **1** to **10** are similar.

1

5

2

6

3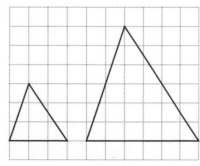

7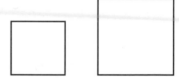

4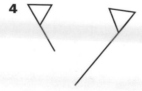

8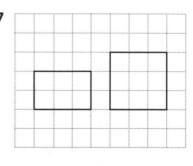

9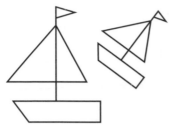

10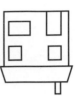

11 Which two rectangles are similar?

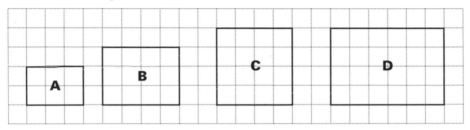

12 Which of the other five shapes are similar to **X**?

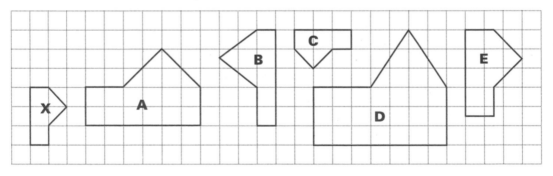

13 Which of the following statements are true?

a All circles are similar.

b All rectangles are similar.

c All squares are similar.

d All triangles are similar.

e All equilateral triangles are similar.

f All isosceles triangles are similar.

**SIMILAR
TRIANGLES**

Some of the easiest similar figures to deal with are triangles. This is because only a small amount of information is needed to prove them to be similar.

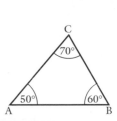

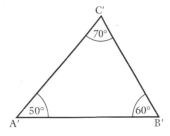

In these triangles the corresponding angles are equal and so the triangles are the same shape. One triangle is an enlargement of the other. These triangles are *similar*.

We can prove that two triangles are similar if we can show that they have the same angles.

EXERCISE 15B

1

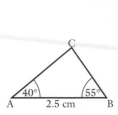

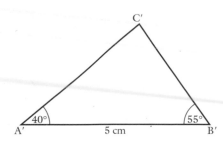

    a Are the triangles similar?

    b Measure the remaining sides.

    c Find $\dfrac{A'B'}{AB}$, $\dfrac{B'C'}{BC}$ and $\dfrac{C'A'}{CA}$

    (as decimals if necessary, correct to 1 d.p.)

    d What do you notice about the answers to part **c**?

Repeat question **1** for the pairs of triangles in questions **2** to **4**.

2

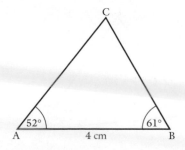

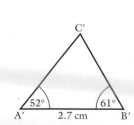

3

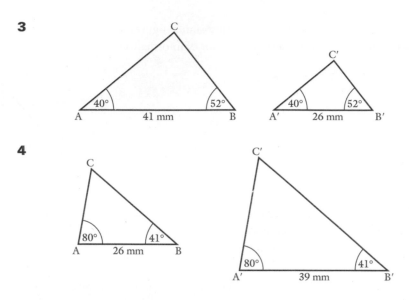

4

Sketch the following pairs of triangles and find the size of each of the missing angles. In each question state whether the two triangles are similar. (One triangle may be turned round or over compared with the other.)

5

6

7

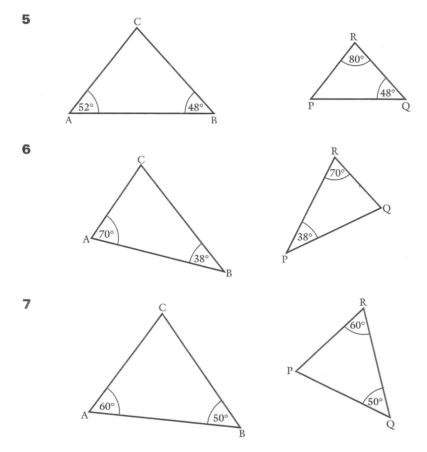

CORRESPONDING
VERTICES

These two triangles are similar and we can see that X corresponds to A, Y to B and Z to C.

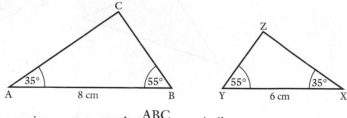

We can write: Δs $\dfrac{ABC}{XYZ}$ are similar

> Make sure that X is written below A, Y below B and Z below C,
> i.e.   that the angles of one triangle are
> below the corresponding angles of the other.

The pairs of corresponding sides are in the same ratio and we can read these from $\dfrac{ABC}{XYZ}$, that is $\dfrac{AB}{XY} = \dfrac{BC}{YZ} = \dfrac{CA}{ZX}$

EXERCISE 15C

> State whether triangles ABC and PQR are similar and, if they are, give the ratios of the sides.
>
>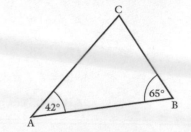
>
> $\hat{Q} = 58°$ (angles of a triangle)
>
> and $\hat{C} = 90°$
>
> so Δs $\dfrac{ABC}{RQP}$ are similar | The angles in each triangle are 32°, 58° and 90°. |
>
> and $\dfrac{AB}{RQ} = \dfrac{BC}{QP} = \dfrac{CA}{PR}$

In questions **1** to **3**, state whether the two triangles are similar and, if they are, give the ratios of the sides.

1

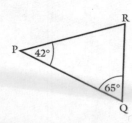

2

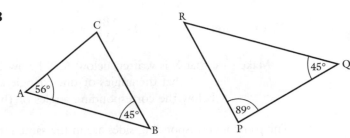

3

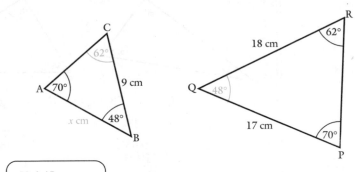

FINDING A
MISSING LENGTH

EXERCISE 15D

State whether the two triangles are similar. If they are, find AB.

Mark AB as x cm.

$\widehat{C} = 62°$ and $\widehat{Q} = 48°$ (angles of a triangle)

so $\quad \Delta s \dfrac{ABC}{PQR}$ are similar $\quad$ and $\quad \dfrac{AB}{PQ} = \dfrac{BC}{QR} = \dfrac{CA}{RP}$

i.e. $\qquad\qquad\qquad \dfrac{x}{17} = \dfrac{9}{18}$

$$\cancel{17} \times \dfrac{x}{\cancel{17}_1} = \dfrac{\cancel{9}'}{\cancel{18}_2} \times 17$$

$$x = \dfrac{17}{2} = 8.5$$

$$AB = 8.5 \text{ cm}$$

In questions **1** to **5**, state whether the pairs of triangles are similar. If they are, find the required side.

1 Find PR.

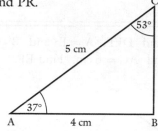

2 Find QR.

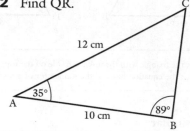

3 Find BC.

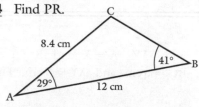

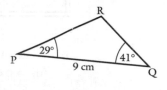

4 Find PR.

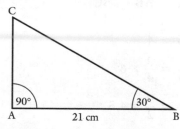

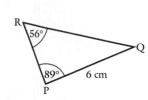

5 Find QR.

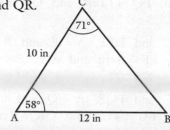

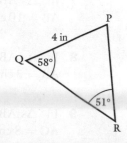

In some cases we do not need to know the sizes of the angles as long as we know that pairs of angles are equal. (Two pairs only are needed as the third pair must then be equal.)

In $\triangle$s ABC and DEF, $\widehat{A} = \widehat{E}$ and $\widehat{B} = \widehat{D}$. AB = 4 cm, DE = 3 cm and AC = 6 cm. Find EF.

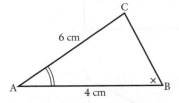

We put the triangle with the unknown side on the top; this makes sure that the equation has x as the numerator of a fraction.

$\triangle$s $\dfrac{\text{EDF}}{\text{ABC}}$ are similar

so $\dfrac{\text{FE}}{\text{CA}} = \dfrac{\text{ED}}{\text{AB}} = \dfrac{\text{DF}}{\text{BC}}$

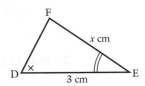

We need $\dfrac{\text{FE}}{\text{CA}}$ and $\dfrac{\text{ED}}{\text{AB}}$.

$\dfrac{x}{6} = \dfrac{3}{4}$

$\cancel{6}^{1} \times \dfrac{x}{\cancel{6}_{1}} = \dfrac{3}{\cancel{4}_{2}} \times \cancel{6}^{3}$

$x = \dfrac{9}{2} = 4.5$

so   EF = 4.5 cm

6 In $\triangle$s ABC and XYZ, $\widehat{A} = \widehat{X}$ and $\widehat{B} = \widehat{Y}$.
AB = 6 cm, BC = 5 cm and XY = 9 cm. Find YZ.

7 In $\triangle$s ABC and PQR, $\widehat{A} = \widehat{P}$ and $\widehat{C} = \widehat{R}$.
AB = 10 cm, PQ = 12 cm and QR = 9 cm. Find BC.

8 In $\triangle$s ABC and DEF, $\widehat{A} = \widehat{E}$ and $\widehat{B} = \widehat{F}$.
AB = 3 cm, EF = 5 cm and AC = 5 cm. Find DE.

9 In $\triangle$s ABC and PQR, $\widehat{A} = \widehat{Q}$ and $\widehat{C} = \widehat{R}$.
AC = 8 cm, BC = 4 cm and QR = 9 cm. Find PR.

a Show that triangles ABC and CDE are similar.

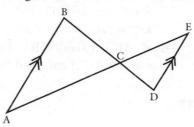

b Given that AC = 15 cm, CE = 9 cm and DE = 8 cm, find AB.

a $\hat{A} = \hat{E}$ (alternate angles, AB||DE)

$\hat{B} = \hat{D}$ (alternate angles, AB||DE)

> Or we could use $B\hat{C}A = E\hat{C}D$ as these are vertically opposite angles.

so $\Delta s \begin{matrix} ABC \\ EDC \end{matrix}$ are similar

b

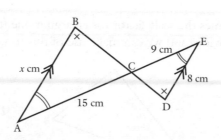

$$\frac{AB}{ED} = \frac{BC}{DC} = \frac{CA}{CE}$$

$$\frac{x}{8} = \frac{15}{9}$$

$$\cancel{8} \times \frac{x}{\cancel{8}_1} = \frac{\cancel{15}^5}{\cancel{9}_3} \times 8$$

$$x = \frac{40}{3}$$

$$= 13\frac{1}{3}$$

$AB = 13\frac{1}{3}$ cm, or 13.3 cm correct to 3 s.f.

10 a Show that Δs ABC and BDE are similar.

b If AB = 6 cm, BD = 3 cm and DE = 2 cm, find BC.

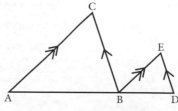

11 a Show that Δs ABC and CDE are similar.

b If AB = 7 cm, BC = 6 cm, AC = 4 cm and CE = 6 cm, find CD and DE.

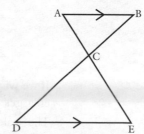

12 a ABCD is a square.
EF is at right angles to BD.
Show that △s ABD and DEF
are similar.

b If AB = 10 cm, DB = 14.2 cm
and DF = 7.1 cm, find EF.

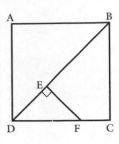

13

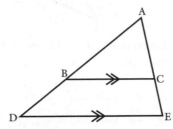

a Show that △s ABC and ADE are similar. (Notice that $\hat{A}$ is
common to both triangles.)

b If AB = 10 cm, AD = 15 cm, BC = 12 cm and
AC = 9 cm, find DE, AE and CE.

**USING THE
SCALE FACTOR
TO FIND THE
MISSING LENGTH**

Sometimes the scale factor for enlarging one triangle into the other is
very obvious and we can make use of this to save ourselves some work.

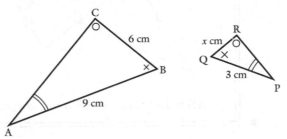

The two triangles above are similar and we can see that the scale factor
for 'enlarging' the first triangle into the second is $\frac{1}{3}$.
We can say straightway that x is $\frac{1}{3}$ of 6.

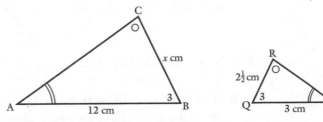

If we wish to find a length in the first triangle, we use the scale factor that
enlarges the second triangle to the first.

The scale factor is 4   so $x = 4 \times 2\frac{1}{2} = 10$.

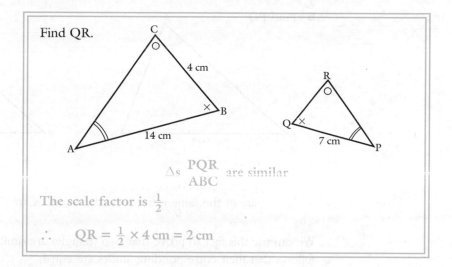

Find QR.

4 cm

14 cm

7 cm

$\triangle$s $\dfrac{\text{PQR}}{\text{ABC}}$ are similar

The scale factor is $\frac{1}{2}$

$\therefore \quad QR = \frac{1}{2} \times 4\,\text{cm} = 2\,\text{cm}$

1 Find BC.

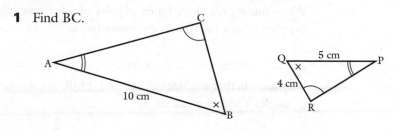

2 Find PR.

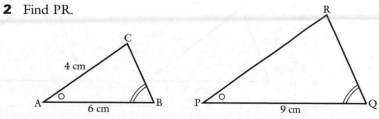

3 Find PR.

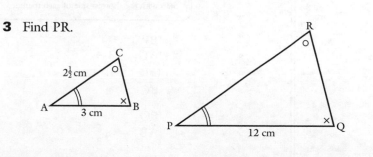

4 Find LN.

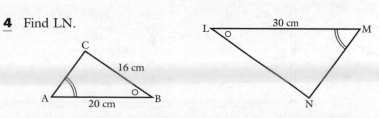

5 Find PQ.

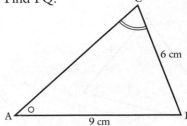
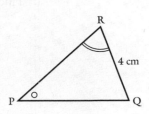

CORRESPONDING SIDES

> If three pairs of corresponding sides of two triangles are in the same ratio, then the triangles are similar.

We can use this fact to prove that two triangles are similar. It then follows that their corresponding angles are equal.

When finding the three ratios of pairs of sides, give each ratio as a whole number or as a fraction in its lowest terms.

EXERCISE 15F

State whether triangles ABC and PQR are similar. Say which angle, if any, is equal to $\widehat{A}$.

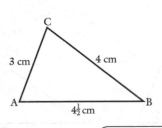
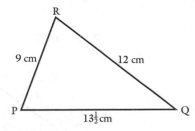

> Start with the shortest side of each triangle.

$$\frac{PR}{AC} = \frac{9}{3} = 3$$

$$\frac{QR}{BC} = \frac{12}{4} = 3$$

$$\frac{PQ}{AB} = \frac{13\frac{1}{2}}{4\frac{1}{2}} = \frac{27}{9} = 3$$

i.e. $$\frac{PR}{AC} = \frac{QR}{BC} = \frac{PQ}{AB}$$

so $$\Delta s \; \frac{PQR}{ABC} \; \text{are similar}$$

∴ $$\widehat{P} = \widehat{A}$$

State whether the following pairs of triangles are similar. In each case say which angle, if any, is equal to $\widehat{A}$.

1

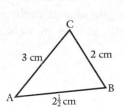
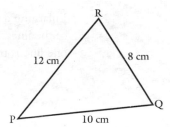

2

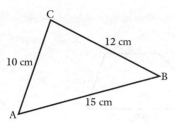
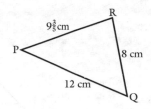

3

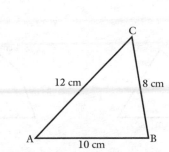
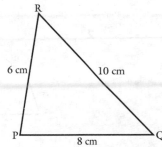

4

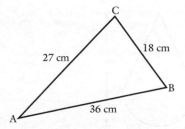
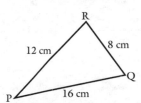

5

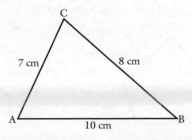
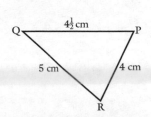

OTHER SHAPES

So far in this chapter we have concentrated on triangles that are similar. However, relationships we have found for similar triangles apply equally well to other shapes.

Enlarging a figure does not alter the angles. It does change the lengths of the lines, but all the lengths change in the same ratio; for example, if one line is trebled in length, all lines are trebled in length.

EXERCISE 15G

In questions **1** to **5** you are given two similar figures. Find each length marked with a letter.

1

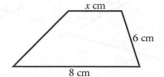
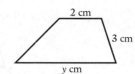

2

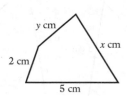
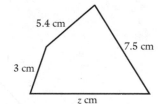

3

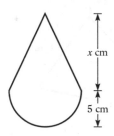
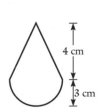

4

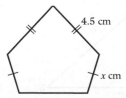
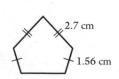

5

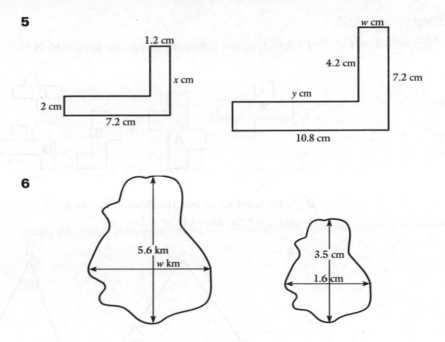

6

The sketches show the length and width of a lake on a map and the corresponding length of the actual lake. What is the width of the lake at this widest point?

7 The sketch shows a field which is planted with two different crops of cereal. Find the length of the common border between the two crops.

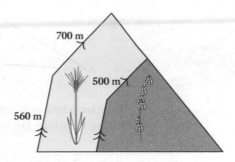

8

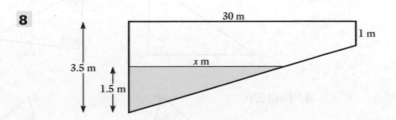

The diagram shows the cross-section of a swimming pool which is 3.5 m deep at one end and 1 m deep at the other. The pool is 30 m long and 10 m wide. Water flows into the pool to a depth of 1.5 m at the deep end. What are the dimensions of the surface of the water?

1 Which of the following shapes are similar to **X**?

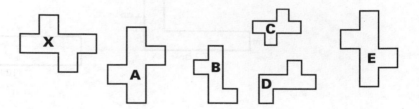

2 State whether or not triangles ABC and DEF are similar. If they are, give the ratio of the sides.

a

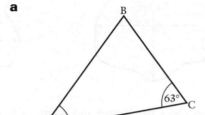

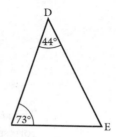

b

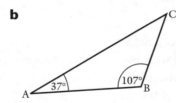

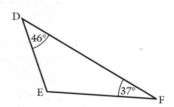

3 State whether triangles ABC and DEF are similar. If they are, find the length of DE.

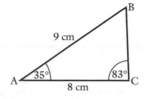

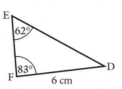

4 Find XY.

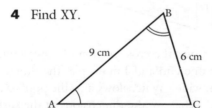

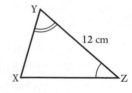

5 State whether or not triangles ABC and XYZ are similar.

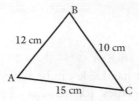
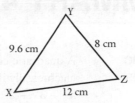

6 The two given shapes are similar. Find the lengths marked with letters.

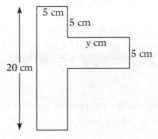
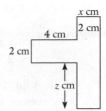

INVESTIGATIONS

1 a Draw a triangle ABC with three unequal sides. Mark a point D in any position on AB. Draw DE parallel to AC so that E lies on BC. Draw EF parallel to AB so that F lies on AC. Join DF. Are triangles ABC and DEF similar?
Mark D in a new position on AB and repeat the process.
Try different positions of D until you have a triangle DEF that is similar to triangle ABC.
(You may find it necessary to draw your original triangle several times.)
Comment on the position of D. Does this position for D result in special positions for E and F?
Justify your answer.

b Draw another triangle ABC. Repeat part **a**.

2 a Draw a quadrilateral ABCD with four unequal sides. Mark the midpoints of the sides AB, BC, CD and DA using the letters P, Q, R and S in that order. Join P, Q, R and S to give a new quadrilateral. Are the two quadrilaterals ABCD and PQRS similar?

b Repeat part **a** for other quadrilaterals including the special quadrilaterals this time. Comment on any quadrilaterals ABCD and PQRS which are found to be similar.

SUMMARY 4

QUADRATIC
EQUATIONS

A quadratic equation, such as $x^2 - 3x + 2 = 0$ can be solved algebraically if the left-hand side can be factorised.

$x^2 - 3x + 2 = 0$ becomes $(x - 2)(x - 1) = 0$

Now we use the fact that if the product of two numbers is zero then one or both of the numbers must be zero.

Therefore as $(x - 2)(x - 1) = 0$

then either $x - 2 = 0$ in which case $x = 2$

or $x - 1 = 0$ in which case $x = 1$

Hence the equation $x^2 - 3x + 2 = 0$ has two solutions;

$x = 2$ and $x = 1$.

A quadratic equation must be rearranged in the order $ax^2 + bx + c = 0$ before factorisation is attempted.

GRAPHS

Straight lines and parabolas

An equation of the form $y = mx + c$ gives a straight line where m is the gradient of the line and c is the intercept on the y-axis.

An equation of the form $y = ax^2 + bx + c$ gives a curve whose shape is called a parabola and looks like this when a is positive.

When the x^2 term is negative, the curve is 'upside down'.

Cubic curves

When the equation of a curve contains x^3 (and maybe terms involving x^2, x and a number), the curve is called a *cubic* curve.

These equations give cubic curves:

$$y = x^3, \quad y = 2x^3 - x + 5, \quad y = x^3 - 2x^2 + 6$$

A cubic curve looks like ⟋ or ⟿ when the x^3 term is positive

and ⟍ or ⟍⟋ when the x^3 term is negative.

Reciprocal curves

The equation $y = \dfrac{a}{x}$ where a is a number,

is called a *reciprocal equation*.

An equation of the form $y = \dfrac{a}{x}$, where
a is a constant (that is, a number), gives
a two-part curve called a *reciprocal curve*
or a *hyperbola*. There are no values for
y when $x = 0$; division by zero is not
possible.

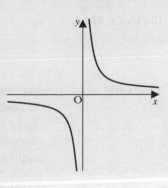

Any two quantities, x and y, that are *inversely proportional*, are related

by the equation $y = \dfrac{k}{x}$ and the graph representing them is a hyperbola.

Gradient

The gradient, or slope, of a curve changes from
point to point.
Gradient gives the rate at which the quantity
on the vertical axis is changing as the quantity
on the horizontal axis increases; the steeper the
graph and therefore the greater the gradient,
the faster y is increasing compared with x.
For graphs of distance plotted against time

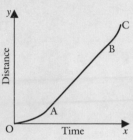

the steeper the graph is, the greater is the speed. In this graph, the
greatest speed is at C; for the section AB, the speed is steady.

AREA

The *area of a trapezium* is equal to
$\frac{1}{2}$ (sum of the parallel sides) × (distance between them)

$$= \tfrac{1}{2}\,(\,a + b\,) \times h$$

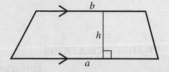

Part of the circumference of a circle is called an *arc*.

Length of the arc AB $= \dfrac{\widehat{AOB}}{360^\circ} \times 2\pi r$

The slice of the circle enclosed by the arc
and the radii is called a *sector*.

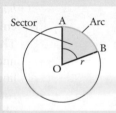

The *area of the sector* AOB $= \dfrac{\widehat{AOB}}{360^\circ} \times \pi r^2$

UNITS OF AREA

$$1 \text{ hectare} = 10\,000\,\text{m}^2 \quad 1 \text{ acre} = 4840 \text{ sq yd}$$
$$1 \text{ hectare} = 2.5 \text{ acres correct to } 1 \text{ d.p.}$$

VECTORS

A *vector* is any quantity that needs to be described by giving both its size (magnitude) and its direction. For example, to explain where one village is in relation to another we need to give both distance and direction, that is, a vector. A quantity that needs only size to describe it is called a *scalar*, for example, time.

A vector can be represented by a straight line with an arrow to show direction, e.g.

When vectors are drawn on squared paper they can be described in terms of the number of squares needed to go across to the right and up.

They are written in the form $\begin{pmatrix} a \\ b \end{pmatrix}$ where a is the number of squares across and b is the number of squares up.

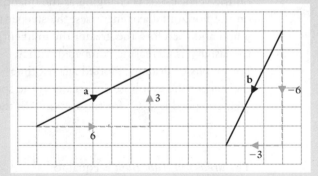

In the diagram, $\mathbf{a} = \begin{pmatrix} 6 \\ 3 \end{pmatrix}$ and $\mathbf{b} = \begin{pmatrix} -3 \\ -6 \end{pmatrix}$

TRANSFORMATIONS

Enlargement by a negative scale factor
When an object is enlarged by a scale factor -2, the guidelines are drawn *backwards* through the centre of enlargement, O, so that $OA' = 2OA$. This produces an image each of whose lines is twice as long as the corresponding line on the object. The image is also rotated by $180°$ with respect to the object.

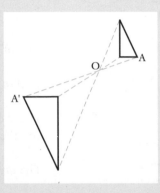

Finding the mirror line

When an object has been reflected, corresponding points on the object and the image are the same distance from the mirror line. Therefore the mirror line can be found by joining a pair of corresponding points on the object and the image, AA′ say, and then finding the line that goes through the midpoint of AA′ and is perpendicular to it.

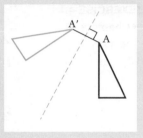

Translation

A *translation* can be described by the vector which gives the movement from the object to the image.

In the diagram, the movement of the point A to A′ is 9 units to the left and 3 units up, so the translation is described by the

vector $\begin{pmatrix} -9 \\ 3 \end{pmatrix}$.

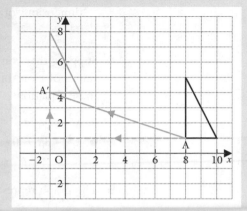

Compound transformation

A compound transformation is the result of one transformation followed by another; for example, the result of reflecting an object in the *y*-axis and then rotating the image obtained by 30° about the origin.

SIMILAR FIGURES

Two figures are similar if they are the same shape but are different in size, that is, one figure is an enlargement of the other. (One figure may be turned over or round with respect to the other.) It follows that the lengths of corresponding sides are all in the same ratio.

Similar triangles

Two triangles can be proved to be similar if it can be shown that either the three angles of one triangle are equal to the three angles of the other

(In practice only two pairs of angles need to be shown to be equal because, since the sum of the three angles in any triangle is 180°, it follows that the third pair must be equal.)

or the three pairs of corresponding sides are in the same ratio.

(Two pairs of sides in the same ratio is not enough to prove that the triangles are similar.)

1 Solve the equations.   **a** $(x+3)(x-4)=0$ **b** $x(x-5)=0$

2 Solve the equations.

a $x^2 - 5x - 14 = 0$ **b** $x^2 - 3x = 0$ **c** $x^2 - 12x + 36 = 0$

3 a Solve the equations.

   i $x^2 - 8x + 15 = 0$ **iii** $x^2 = 8x + 20$

  ii $3x^2 - x = 0$ **iv** $(x+4)(x-4)+12=0$

b A rectangle is x cm wide and is 5 cm longer than it is wide. Its area is 24 cm^2. Form an equation in x and solve it to find the dimensions of the rectangle.

4 Suppose that you have drawn the graph of $y = x^2 - 3x - 5$. What values of y on this graph give the values of x that are solutions to these equations.

a $x^2 - 3x - 7 = 0$ **c** $x^2 - 3x + 1 = 0$

b $x^2 - 3x = 0$ **d** $2x^2 - 6x - 11 = 0$?

5 The graph representing $y = x^2 - x$ could be

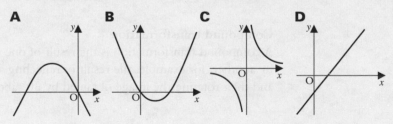

A　　　　**B**　　　　**C**　　　　**D**

6 Liquid is poured at a constant rate into each of the containers whose cross-sections are shown below. For each shape sketch the graph showing how the depth of liquid in the container increases as the liquid is poured in.

a　　　　**b**　　　　**c**　　　　**d**

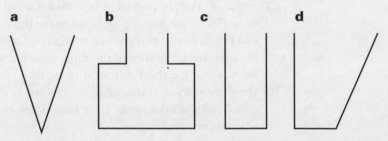

7 Find the area of each shape.

a

6.4 cm

3.5 cm

4.8 cm

b

32 cm

12 cm

27 cm

8 Find **a** the length of the minor arc AB

b the area of the sector

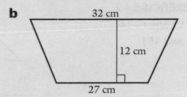

A

150°

12 cm

B

9 a The volume of this solid is 234 cm³.

Find **i** the area of cross-section

ii the length of the solid.

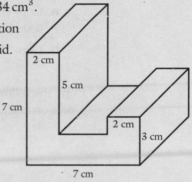

2 cm

5 cm

7 cm

2 cm

3 cm

7 cm

b Bourbon pâté is sold in tins like this.
The top has an area of 32 cm²
and the tin is 3 cm deep.

It is decided to change to cylindrical tins
that are 4 cm deep, but hold the same
quantity of pâté. Find the diameter
of the top of the cylindrical tin.

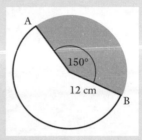

Pâté

3 cm

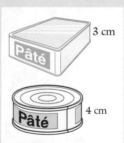

Pâté

4 cm

10 a State whether each of the following quantities should be
measured in units of length or area or volume.

  i The space inside a cubical box

  ii The distance round the edge of a lake

  iii The region inside a trapezium

  iv The amount of land needed to park 100 cars.

b Water issues from a pipe of cross-section 4.3 cm² at a speed of
0.6 m/s. How much water issues from the pipe in one second?

1

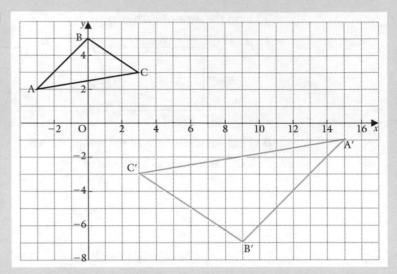

a Copy the diagram and find the centre of enlargement and the scale factor when $\triangle ABC$ is enlarged to give $\triangle A'B'C'$.

b Draw the reflection of $\triangle ABC$ in the line $x = 5$.

Copy the following diagram onto 5 mm squared paper using 1 square as 1 unit on both axes, and use it to answer questions **2** to **5**.

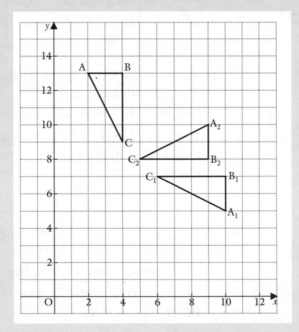

2 Draw the mirror line so that $\triangle A_1B_1C_1$ is the image of $\triangle ABC$. Write down its gradient and y-intercept.

3 △ABC is mapped onto △$A_2B_2C_2$ by a rotation. Find the centre and angle of rotation.

4 △$A_1B_1C_1$ is reflected in the line $y = 4$ to give △$A_3B_3C_3$. Mark △$A_3B_3C_3$ on your diagram and write down the coordinates of C_3.

5 a Write down the vector that defines the translation that maps

    i C_3 onto C_2 **i** C_2 onto C_3.

  b Describe the translation that maps △$A_2B_2C_2$ onto △$A_3B_3C_3$.

6 a State, with reasons, whether or not the following triangles are similar.

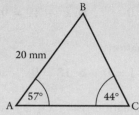 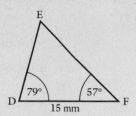

  b If they are, give the ratio of corresponding sides.

7 In triangles ABC and PQR, $\widehat{A} = \widehat{P}$ and $\widehat{C} = \widehat{R}$. AB = 12 cm, PQ = 15 cm and QR = 10 cm. Find BC.

8 a Show that triangles ABC and CDE are similar.

  b If AB = 12 cm, BC = 10 cm, AC = 9 cm and CE = 6 cm find DE and CD.

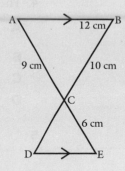

9

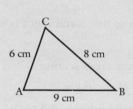 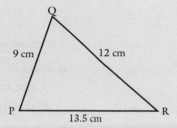

State whether or not triangles ABC and PQR are similar. Say which angle, if any, is equal to $\widehat{A}$.

10

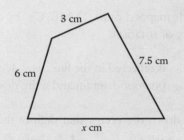 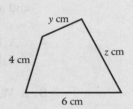

These two quadrilaterals are similar. Find the values of x, y and z.

**REVISION
EXERCISE 4.3
(Chapters 11
to 15)**

1 Solve the equations.

   a $x^2 + 7x = 0$ **b** $x^2 - 49 = 0$ **c** $x^2 + 3x - 28 = 0$

2 Solve the equations.

   a $x^2 - 2x + 1 = 0$ **b** $18 = 9x - x^2$ **c** $3x^2 + 12x + 9 = 0$

3 A curve has equation $y = (x + 2)(x - 1)(x - 5)$.

   a What name do we give to this type of equation?

   b How many values of x satisfy the equation
      $(x + 2)(x - 1)(x - 5) = 0$? What are they?

4 The equation of this curve could be

   A $y = \dfrac{12}{x}$

   B $y = x^3 - 4x$

   C $y = 4x - x^3$

   D $y = x^2 + 4x - 4$

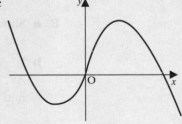

5

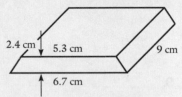

The cross-section of this metal ingot is a trapezium.

   a Find

     i the area of the cross-section

     ii the volume of the ingot.

   b The ingot is melted down and recast, without any change in
     volume, into a cube. What is the length of an edge of this cube?
     Give your answer correct to 3 significant figures.

6 On squared paper draw axes for x and y using the ranges
$-6 \leqslant x \leqslant 6$ and $-6 \leqslant y \leqslant 6$ and a scale of one square to 1 unit.
Plot the points A$(-5, 5)$, B$(4, 5)$, C$(2, -3)$, D$(-1, -3)$
and join them in alphabetical order to form a closed quadrilateral.
Find, in square units, the area of the resulting shape.

7 Copy the diagram.

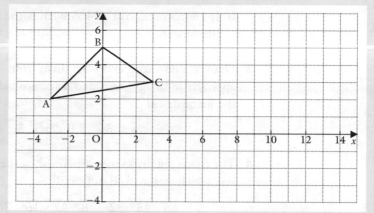

a Draw the image of $\triangle$ABC when it is rotated through $90°$
clockwise about the point $(-2, 1)$. Mark it $\triangle A_1 B_1 C_1$.

b Draw the reflection of $\triangle A_1 B_1 C_1$ in the line $x = 3$.
Mark it $\triangle A_2 B_2 C_2$.

c Draw the image of $\triangle$ABC when it undergoes a translation
defined by the vector $\begin{pmatrix} 10 \\ 0 \end{pmatrix}$. Mark it $\triangle A_3 B_3 C_3$.

d Describe the single transformation that maps $\triangle A_3 B_3 C_3$ onto
$\triangle A_2 B_2 C_2$.

8 a One rectangle measures 12 cm by 8 cm and another rectangle
measures 10 cm by 6 cm. Are these rectangles similar? Justify
your answer.

b Which of the following statements are true and which are false?

   i Any two isosceles triangles are similar.

   ii Two regular octagons are always similar.

   iii Any two parallelograms are similar.

   iv All squares are similar.

9 a Show that triangles ABC and CDE
are similar.

b Given that AC $= 4$ cm, CE $= 10$ cm
and AB $= 5$ cm find DE.

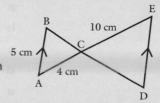

1 a Express each fraction as a decimal.

    i $\frac{1}{200}$ **ii** $5\frac{5}{8}$ **iii** $\frac{8}{25}$ **iv** $\frac{39}{50}$ **v** $\frac{27}{500}$ **vi** $\frac{9}{8}$

b The mass, w kg, of a bag of apples is given as 5 kg correct to the nearest whole number. Find the range of values in which w lies.

2 a Express $9 : 4$ in the form $n : 1$.

b If 15 machines cost £9465 what is the cost of 18 similar machines?

3 Draw x- and y-axes and scale them from -6 to 4. Draw the appropriate straight lines and leave unshaded the region defined by the inequalities

$$-5 < x \leqslant 3, \qquad 0 \leqslant y < 3, \qquad y \leqslant \tfrac{1}{2}x + 3$$

4 The following marks were obtained by the 80 candidates in a history test which was marked out of 50.

```
48  43  32   8  47  15  36  46  43  30
41  25  36  29  17  20  22  35  23   9
30   7  38  44  35  31  36  44  10  34
46  19  31  28  39  37  48  17  28  33
23  45  39   7  22  33  24  40  36  23
35  31  36  12  27  12  37  33  33  46
27  26  37  32  33  26  24  42  23  10
14  30  38  34  34  31  26  11  26  48
```

Use this data to complete the following table.

Interval	Tally	Frequency	Mark	Cumulative frequency
0–9			$\leqslant 9$	
10–19			$\leqslant 19$	
20–29			$\leqslant 29$	
30–39				
40–50				

Use the information in your table to draw a cumulative frequency curve and from it estimate

a the median mark

b the upper and lower quartiles

c the number of candidates who passed if the pass mark was 30

d the pass mark if 75% of the candidates passed

e the probability that a pupil selected at random scored less than 20.

5 a Make the letter in brackets the subject of the formula.

i $p = 5 + q$ (q) **ii** $x = \frac{1}{3}t$ (t) **iii** $z = \frac{x}{y}$ (x)

b Given that $V = 3b + \frac{8a}{b}$ and that $a = \frac{b^2}{2}$ find an expression for V in terms of b.

6 Solve the simultaneous equations.

a $4x + 3y = 3$ **b** $2x + 5y = -14$
$2x - y = 9$ $5x + 4y = -1$

7 Solve the equations.

a $x^2 + 10x + 24 = 0$ **b** $x^2 = 3x + 28$ **c** $x(x+2) = 24$

8 The equation of this curve could be

A $y = x^2$

B $y = 9 - x^2$

C $y = \frac{5}{x} + 4$

D $y = x^2 + 3x + 3$

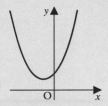

9 a On x- and y-axes scaled from 0 to 12 plot the points $A(6, 3)$, $B(12, 12)$, $C(12, 3)$.

b An enlargement, centre the origin, with scale factor $\frac{1}{3}$ maps $\triangle ABC$ onto $\triangle A_1B_1C_1$. Draw $\triangle A_1B_1C_1$.

c A translation of $\triangle A_1B_1C_1$ onto $\triangle A_2B_2C_2$ is defined by the vector $\begin{pmatrix} 6 \\ 2 \end{pmatrix}$. Draw $\triangle A_2B_2C_2$. What are the coordinates of C_2?

d $\triangle ABC$ can be transformed onto $\triangle A_2B_2C_2$ by an enlargement. Give the centre of enlargement and the scale factor.

10

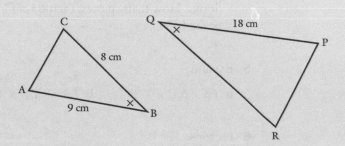

In triangles ABC and PQR, $\widehat{A} = \widehat{P}$ and $\widehat{B} = \widehat{Q}$, $AB = 9\,cm$, $BC = 8\,cm$ and $PQ = 18\,cm$. Find QR.

1 a Illustrate the following inequalities on a number line.

      i $0 < x < 8$ **ii** $-3 < x \leqslant 4$

b Express the following decimals as fractions.

      i 0.55 **ii** 0.56 **iii** 0.056 **iv** 0.0055

c Simplify $\dfrac{2\frac{1}{9} - \frac{2}{3} \times 1\frac{5}{6}}{\frac{1}{5} \times 3\frac{1}{3} + 1\frac{5}{9}}$

2 The school bus never leaves early in the morning. The probability that it leaves on time is $\frac{2}{5}$.

a What is the probability that it is late leaving?

b Copy and complete the following tree diagram by writing the probabilities on the branches.

                                 Monday   Tuesday

What is the probability that

c the bus is late on both days

d the bus is late on one day
and on time on the other?

3 a George Sharp buys a greyhound for £360 and sells it at a loss of $12\frac{1}{2}\%$. How much does he lose?

b A small business made a profit of £8909 this year. This is an increase of 18% on the profit last year. How much profit was made last year?

4 a Divide 88 cm in the ratio $4 : 7$.

b A ball of string can be cut into 45 pieces if each piece is 28 cm long. How many pieces, each 63 cm long, could be cut from the same ball?

5 Expand

    a $(a - b)(a - 2b)$ **b** $(3x + 1)(x + 2)$ **c** $(p + q)^2$

6 Factorise

    a $a^2 - 3a$ **c** $x^2 - 49$

    b $x^2 + 6x - 7$ **d** $14 + 5x - x^2$

7 a Given that $p = q + 2r$ find r when $p = 7.8$ and $q = 3$.

b Write down the first four terms and the tenth term of the sequence for which the nth term is given by $(n+1)(n+2)$.

c Find, in terms of n, an expression for the nth term of the sequence $0, 5, 10, 15, \ldots$

d $P = x^2 + 3xy$ and $x = y + 2$. Find P in terms of y.

8 Solve the equations.

a $5x^2 + 4x = 0$ **c** $x^2 - 5x = 24$

b $x^2 - x - 56 = 0$ **d** $(x+5)(x-4) + 8 = 0$

9 The diagram shows a lean-to workshop.

Find   **a** the area of cross-section

       b the amount of space it occupies

       c the number of litres of rain that falls vertically on it in a storm when the rainfall is 8 mm.

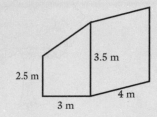

10

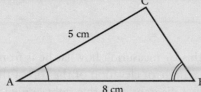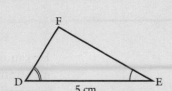

In triangles ABC and DEF $\widehat{A} = \widehat{E}$ and $\widehat{B} = \widehat{D}$. AB=8 cm, DE $= 5$ cm and AC $= 5$ cm. Find EF.

11 In the first round of a golf tournament the following scores were recorded.

70	68	71	67	74	69	69	71	68	70
71	70	72	69	69	68	71	70	70	72
72	69	68	70	68	69	67	71	69	70
68	67	70	70	73	69	71	67	69	68

a Construct a cumulative frequency table for these scores.

b How many rounds of less than 70 were there?

c How many rounds of more than 69 were there?

d Find the median score.

e Explain why you do not need a cumulative frequency curve to estimate the median.

TRIGONOMETRY: TANGENT OF AN ANGLE

Sam wants to know the height of the tree.

He can find this by measuring how far he is from the tree and then measuring the angle of elevation of the top of the tree. Showing this information in a diagram gives this right-angled triangle.

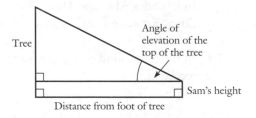

The diagram can be drawn to scale, and the height of the tree measured from the scale drawing.

- The disadvantage of this method is that it takes time and precision to get a reasonably accurate result. In this chapter we find how to calculate such a length by a method which, with the help of a scientific calculator, is fast and whose accuracy depends only on the accuracy of the initial measurements.

We start by investigating the relationship between the size of an angle and the lengths of two sides in a right-angled triangle.

EXERCISE 16A

1 a Draw the given triangle accurately using a protractor and a ruler.

b Measure $\widehat{A}$.

c Find $\dfrac{BC}{AB}$ as a decimal.

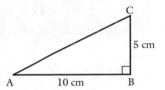

Repeat question **1** for the triangles in questions **2** to **5**.

2
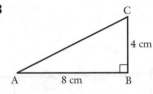

4
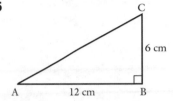

3
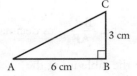

5
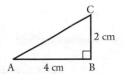

6 Are the triangles in questions **1** to **5** similar?

Repeat question **1** for the triangles in questions **7** to **10**.

7
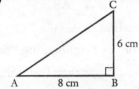

9
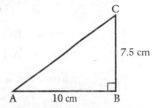

8
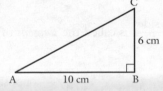

10
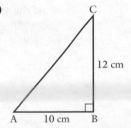

11

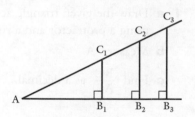

Similar triangles can be drawn so that they overlap as in this diagram. Copy the diagram above onto squared paper. Choose your own measurements but make sure that the lengths of the horizontal lines are whole numbers of centimetres. Measure $\widehat{A}$.

Find $\dfrac{B_1C_1}{AB_1}$, $\dfrac{B_2C_2}{AB_2}$ and $\dfrac{B_3C_3}{AB_3}$ as decimals.

12 Copy and complete the table using the information from questions **1** to **5**.

$\widehat{A}$	$\dfrac{BC}{AB}$
26.6°	0.5

TANGENT OF AN ANGLE

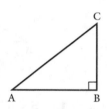 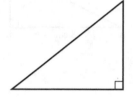

If we consider the set of all triangles that are similar to $\triangle ABC$ then, for every triangle in the set,

the angle corresponding to $\widehat{A}$ is the same

the ratio corresponding to $\dfrac{BC}{AB}$ is the same

where BC is the side *opposite* to $\widehat{A}$

and AB is the *adjacent* (or neighbouring) side to $\widehat{A}$.

From the last exercise you can see that, in a right-angled triangle the ratio $\dfrac{\text{opposite side}}{\text{adjacent side}}$ is always the same for a given angle whatever the size of the triangle.

The ratio $\dfrac{\text{opposite side}}{\text{adjacent side}}$ is called the *tangent* of the angle.

$$\text{tangent of the angle} = \dfrac{\text{opposite side}}{\text{adjacent side}}$$

More briefly,

$$\tan (\text{angle}) = \frac{\text{opp}}{\text{adj}}$$

The information about this ratio is used so often that we need a more complete and more accurate list than the one made in the last exercise. The complete list is stored in scientific calculators.

FINDING TANGENTS OF ANGLES

To find the tangent of 33°, press **tan** **3** **3** **=**. You will obtain a number which fills the display. Write down one more figure than the accuracy required,

e.g. $\tan 33° = 0.64940\ldots$

            $= 0.6494$ correct to 4 significant figures.

If you do not get the correct answer, one reason could be that your calculator is not in 'degree mode'. For all trigonometric work at this stage, angles are measured in degrees, so make sure that your calculator is in the correct mode. Calculators also vary in the order in which buttons have to be pressed; if the order given above does not work, try

3 **3** **tan** .

EXERCISE 16B

> Find the tangent of 56° correct to 3 significant figures.
>
> $\tan 56° = 1.482\ldots$
>
> $= 1.48$ (correct to 3 significant figures)

Find the tangents of the following angles correct to 3 significant figures.

1 20°   **2** 28°   **3** 72°   **4** 53°   **5** 59°

6   Find the tangents of the angles listed in question **12** in **Exercise 16A**. How do the answers you now have compare with the decimals you worked out? If they are different, give a reason for this.

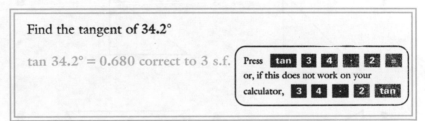

Find the tangent of **34.2°**

$\tan 34.2° = 0.680$ correct to 3 s.f.   Press **tan** **3** **4** **.** **2** **=**
or, if this does not work on your
calculator, **3** **4** **.** **2** **tan**

Find the tangents of the following **angles correct to 3 significant figures**.

7 15.5°   **8** 29.6°   **9** 11.4°   **10** 60.1°   **11** 3.8°

THE NAMES OF THE SIDES OF A RIGHT-ANGLED TRIANGLE

Before we can use the tangent for finding sides and angles we need to know which is the side opposite to the given angle and which is the adjacent side.

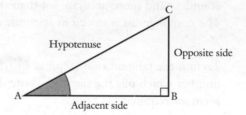

- The longest side, that is, the side opposite to the right angle, is called the *hypotenuse*.
- The side next to the angle A (not the hypotenuse) is called the *adjacent side*. (Adjacent means 'next to'.)
- The third side is the *opposite side*. It is opposite to the particular angle A that we are concerned with.

Sometimes the triangle is in a position different from the one we have been using.

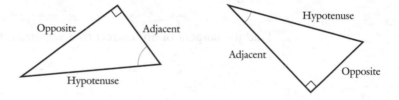

EXERCISE 16C

Sketch the following triangles. The angle we are concerned with is marked with a green arc like this ⟋. Label the sides 'hypotenuse', 'adjacent' and 'opposite'. If necessary, turn the page round so that you can see which side is which.

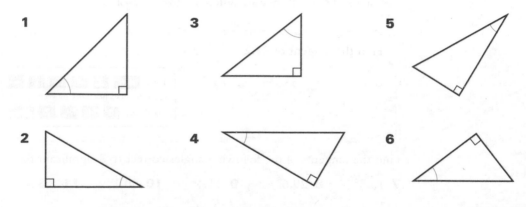

FINDING A SIDE
OF A TRIANGLE

We can now use the tangent of an angle to find the length of the opposite side in a right-angled triangle provided that we know an angle and the length of the adjacent side.

EXERCISE 16D

Use a calculator. Give your answers correct to 3 significant figures.

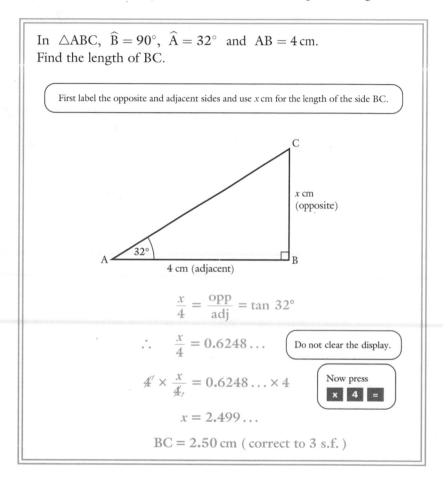

In $\triangle$ABC, $\widehat{B} = 90°$, $\widehat{A} = 32°$ and AB = 4 cm.
Find the length of BC.

First label the opposite and adjacent sides and use x cm for the length of the side BC.

$\dfrac{x}{4} = \dfrac{\text{opp}}{\text{adj}} = \tan 32°$

$\therefore \quad \dfrac{x}{4} = 0.6248\ldots$ Do not clear the display.

$4 \times \dfrac{x}{4} = 0.6248\ldots \times 4$ Now press

$x = 2.499\ldots$

BC = 2.50 cm (correct to 3 s.f.)

Find the length of BC in questions **1** to **8**.

1

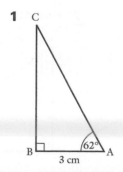

2

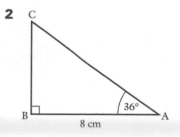

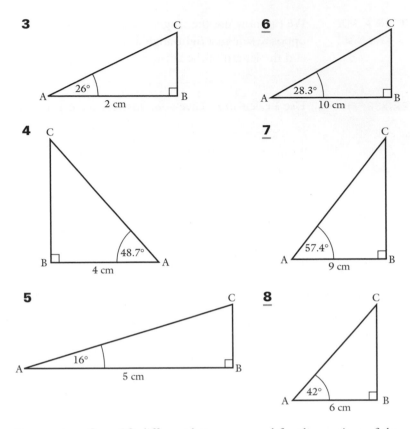

3

4

5

6

7

8

In questions **9** to **12** different letters are used for the vertices of the triangles. In each case find the side required.

9 Find PQ.

11 Find AC.

10 Find YZ.

12 Find AC.

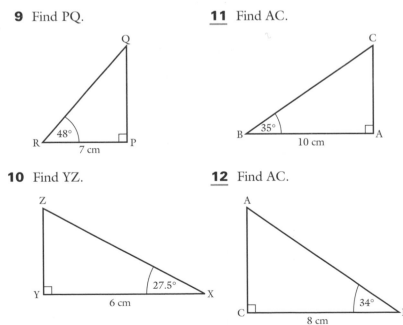

Find BC in questions **13** to **20**. Turn the page round if necessary to identify the opposite and adjacent sides.

13

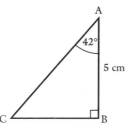

17

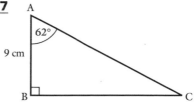

14

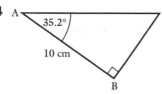

18

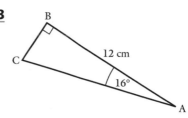

15

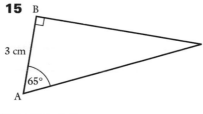

19

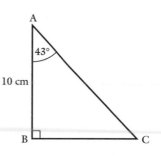

16

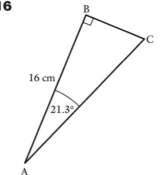

20

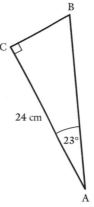

21 In $\triangle ABC$, $\widehat{B} = 90°$, $AB = 6\,cm$ and $\widehat{A} = 41°$. Find BC.

22 In $\triangle PQR$, $\widehat{Q} = 90°$, $PQ = 10\,m$ and $\widehat{P} = 16.7°$. Find QR.

23 In $\triangle DEF$, $\widehat{F} = 90°$, $DF = 12\,cm$ and $\widehat{D} = 56°$. Find EF.

24 In $\triangle XYZ$, $\widehat{Z} = 90°$, $YZ = 11\,cm$ and $\widehat{Y} = 40°$. Find XZ.

**FINDING A SIDE
ADJACENT TO
THE GIVEN
ANGLE**

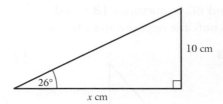

x cm

Sometimes the side whose length we are asked to find is adjacent to the given angle instead of opposite to it. Using $\dfrac{10}{x}$ instead of $\dfrac{x}{10}$ can lead to an awkward equation so we work out the size of the angle opposite x and use it instead. In this case the other angle is $64°$ and we label the sides 'opposite' and 'adjacent' to this angle.

Using $64°$,

$$\frac{x}{10} = \frac{\text{opposite}}{\text{adjacent}} = \tan 64°$$

so $\dfrac{x}{10} = 2.05$ giving $x = 20.5$.

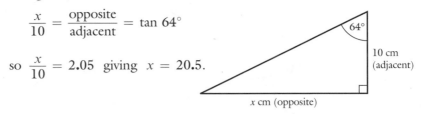

EXERCISE 16E

Use a calculator. Give your answers correct to 3 significant figures.

In $\triangle PQR$, $\hat{P} = 90°$, $\hat{Q} = 51°$ and $PR = 4\,\text{cm}$.
Find the length of PQ.

First find the other angle, i.e., $\hat{R}$.

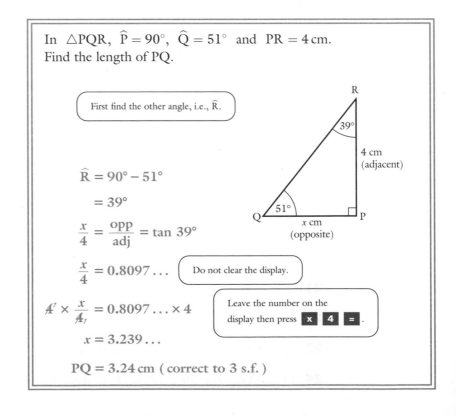

$\hat{R} = 90° - 51°$

$\quad = 39°$

$\dfrac{x}{4} = \dfrac{\text{opp}}{\text{adj}} = \tan 39°$

$\dfrac{x}{4} = 0.8097\ldots$ Do not clear the display.

$\cancel{4} \times \dfrac{x}{\cancel{4}} = 0.8097\ldots \times 4$ Leave the number on the display then press ⊠ **4** **=**.

$x = 3.239\ldots$

$PQ = 3.24\,\text{cm}$ (correct to 3 s.f.)

1 Find ZY.

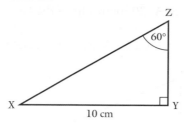

4 Find FD.

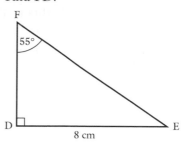

2 Find QP.

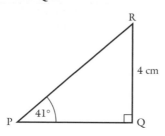

5 Find BC.

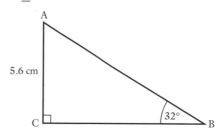

3 Find XZ.

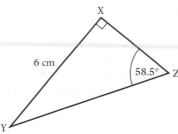

6 Find AB.

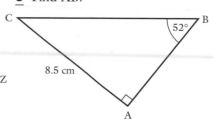

7 In $\triangle$PQR, $\widehat{Q} = 90°$, $\widehat{R} = 31°$ and PQ = 6 cm. Find RQ.

8 In $\triangle$XYZ, $\widehat{Z} = 90°$, $\widehat{Y} = 38°$ and ZX = 11 cm. Find YZ.

9 In $\triangle$DEF, $\widehat{D} = 90°$, $\widehat{E} = 34.8°$ and DF = 24 cm. Find DE.

10 In $\triangle$ABC, $\widehat{C} = 90°$, $\widehat{A} = 42.4°$ and CB = 3.2 cm. Find AC.

11 In $\triangle$LMN, $\widehat{L} = 90°$, $\widehat{N} = 15°$ and LM = 4.8 cm. Find LN.

12 In $\triangle$STU, $\widehat{U} = 90°$, $\widehat{S} = 42.2°$ and TU = 114 cm. Find SU.

Rita is at a point A, 20 metres from the base of a tree which is standing on level ground. From A the angle of elevation of the top, C, of the tree is 23°. What is the height of the tree?

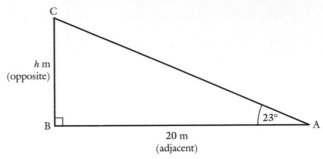

$$\frac{h}{20} = \frac{\text{opp}}{\text{adj}} = \tan 23°$$

$$\frac{h}{20} = 0.4244\ldots$$

$$20 \times \frac{h}{20} = 0.4244\ldots \times 20$$

$$h = 8.489\ldots$$

The height of the tree is 8.49 m correct to 3 s.f.

13   A pole BC stands on level ground. A is a point on the ground 10 metres from the foot of the pole. The angle of elevation of the top C from A is 27°. What is the height of the pole?

14   ABCD is a rectangle. $AB = 42\,\text{m}$ and $\widehat{BAC} = 59°$.
Find the length of BC.

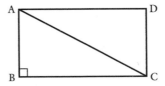

15   In rectangle ABCD, the angle between the diagonal AC and the side AB is 22°, $AB = 8\,\text{cm}$.
Find the length of BC.

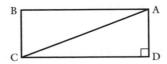

16 In △PQR, PQ = QR. From symmetry,
S is the midpoint of PR.
$\widehat{P} = 72°$, PR = 20 cm.
Find the height QS of the triangle.

17 A point R is 14 m from the foot
of a flagpole PQ. The angle of
elevation of the top of the pole
from R is 22°.
Find the height of the pole.

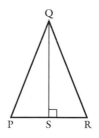
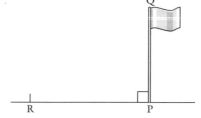

18 A ladder leans against a vertical wall
so that it makes an angle of 35° with
the wall. The top of the ladder is
2 m up the wall. How far out from
the wall is the foot of the ladder?

19 A dinghy B is 60 m out to sea
from the foot A of a vertical
cliff AC. From C the angle
of depression of B is 16°.

a Find $\widehat{B}$.

b Find the height of the cliff.

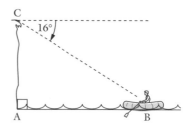

20 From a yacht, the angle of
elevation of the top of a
lighthouse is 26° and the
angle of elevation of the base
of the lighthouse is 20°. The
lighthouse is standing on top
of a 40 metre high cliff. Find

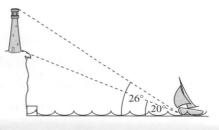

a how far the yacht is from the base of the cliff

b the height of the lighthouse.

If we are given the value of the tangent of an angle, we can use a calculator to find that angle.

For example, if the tangent of angle A is 0.732,

i.e. $\tan \widehat{A} = 0.732,$

we find $\widehat{A}$ by pressing shift tan then entering $0.732=$ in the calculator. (The 'shift' key or '2nd' key accesses the '$\tan^{-1}$' function which means 'the angle whose tan is'.)

This gives the size of the angle in degrees.

So when $\tan \widehat{A} = 0.732$
            $\widehat{A} = 36.20\ldots^\circ = 36.2^\circ$ correct to 1 decimal place.

Check this on your calculator; if it does not give this result first check that you are using degree mode then, if it is still not correct, consult your manual.

Find, correct to 1 decimal place, the angles whose tangents are given.

1 2.2	**4** 4.1	**7** 0.6752	**10** 2.0879
2 0.36	**5** 1.4	**8** 0.99293	**11** 1
3 0.41	**6** 0.31	**9** 0.37624	**12** 0.33333

Find the angle whose tangent is $\frac{3}{4}$

$\tan \widehat{A} = \frac{3}{4} = 0.75$ First express $\frac{3}{4}$ as a decimal.

   $\widehat{A} = 36.86\ldots^\circ = 36.9^\circ$ (correct to 1 d.p.)

Find correct to 1 decimal place, the angles whose tangents are given below.

13 $\frac{3}{5}$	**15** $\frac{1}{2}$	**17** $\frac{3}{20}$	**19** $\frac{3}{8}$	**21** $\frac{3}{25}$
14 $\frac{4}{5}$	**16** $\frac{7}{10}$	**18** $\frac{5}{4}$	**20** $2\frac{1}{4}$	**22** $2\frac{2}{5}$

Find the angle whose tangent is $\frac{2}{3}$

$\tan \widehat{A} = \frac{2}{3} = 0.6666\ldots$ Angle A can be found in one step on the calculator; press
   $\widehat{A} = 33.69\ldots^\circ = 33.7^\circ$
   (correct to 1 d.p.)   tan⁻¹ (2 ÷ 3) =

Find, correct to 1 decimal place, the angles whose tangents are

23 $\frac{1}{3}$	**25** $\frac{1}{6}$	**27** $\frac{5}{3}$	**29** $\frac{2}{9}$	**31** $\frac{4}{9}$
24 $\frac{1}{7}$	**26** $\frac{5}{6}$	**28** $\frac{3}{7}$	**30** $\frac{7}{3}$	**32** $\frac{4}{3}$

FINDING AN
ANGLE IN A
RIGHT-ANGLED
TRIANGLE

We can now use tangents of angles to calculate the sizes of the angles in the triangle. The next worked example shows the method.

EXERCISE 16G

Give angles correct to 1 decimal place.

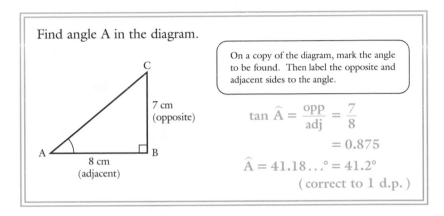

Find angle A in the diagram.

On a copy of the diagram, mark the angle to be found. Then label the opposite and adjacent sides to the angle.

C

7 cm
(opposite)

A

8 cm
(adjacent)

B

$$\tan \widehat{A} = \frac{\text{opp}}{\text{adj}} = \frac{7}{8}$$

$$= 0.875$$

$$\widehat{A} = 41.18\ldots° = 41.2°$$

(correct to 1 d.p.)

Find $\widehat{A}$ in questions **1** to **10**.

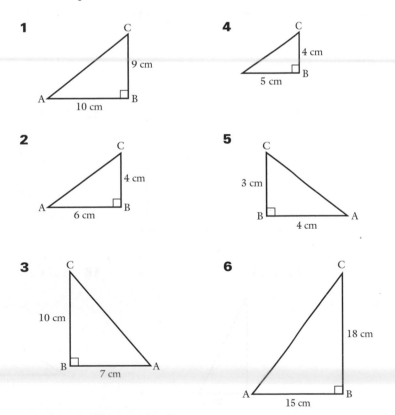

1

C

9 cm

A

10 cm

B

2

C

4 cm

A

6 cm

B

3

C

10 cm

B

7 cm

A

4

C

4 cm

B

5 cm

5

C

3 cm

B

4 cm

A

6

C

18 cm

A

15 cm

B

7

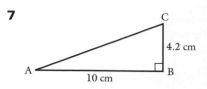

9

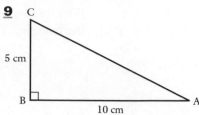

8

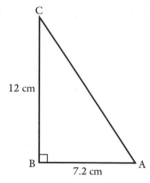

10

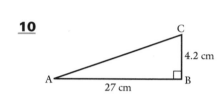

11 Find $\widehat{P}$.

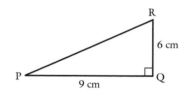

14 Find $\widehat{N}$.

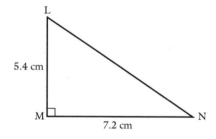

12 Find $\widehat{B}$.

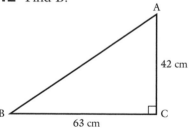

15 Find $\widehat{D}$.

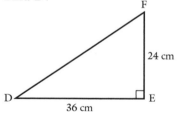

13 Find $\widehat{Y}$.

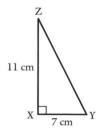

16 Find $\widehat{C}$.

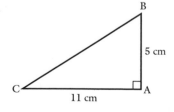

Find $\widehat{A}$ in questions **17** to **26**. Turn the page round if necessary before labelling the sides.

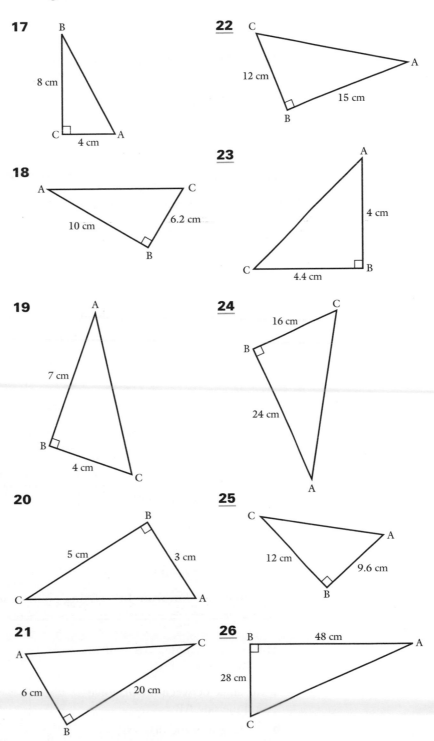

17

B

8 cm

C 4 cm A

18

A

10 cm

C

6.2 cm

B

19

A

7 cm

B

4 cm

C

20

B

5 cm 3 cm

C A

21

A C

6 cm 20 cm

B

22

C

12 cm

B 15 cm

A

23

A

4 cm

C B

4.4 cm

24

C

16 cm

B

24 cm

A

25

C A

12 cm 9.6 cm

B

26

B 48 cm A

28 cm

C

27 In △ABC, $\widehat{B} = 90°$, AB = 12 cm, BC = 11 cm. Find $\widehat{A}$.

28 In △PQR, $\widehat{P} = 90°$, PQ = 3.2 m, PR = 2.8 m. Find $\widehat{Q}$.

29 In △DEF, $\widehat{D} = 90°$, DE = 108 m, DF = 72 m. Find $\widehat{F}$.

If we know two sides in a right-angled triangle we can use Pythagoras' Theorem to find the remaining side.

A man walks due north for 5 km from A to B, then 4 km due east to C.

a What is the bearing of C from A?

b How far is he from his starting point?

a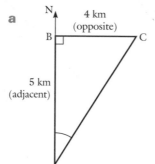

We start by drawing a diagram showing all the information given. The bearing of C from A is angle A in the triangle, so we mark this angle.

$$\tan \widehat{A} = \frac{\text{opp}}{\text{adj}} = \frac{4}{5}$$
$$= 0.8$$
$$\widehat{A} = 38.65\ldots° = 38.7°$$
$$(\text{correct to 1 d.p.})$$

The bearing of C from A is 038.7°

Notice that we add '0' to make a 3-figure bearing.

b The length of AC in the diagram gives the distance required. Tangents will not help to find this, but as we know the other two sides in the right-angled triangle we can use Pythagoras' theorem.

$$AC^2 = AB^2 + BC^2 \ (\text{Pythagoras})$$
$$= 25 + 16 = 41$$
$$AC = \sqrt{41} = 6.403\ldots$$

The man is 6.40 km from his starting point (correct to 3 s.f.)

---

30 ABCD is a rectangle. AB = 60 m and BC = 36 m. Find

a the angle between the diagonal and the side AB

b the length of the diagonal AC.

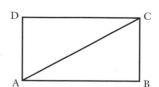

31 A flagpole PQ is 10 m high.
R is a point on the ground 20 m
from the foot of the pole.
Find the angle of elevation of
the top of the pole from R
(that is, $\widehat{R}$).

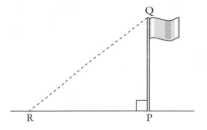

32 In △ABC, AB = BC. AC = 12 cm.
D is the midpoint of AC. The height
BD of the triangle is 10 cm. Find

a $\widehat{C}$ and the other angles of the
triangle

b the lengths of AB and BC.

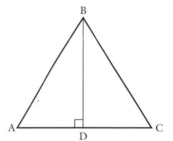

33 A ladder leans against a vertical wall.
Its top, Q, is 3 m above the ground and its
foot, P, is 2 m from the foot of the wall.
Find the angle of slope of the ladder
(that is, $\widehat{P}$).

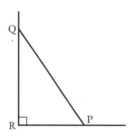

34 The bearing of town A from town B is 032.4°.
A is 16 km north of B.

a How far east of B is A?

b How far is A from B?

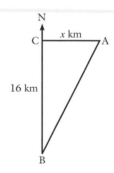

35 In a square ABCD of side 8 cm, A is joined to the midpoint E of BC.
Find EÂB, CÂB and CÂE. Notice that AE does *not* bisect CÂB.

36 A ladder leans against a vertical wall. It makes an angle of 72° with
the horizontal ground and its foot is 1 m from the foot of the wall.
How high up the wall does the ladder reach?

37 Sketch axes for x and y from 0 to 5. A is the point (1, 0) and B is
(5, 2). What angle does the line AB make with the x-axis?

38 In a rhombus the two diagonals are of lengths 6.2 cm and 8 cm.
Find **a** the angles of the rhombus

b the lengths of the sides of the rhombus.

39 In rectangle ABCD, AB = 24 cm and BC = 11 cm. Find CÂB
and hence find the obtuse angle between the diagonals.

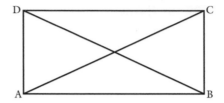

40 In △ABC, AB = BC, CA = 10 cm and Ĉ = 72°. Find the
height BD of the triangle.

41 The diagram shows a section through an underground railway
tunnel.

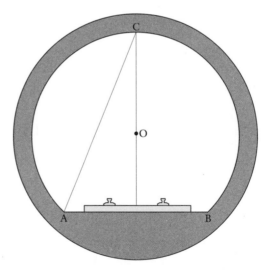

The floor of the tunnel is 2 m wide and the angle of elevation of the
highest point C, of the tunnel from the edge A of the floor is 69°.

a Draw a diagram showing just the inside wall and floor of the
tunnel. Mark the points A, B, C and O which is the centre of the
circle part of which forms the inside wall.

b Calculate the height of the tunnel.

42

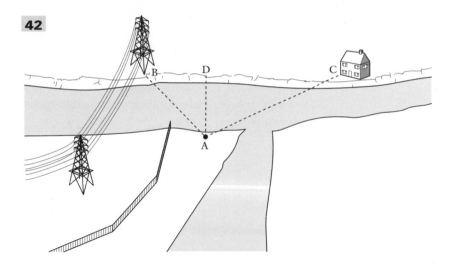

Sula is at the point A on the bank of the river and needs to find the distance between the pylon at B and the house at C, both of which are on the opposite bank. She cannot cross the rivers nor can she stand directly opposite the pylon at B.

From a map she knows that the width, AD, of the river is 60 metres. She measures the angle BAD as 35° and the angle DAC as 58°. Calculate the distance between the pylon and the house.

Assume that BDC is a straight line.

43

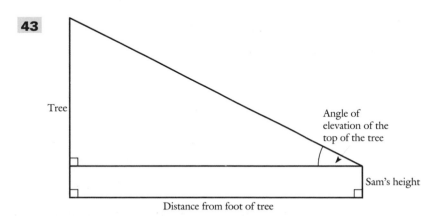

We started the chapter with Sam wanting to find the height of a tree. Sam is 1.75 m tall. He paces out a distance of 40 m from the base of the tree and then measures the angle of elevation of the top of the tree as 32°. Calculate the height of the tree.

PRACTICAL
WORK

Use what you have learnt in this chapter to find the height of a chimney on a building near you. You will need a long measuring tape and a theodolite to measure angles. You can make a rough theodolite from a protractor, a length of string, some blu-tack and a small weight such as a pebble.

INVESTIGATION

John has a problem. For safety reasons he has to find the inclination of the ladder to the vertical. All he has with him is a scientific calculator and a straight stick. He does not have a ruler or a measuring tape or a protractor and he does not know how long the ladder is.

Investigate how John can calculate the angle that the ladder makes with the wall and how accurate an answer he can expect.

SINE AND COSINE OF AN ANGLE

The tangent of an angle is useful when the opposite and adjacent sides of a right-angled triangle are involved, but there are many problems where the hypotenuse and one of the other sides are involved instead. Consider this example.

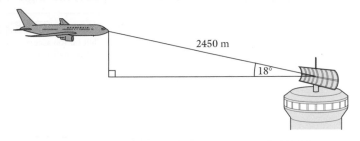

The reading from a radar instrument shows that a plane is 2450 metres from the radar and at an angle of elevation of 18°.

To find the height at which the plane is flying, we need a relationship between the side opposite the 18° angle and the hypotenuse of the triangle.

SINE OF AN ANGLE

In a right-angled triangle, the ratio formed by the side opposite an angle and the hypotenuse is called the *sine* of the angle.

In this diagram

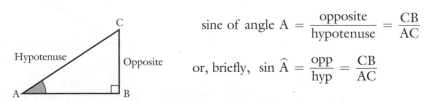

$$\text{sine of angle A} = \frac{\text{opposite}}{\text{hypotenuse}} = \frac{CB}{AC}$$

$$\text{or, briefly,} \quad \sin \widehat{A} = \frac{\text{opp}}{\text{hyp}} = \frac{CB}{AC}$$

All right-angled triangles containing, say, an angle of 40° are similar, so the sine ratio, $\frac{\text{opp}}{\text{hyp}}$, always has the same value.

The value of the sine ratio of every acute angle is stored in scientific calculators and methods of calculation are similar to those involving tangents.

EXERCISE 17A

Find , correct to 3 significant figures, the sine of

a 72° **b** 38.2°

a $\sin 72° = 0.951$ (correct to 3 s.f.)

Press

b $\sin 38.2° = 0.6184\ldots = 0.618$ (correct to 3 s.f.)

Find, correct to 3 significant figures, the sines of the following angles.

1 26° **2** 84° **3** 25.4° **4** 37.1° **5** 78.9°

Find, correct to 1 decimal place, the angle whose sine is 0.909

$\sin \widehat{A} = 0.909$

Press SHIFT sin 0 · 9 0 9 =
(The shift button accesses the $\sin^{-1}$ function which means 'the angle whose sin is')

$\widehat{A} = 65.36\ldots°$

$= 65.4°$ (correct to 1 d.p.)

Find, correct to 1 decimal place, the angles whose sines are given below.

6 0.834 **7** 0.413 **8** 0.639 **9** 0.704 **10** 0.937

USING THE SINE RATIO

The sine ratio can be used, in a similar manner to the tangent ratio, to find angles and sides in right-angled triangles.

EXERCISE 17B

Find the length of BC.

On a copy of the diagram, label the sides in relation to the angle.

7 cm (hypotenuse)

x cm (opposite)

28°

A B C

$\sin 28° = \dfrac{\text{opp}}{\text{hyp}} = \dfrac{x}{7}$

We write the equation the other way round as it is easier to handle when the term containing x is on the l.h.s.

$\dfrac{x}{7} = 0.4694\ldots$

$x = 7 \times 0.4694\ldots = 3.286\ldots$

$BC = 3.29 \text{ cm}$ (correct to 3 s.f.)

Give answers correct to 3 significant figures.

1 Find BC.

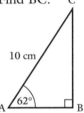

2 Find BC.

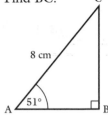

3 Find AC.

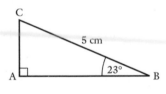

4 Find BC.

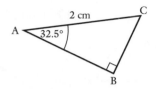

5 Find QR.

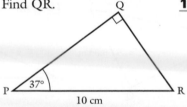

6 Find RQ.

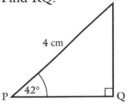

7 Find PQ.

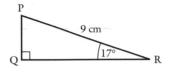

8 Find XY.

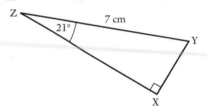

9 Find LM.

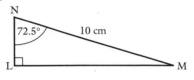

10 Find PQ.

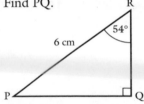

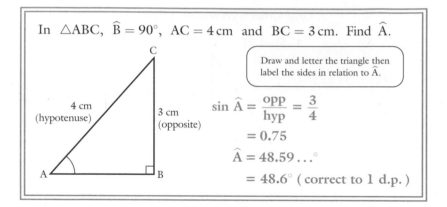

In △ABC, $\widehat{B} = 90°$, AC = 4 cm and BC = 3 cm. Find $\widehat{A}$.

Draw and letter the triangle then label the sides in relation to $\widehat{A}$.

4 cm (hypotenuse)

3 cm (opposite)

$$\sin \widehat{A} = \frac{opp}{hyp} = \frac{3}{4}$$

$$= 0.75$$

$$\widehat{A} = 48.59\ldots°$$

$$= 48.6° \ (\text{correct to 1 d.p.})$$

Give angles correct to 1 decimal place.

11 Find $\widehat{A}$.

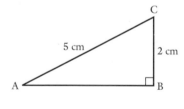

13 Find $\widehat{P}$.

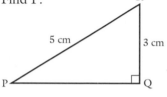

12 Find $\widehat{A}$.

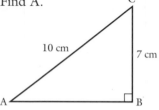

14 Find $\widehat{Q}$.

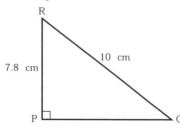

15 Find $\widehat{Y}$.

17 Find $\widehat{M}$.

16 Find $\widehat{P}$.

18 Find $\widehat{A}$.

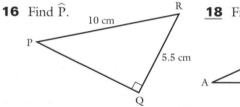

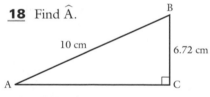

19 In $\triangle ABC$, $\widehat{B} = 90°$, $\widehat{C} = 36°$ and $AC = 3.5$ cm. Find AB.

20 In $\triangle PQR$, $\widehat{Q} = 90°$, $PQ = 2.6$ cm and $PR = 5.5$ cm. Find $\widehat{R}$.

COSINE OF AN ANGLE

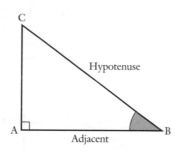

If we are given the adjacent side and the hypotenuse, then we can use a third ratio, $\dfrac{\text{adjacent side}}{\text{hypotenuse}}$.

This is called the *cosine* of the angle (cos for short).

$$\cos \widehat{B} = \frac{\text{adj}}{\text{hyp}} = \frac{AB}{BC}$$

Cosines of acute angles are stored in scientific calculators.

EXERCISE 17C

> Find, correct to 3 significant figures, the cosine of
>
> **a** 41°   **b** 28.7°
>
> **a** cos 41° = 0.755   **b** cos 28.7° = 0.877

Find, correct to 3 significant figures, the cosines of the following angles.

1 59°   **2** 48°   **3** 4°   **4** 44.9°   **5** 60.1°

> Find, correct to 1 decimal place, the angle whose cosine is 0.493
>
> cos $\widehat{A}$ = 0.493
>
> $\widehat{A} = 60.46\ldots° = 60.5°$ (correct to 1 d.p.)

In questions **6** to **10**, cos $\widehat{A}$ is given. Find $\widehat{A}$ correct to 1 decimal place.

6 0.435   **7** 0.943   **8** 0.012   **9** 0.7   **10** 0.24

USING THE COSINE RATIO

The cosine ratio can be used in the same manner as the sine and tangent ratios to find angles and sides in right-angled triangles.

EXERCISE 17D

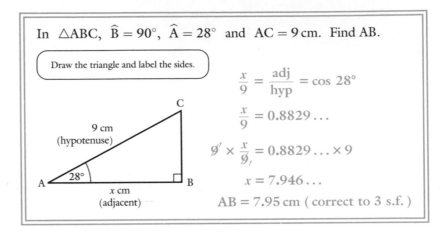

In △ABC, $\widehat{B} = 90°$, $\widehat{A} = 28°$ and $AC = 9$ cm. Find AB.

Draw the triangle and label the sides.

9 cm (hypotenuse)

28°

A

x cm (adjacent)

B

C

$$\frac{x}{9} = \frac{adj}{hyp} = \cos 28°$$

$$\frac{x}{9} = 0.8829\ldots$$

$$\cancel{9} \times \frac{x}{\cancel{9}} = 0.8829\ldots \times 9$$

$$x = 7.946\ldots$$

$$AB = 7.95 \text{ cm (correct to 3 s.f.)}$$

In the following triangles find the required lengths, correct to 3 significant figures.

1 Find AB.

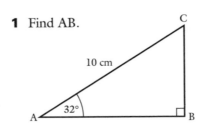

10 cm

32°

A

B

C

4 Find AB.

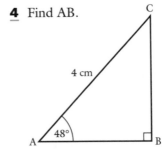

4 cm

48°

A

B

C

2 Find PQ.

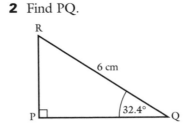

R

6 cm

32.4°

P

Q

5 Find AC.

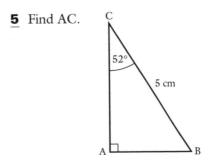

52°

5 cm

A

B

C

3 Find PQ.

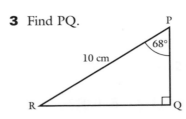

P

68°

10 cm

R

Q

6 Find XZ.

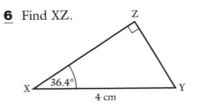

Z

36.4°

X

4 cm

Y

7 Find PQ.

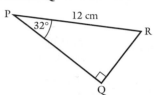

12 cm

8 Find PR.

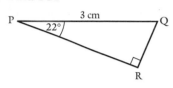

3 cm

In $\triangle ABC$, $\widehat{B} = 90°$, $AB = 4\,cm$ and $AC = 6\,cm$. Find $\widehat{A}$.

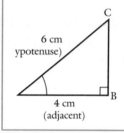

6 cm
ypotenuse)

4 cm
(adjacent)

$$\cos \widehat{A} = \frac{adj}{hyp} = \frac{4}{6} = 0.6666\ldots$$

$$\widehat{A} = 48.18\ldots°$$

$$= 48.2° \ (\text{correct to 1 d.p.})$$

Give angles correct to 1 decimal place.

9 Find $\widehat{A}$.

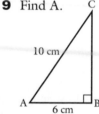

10 cm

6 cm

13 Find $\widehat{A}$.

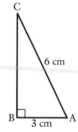

6 cm

3 cm

10 Find $\widehat{A}$.

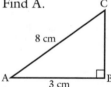

8 cm

3 cm

14 Find $\widehat{Q}$.

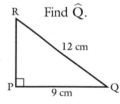

12 cm

9 cm

11 Find $\widehat{Y}$.

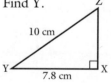

10 cm

7.8 cm

15 Find $\widehat{C}$.

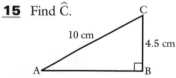

10 cm

4.5 cm

12

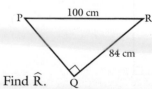

100 cm

84 cm

Find $\widehat{R}$.

16 Find $\widehat{X}$.

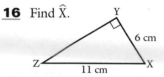

6 cm

11 cm

USE OF ALL
THREE RATIOS

To remember which ratio is called by which name, some people use the word SOHCAHTOA.

$$\text{Sin } \widehat{A} = \frac{\text{Opposite}}{\text{Hypotenuse}} \qquad \text{SOH}$$

$$\text{Cos } \widehat{A} = \frac{\text{Adjacent}}{\text{Hypotenuse}} \qquad \text{CAH}$$

$$\text{Tan } \widehat{A} = \frac{\text{Opposite}}{\text{Adjacent}} \qquad \text{TOA}$$

EXERCISE 17E

State whether sine, cosine or tangent should be used for the calculation of the marked angle.

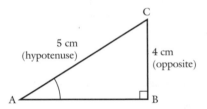

The opposite side and the hypotenuse are given so we should use sin $\widehat{A}$.

In questions **1** to **6** label the sides whose lengths are known, as 'hypotenuse', 'opposite' or 'adjacent'. Then state whether sine, cosine or tangent should be used for the calculation of the marked angle.

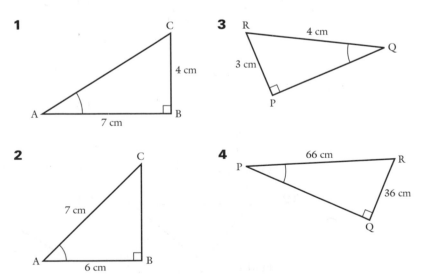

5

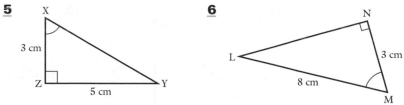

6

State whether sine, cosine or tangent should be used for the calculation to find x.

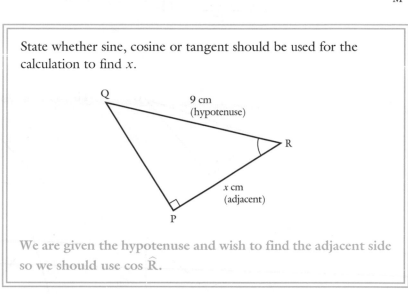

We are given the hypotenuse and wish to find the adjacent side so we should use $\cos \hat{R}$.

In questions **7** to **12**, use 'opposite', 'adjacent' or 'hypotenuse' to label the side whose length is given and the side whose length is to be found. Then state whether sine, cosine or tangent should be used for the calculation to find x.

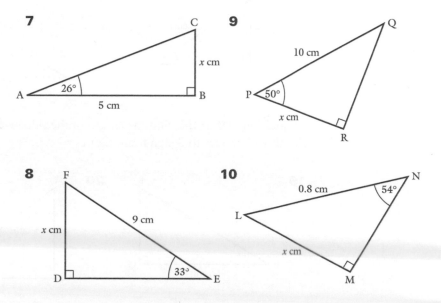

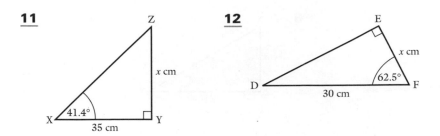

11

12

In questions **13** to **18** find the marked angle correct to 1 decimal place.

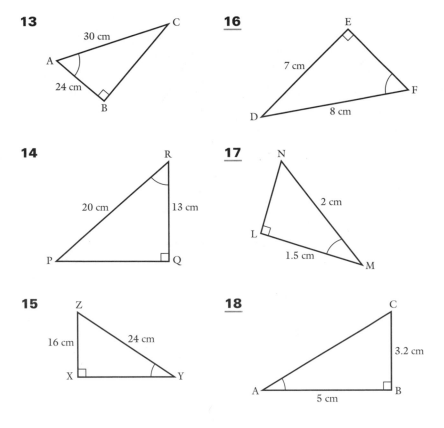

13

16

14

17

15

18

In questions **19** to **24**, find the length of the side marked x cm, giving the answer correct to 3 significant figures.

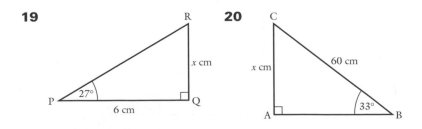

19

20

21

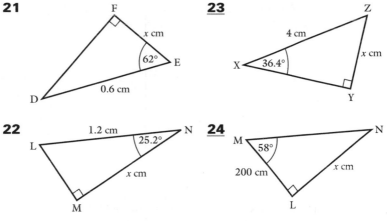

23

22

24

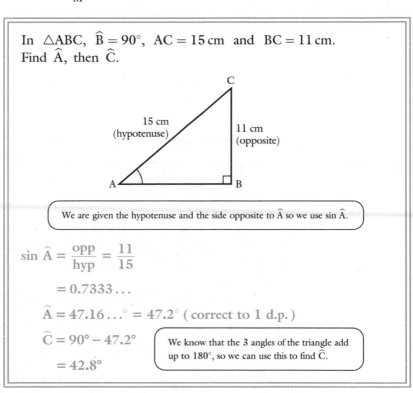

In △ABC, $\widehat{B} = 90°$, AC = 15 cm and BC = 11 cm.
Find $\widehat{A}$, then $\widehat{C}$.

We are given the hypotenuse and the side opposite to $\widehat{A}$ so we use sin $\widehat{A}$.

$$\sin \widehat{A} = \frac{\text{opp}}{\text{hyp}} = \frac{11}{15}$$

$$= 0.7333\ldots$$

$$\widehat{A} = 47.16\ldots° = 47.2° \text{ (correct to 1 d.p.)}$$

$$\widehat{C} = 90° - 47.2°$$

We know that the 3 angles of the triangle add up to 180°, so we can use this to find $\widehat{C}$.

$$= 42.8°$$

Give angles correct to 1 decimal place and lengths correct to 3 significant figures.

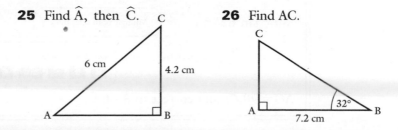

25 Find $\widehat{A}$, then $\widehat{C}$.

26 Find AC.

27 Find $\widehat{X}$, then $\widehat{Z}$.

28 Find PQ.

29 Find $\widehat{C}$, then AB.

30 Find AB.

31 Find XZ.

32 Find $\widehat{A}$.

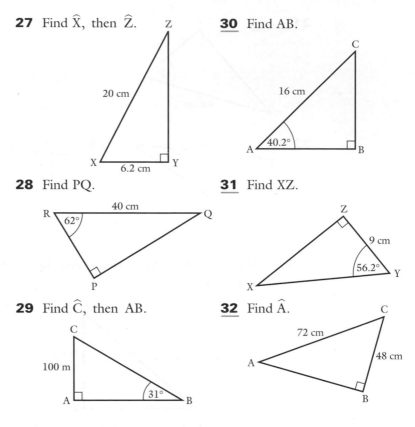

FINDING THE
HYPOTENUSE

So far when finding the length of a side we have been able to form an equation in which the unknown length is on the top of a fraction. This is not possible when the hypotenuse is to be found, and the equation we form takes slightly longer to solve.

EXERCISE 17F

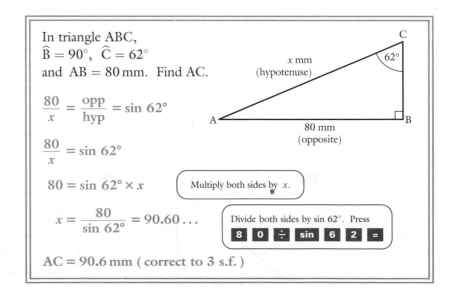

In triangle ABC,
$\widehat{B} = 90°$, $\widehat{C} = 62°$
and AB = 80 mm. Find AC.

$\dfrac{80}{x} = \dfrac{\text{opp}}{\text{hyp}} = \sin 62°$

$\dfrac{80}{x} = \sin 62°$

$80 = \sin 62° \times x$ (Multiply both sides by x.)

$x = \dfrac{80}{\sin 62°} = 90.60\ldots$ Divide both sides by sin 62°. Press

$\boxed{8}\ \boxed{0}\ \boxed{\div}\ \boxed{\sin}\ \boxed{6}\ \boxed{2}\ \boxed{=}$

AC = 90.6 mm (correct to 3 s.f.)

In questions **1** to **6**, use the information given in the diagram to find the hypotenuse, correct to 3 significant figures.

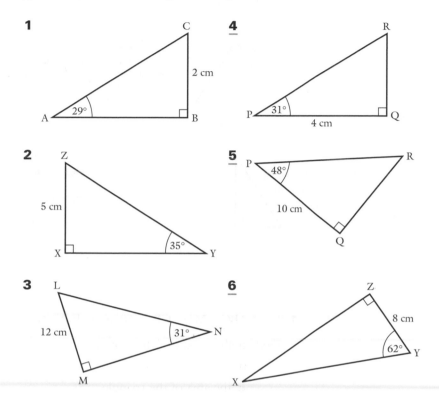

7 In triangle ABC, $\widehat{B} = 90°$, $\widehat{A} = 43°$ and BC = 3 cm. Find AC.

8 In triangle PQR, $\widehat{P} = 90°$, $\widehat{Q} = 28°$ and PR = 7 cm. Find QR.

9 In triangle LMN, $\widehat{L} = 90°$, $\widehat{M} = 14°$ and LN = 8 cm. Find MN.

10 In triangle XYZ, $\widehat{Z} = 90°$, $\widehat{Y} = 62°$ and ZY = 20 cm. Find XY.

APPLICATIONS

Trigonometry has many applications. These range from finding unknown angles and lengths in geometrical figures to solving problems in surveying and navigation. The following exercise gives a variety of problems where angles or sides have to be found.

Remember that the sine, cosine and tangent ratios have been defined in right-angled triangles, so if your diagram does not contain a right-angled triangle you will have to find one by adding a suitable line.

Remember also that Pythagoras' theorem can be used in a right-angled triangle to find one side when the other two sides are known.

EXERCISE 17G

In an isosceles triangle PQR, PQ = QR = 5 cm and PR = 6 cm.
Find the angles of the triangle.

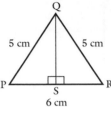

By dividing the triangle down the middle with the line QS, we create two identical right-angled triangles,. We can then draw one of these triangles to work with.

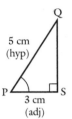

$$\cos \widehat{P} = \frac{\text{adj}}{\text{hyp}} = \frac{PS}{PQ} = \frac{3}{5} \qquad \boxed{PS = \tfrac{1}{2} PR = 3 \text{ cm}}$$

$\widehat{P} = 53.1°$ (correct to 1 d.p.)

$\widehat{R} = 53.1°$ (isosceles △; base angles equal)

$P\widehat{Q}R = 73.8°$ (angles of a △ add up to 180°)

1 In △ABC, AC = CB = 10 m and $\widehat{A} = 64°$.

Find **a** the height of the triangle

    b the length of AB

    c the area of the triangle.

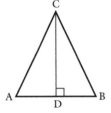

2 An aeroplane climbs at a steady angle. When it has been climbing for 2 minutes, the cockpit instruments show that it has reached an altitude of 1000 m and has travelled a distance of 3 km. At what angle is the plane climbing?

3

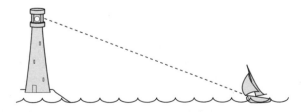

An observer at the top of the lighthouse measures the angle of depression of a yacht as 15°. The lighthouse is 50 m high. How far is the yacht from the base of the lighthouse?

4

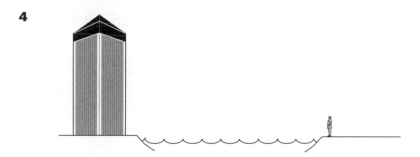

The diagram shows an office tower which is 100 m tall (to the top of the roof). The tower is on one bank of a river. Peter is standing on the opposite bank, 150 m away from the base of the tower. What is the angle of elevation of the top of the tower from Peter? (Neglect Peter's height.)

5 A girl walks 600 m along a road that slopes up uniformly at 5° to the horizontal. How far, to the nearest metre, has she risen?

6 A ski run is 1500 m long and slopes uniformly at 8° to the horizontal. How far, to the nearest metre, will a skier descend vertically when making this run?

7 An escalator is to be installed in a shopping mall to raise shoppers through a height of 8 metres. The escalator must be inclined at 20° to the horizontal.

a How long must the escalator be?

b What is the least horizontal distance needed to install it?

8 A treasure hunter locates two large trees, A and B, which are 250 m apart. He then tries to find a point C such that angle ABC is 90° and angle BAC is 64°.

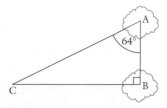

a How far is C from B?

b If he makes an error in measuring BÂC, using 63° instead of 64°, and sets out to walk from A to C, how far will he be from the true position of C when he thinks he has arrived? Give your answer to the nearest metre.

9 ABCD is a rectangle in which AB = 360 mm and BC = 184 mm.

   a Find the angle that the diagonal AC makes with the side BC.

   b Hence find the acute angle between the diagonals.

10 The drawing shows one of a series of roof trusses which are to be made to construct a factory roof.

Each length in the truss is to be made from angle iron. What length of angle iron is required to make one truss?

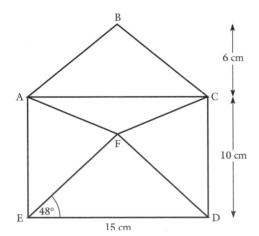

11 The diagram shows a rectangular envelope viewed from the back with the flap opened flat.

How far is B from F when the envelope is sealed?

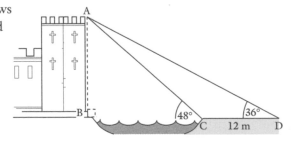

12 The diagram shows a castle tower and a moat. The ground beyond the moat is horizontal.

   a Draw a diagram showing just triangles ABC and ABD. Mark the given information on your diagram and mark AB as *x* m and BC as *y* m.

   b Using triangle ABC, find an equation relating *x* and *y*. Express the equation in the form $x = \ldots$

   c Using triangle ABD, find another equation relating *x* and *y*. Express this equation also in the form $x = \ldots$

   d Use the equations found in parts **b** and **c** to form an equation in *y* only. Solve this equation and hence find the height of the tower and the width of the moat.

13 An aeroplane flies 120 km on a bearing of 130°. It then alters course and flies 250 km on a bearing of 035°.

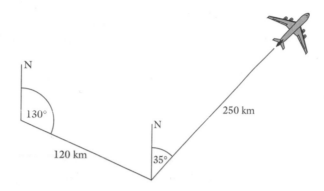

a How far east is the plane from its starting point?

b How far north is the plane from its starting point?

c How far is the plane from its starting point?

14

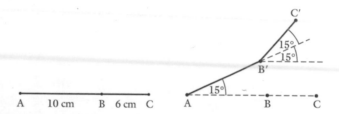

The diagram shows a section through a lift-up warehouse platform. The platform is hinged to the floor at A. The section BC is hinged to AB at B. As the platform is raised, the section BC turns through twice the angle turned through by the section AB.

When the section AB is inclined at 15° to the floor, find the height of C above the floor.

15

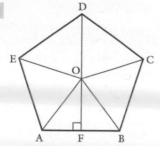

ABCDE is a regular pentagon of side 10 cm.

a Find the height of triangle AOB.

b Hence find the area of the pentagon.

16 ACB is a sector of a circle of radius 12 cm. The arc CB subtends an angle of 28° at the centre A of the circle. Find

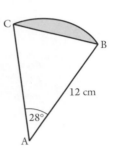

a the area of the sector

b the distance of A from the base BC of triangle ABC

c the area of the triangle ABC

d the area enclosed by the arc BC and the line BC. (This is the region shaded in the diagram and it is called a *segment* of a circle.)

INVESTIGATION

Surveyors use a process called 'triangulation' to find distances and heights. Firstly two stations, A and B, are established on the same level and the distance between them is measured. The position of any other point C, such as a point on the opposite side of a river, is then found by measuring the angles between AB and the lines of sight from A to C and from B to C.

This gives a triangle in which one side and the angles at each end of the side are known. The lengths of the two other sides can then be calculated from this information.
Investigate how this can be done.

LOCI

There are many situations in everyday life where the position in which an object is placed is important.

Jim wants to install security lighting on the front of his house. He buys a light that will come on if there is movement within five metres of it. It is recommended that it should be at least three metres above the ground. He has to decide where to fix the light if the largest possible area around his front door is to be protected.

- Jim can try different positions for the light until he finds one that is satisfactory. This is time-consuming and presents some risks because it involves using ladders and carrying a 'live' electricity cable.
- Jim can try using pencil and paper methods to work out where to fix the light so that it comes on when anyone approaches the front of the house in the dark.

This chapter covers the possible positions of an object if it is to obey certain rules, such as the position of a security light if it is to be useful.

EXERCISE 18A In questions **1** to **4** discuss how a knowledge of where an object can be, and what rules it has to obey, would be useful to solve each problem.

1 Do I need an extension cable for the electric hedgetrimmer if I am to cut my hedge?

2 Natalie wants to tether a goat in her field so that the area it can feed from is not more than 40 square metres.

3 Sally wishes to position a new seat in her garden to maximise the amount of sun that falls on it.

4 A developer is laying water pipes on a new housing estate. He wants the total length of piping to be as small as possible.

LOCI IN TWO DIMENSIONS

A *locus* is the set of all the points whose positions satisfy a given rule.

When the locus is a straight or curved line it is convenient to think of the locus as the path that is traced out by a single moving point.

Remember that every point on a locus must obey the given conditions or law, and that every point which obeys the law must lie on the locus.

The plural of locus is *loci*.

EXERCISE 18B

Describe, and illustrate with a sketch, the locus of the tip of the minute hand of a clock as it moves between 12 noon and 12.30 p.m.

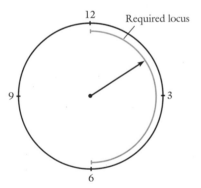

The required locus is a semicircle, centre at the centre of the clockface, radius the length of the minute hand.

In questions **1** to **10** describe, and illustrate with a sketch, the given locus.

1 The tip of the minute hand of a clock as it moves between 1 a.m. and 2 a.m.

2 The tip of the hour hand of a clock as it moves between 1 a.m. and 2 a.m.

3 A cricket ball when bowled at the wicket.

4 A cricket ball when hit along the ground for four.

5 A cricket ball when hit for six.

6 A goat on the end of a rope winding it around a tree.

7 The centre of the wheel of a bicycle as the bicycle

    a travels in a straight line

    b travels around a bend.

8 The number at the top of this page as you turn the page over.

9 A satellite circling the Earth.

10 The Earth moving around the Sun.

A rod, OA, turns through a complete revolution about one end, O, which is fixed. Describe the locus of A.

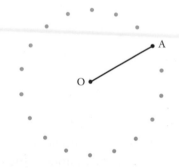

> Mark several possible positions for A until the overall shape of the path of A becomes clear.

The locus of A is the circumference of a circle, centre O radius OA.

---

11 The minute hand of a clock is 80 cm long. Describe the locus of its tip   **a** from 5 a.m. to 6 a.m.   **b** from 2.15 p.m. to 2.45 p.m.

12 Draw a straight line AB, about 10 cm long, on a page of your exercise book. Describe the locus of a point X on the page which moves so that it is always 3 cm away from AB.

13

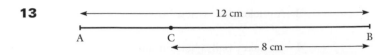

The rod AB is rotated about C. Describe the locus of

a the point A **b** the point B.

14

A and B are two fixed points. Describe the locus of the points on this page that are equidistant (that is, the same distance) from A and B.

15

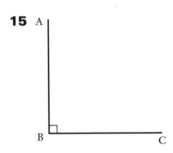

If $A\widehat{B}C = 90°$ describe the locus of points that are equidistant from AB and BC.

16 ABCD is a square of side 10 cm.

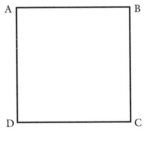

a Find the locus of points within the square equidistant from

i AB and BC   **ii** AB and AD.

b Is there any point that is equidistant from all three lines AB, BC and AD? If so, where is it?

17

A ————————————> B

C ————————————> D

AB and CD are two parallel lines. Describe the locus of a point, between the two lines, that is always twice as far from AB as it is from CD.

18

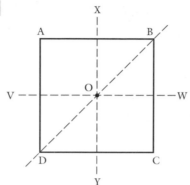

ABCD is a square, centre O, of side 4 cm. VW is parallel to AB and XY is parallel to AD. Describe the locus of A as the square is rotated about

a XY

b VW

c DB

d the axis through O perpendicular to ABCD.

SPECIAL LOCI

In question **11** from **Exercise 18B** the locus is a set of points traced out by a particular point, while in questions **14–17** the locus is a set of position points, all of which exist at the same time.

These loci have introduced us to the four most important loci in two-dimensional work.

1 The locus of a point that moves in such a way that it is always at a fixed distance from a fixed point is called a circle. The fixed point is the centre of the circle, and the fixed distance is its radius.

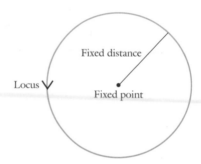

2 The locus of a point that moves in such a way that it is at a constant distance *d* from a line through two fixed points A and B, is the pair of straight lines drawn parallel to AB and distant *d* from it.

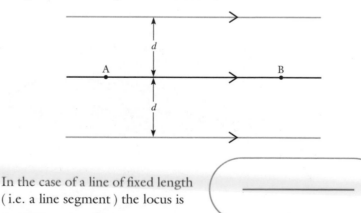

In the case of a line of fixed length (i.e. a line segment) the locus is

3 For points that are equidistant from two fixed points A and B, the locus is the line that bisects AB and is perpendicular to AB. This line is called the *perpendicular bisector* of AB.

4 For points that are equidistant from two intersecting straight lines AXB and CXD, the locus is the pair of bisectors of the angles between the given lines. These bisectors are always at right angles to each other.

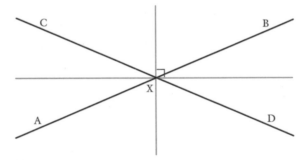

EXERCISE 18C In this exercise copy the diagram and mark several possible positions of the point in question in order to see the shape that the locus is taking.

1 M is the midpoint of a chord AB of fixed length in a circle, centre O. Describe the locus of M as AB moves around the circle.

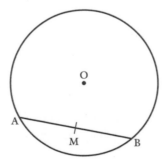

2 A and B are two fixed points where AB = 6 cm. Describe the locus of a third point C if the area of triangle ABC is 12 cm².

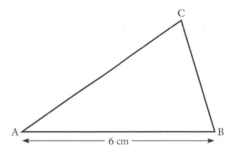

3 A is a fixed point. Describe the locus of the centres of circles which pass through A and have a radius of 5 cm.

4

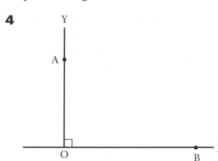

A is a point on OY such that OA = 4 cm and B is a point on OX.
If XÔY = 90° describe the locus of the midpoint of AB as B moves along OX.

5

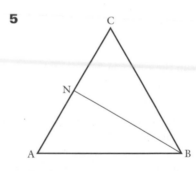

ABC is an equilateral triangle. The triangle is rotated clockwise about B until BC becomes parallel to the lower edge of the page.

a Sketch the locus of
  i C
  ii the foot N of the perpendicular from B to AC.

b What angle has BA turned through?

6

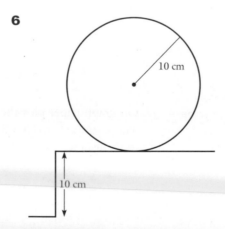

A wheel of radius 10 cm rolls across a horizontal path, and then down a step 10 cm deep, before continuing to roll horizontally. Sketch the locus of the centre of the wheel.

7 Describe the locus of the centre of a coin of diameter 2 cm if it

  a rolls around the inside of a circle with centre C and radius 5 cm

  b rolls around the outside of a circle of radius 5 cm.

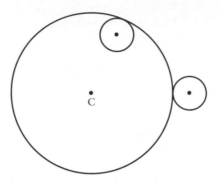

8

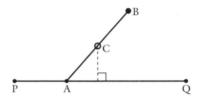

A metal rod AB passes through a ring C which is fixed above the midpoint of a groove PQ. The end A slides in the groove. When A is at P, B rests in the ring. Make a sketch to show the locus of B as A moves from P to Q.

9 Describe the locus of the centre of a circle, of variable radius, which passes through two fixed points A and B.

10 Draw any triangle ABC such that $A\widehat{B}C = 90°$. Draw the locus of points equidistant from   **a** A and B   **b** B and C.
  What do you notice about the point of intersection of the two loci?

11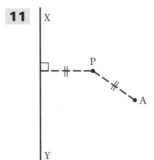

On graph paper plot the locus of a point P whose distance from a fixed point A is equal to its distance from a fixed line XY.

**IMPORTANT
CONSTRUCTIONS**

The easiest way to solve some problems concerning loci is by making accurate drawings. Three constructions that are particularly useful are considered next.

**BISECTING
ANGLES**

Bisect means 'cut exactly in half'.

The construction for bisecting an angle makes use of the fact that, in an isosceles triangle the line of symmetry cuts $\widehat{A}$ in half.

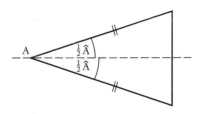

To bisect $\widehat{A}$, open your compasses to a radius of about 6 cm.

With the point on A, draw an arc to cut both arms of $\widehat{A}$ at B and C. (If we joined BC, $\triangle$ABC would be isosceles.)

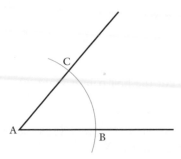

With the point on B, draw an arc between the arms of $\widehat{A}$.

Move the point to C (being careful not to change the radius) and draw an arc to cut the other arc at D.

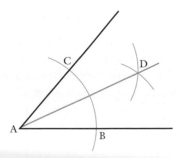

Join AD.

The line AD then bisects $\widehat{A}$.

CONSTRUCTION OF THE PERPENDICULAR BISECTOR OF A LINE

The perpendicular bisector of the line segment XY cuts XY in half at right angles. To find this we construct a rhombus with the given line as one diagonal, but we do not join the sides of the rhombus.

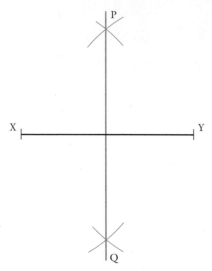

To bisect XY, open your compasses to a radius that is about $\frac{3}{4}$ of the length of XY.

With the point on X, draw arcs above and below XY.

Move the point to Y (being careful not to change the radius) and draw arcs to cut the first pair at P and Q.

Join PQ.

The point where PQ cuts XY is the midpoint of XY, and PQ is perpendicular to XY.

(XPYQ is a rhombus since the same radius is used to draw all the arcs, that is, XP = YP = YQ = XQ. PQ and XY are the diagonals of the rhombus. The diagonals of a rhombus intersect at right angles so PQ is the perpendicular bisector of XY.)

Note that this construction can also be used to bisect a line.

DROPPING A PERPENDICULAR FROM A POINT TO A LINE

If you are told to drop a perpendicular from a point C to a line AB, this means that you have to draw a line through C which is at right angles to the line AB.

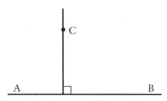

To drop a perpendicular from C to AB, open your compasses to a radius that is about $1\frac{1}{2}$ times the distance of C from AB.

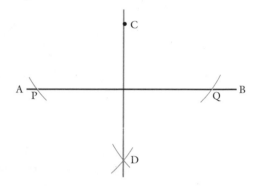

With the point on C, draw arcs to cut the line AB at P and Q.

Move the point to P and draw an arc on the other side of AB. Move the point to Q and draw an arc to cut the last arc at D.

Join CD.

CD is then perpendicular to AB.

Remember to keep the radius unchanged throughout this construction: you then have a rhombus, PCQD, of which CD and PQ are the diagonals.

EXERCISE 18D

Remember to make a rough sketch before you start each construction. Use suitable instruments including a *sharp* pencil.

1 a Construct an equilateral triangle of side 8 cm. What is the size of each angle in this triangle?

b Use what you have learnt in part **a** to construct an angle of 60°. Now bisect this angle. What size should each new angle be? Measure both of them.

c Use what you have learned in part **b** to construct an angle of 30°.

2 a Draw a straight line about 10 cm long and mark a point A near the middle.

You have an angle of 180° at A. Now bisect angle A. What is the size of each new angle? Measure both of them.

b Use what you have learnt in part **a** to construct an angle of 90°.

c Construct an angle of 45°.

3 Construct the following figures using only a ruler and a pair of compasses.

a

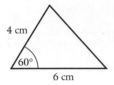

b

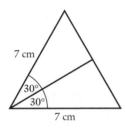

c

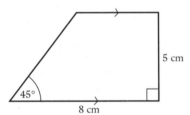

4 Construct a triangle ABC, in which AB = 6 cm, BC = 8 cm and CA = 10 cm. Using a ruler and compasses only, drop a perpendicular from B to AC.

5 Construct a triangle PQR in which PQ = 6 cm, PR = 6 cm and RQ = 10 cm. Using a ruler and compasses only, drop a perpendicular from R to QP; if necessary extend QP.

6 Construct the isosceles triangle LMN in which LM = 6 cm, LN = MN = 8 cm. Construct the perpendicular bisector of the side LM. Explain why this line is a line of symmetry of △LMN.

7 Draw a circle of radius 6 cm and mark the centre C. Draw a chord AB about 9 cm long. Construct the line of symmetry of this diagram.

8 Construct a triangle ABC, in which AB = 8 cm, BC = 10 cm and AC = 9 cm. Construct the perpendicular bisector of AB. Construct the perpendicular bisector of BC. Mark G where these two perpendicular bisectors intersect (that is, cross). With the point of your compasses on G and with a radius equal to the length of GA, draw a circle.

This circle should pass through B and C, and it is called the *circumcircle* of △ABC.

9 Construct an angle $\widehat{ABC}$ of 60°. Construct the locus of points equidistant from AB and BC.

10 A and B are two points such that AB = 9 cm. Construct the locus of points equidistant from A and B.

11 Draw a line AB that is 10 cm long. Construct the locus of a point P such that △ABP is isosceles with AP = BP.

12 Draw a line AB that is 8 cm long. Construct the locus of a point P such that the area of △ABP is 24 cm².

The remaining questions require the accurate drawing of a line parallel to a given line and at a given distance from it. A straightforward method uses a ruler and a set-square.

To draw a line parallel to AB and 2 cm from AB, place the set-square on AB with the right-angled corner on the line as shown.

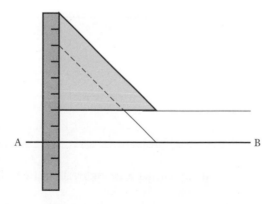

Then place a ruler against the edge of the set-square on the other arm of the right angle. Hold the ruler firmly and slide the set-square along the ruler until the lower edge is 2 cm above the line. Draw the line against this edge.

Construct a triangle ABC in which AB = 9.5 cm, BC = 7 cm and AB̂C = 60°. Find the point D, within the triangle, that is 2 cm from AB and 5 cm from C. Measure BD.

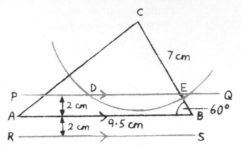

Points that are 2 cm from AB lie on one or other of the two lines, PQ and RS, that are shown parallel to AB. Points that are 5 cm from C lie on the circle, centre C, radius 5 cm.

This circle cuts PQ at D and E but cannot cut RS. Therefore RS need not be drawn in the accurate construction.

From the sketch, D satisfies the given conditions but we cannot be certain whether E lies inside or outside the triangle until we do the construction.

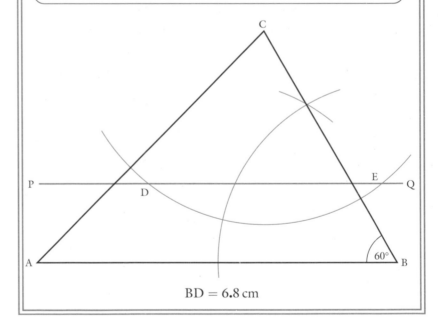

BD = 6.8 cm

13 a Construct a triangle ABC in which AB = 13 cm, AB̂C = 45° and BÂC = 30°.

b Draw the locus of points that are 2.5 cm from BC.

c Draw the locus of points that are 1.5 cm from AB.

d Hence find the point, P, within the triangle, that is 2.5 cm from BC and 1.5 cm from AB. Measure AP.

14 ABCD is a rectangle with AB = 8 cm and BC = 5 cm. Construct this rectangle and find the point P which is 2 cm from AB and equidistant from AD and BC. Measure PB.

15 Construct a triangle ABC in which AB = 9.5 cm, BC = 8 cm and $\widehat{ABC} = 60°$. Find the point D, on the opposite side of AB from C, that is 7 cm from BC and 4.5 cm from AC. Measure CD.

16 Construct a rectangle ABCD with AB = 6.5 cm and AD = 8 cm. Find the point X which is 3 cm from AD and equidistant from AD and DC. Measure DX.

17 ABC is a triangle in which AB = 12 cm, BC = 9 cm and $\widehat{ABC} = 90°$. Construct this triangle and find a point D that is 4.5 cm from BC and equidistant from A and C. Measure AD.

18 Construct a triangle with the given measurements. In each case say how many different triangles can be constructed.

    a $\triangle ABC$ with AB = 5 cm, BC = 6 cm and AC = 5 cm

    b $\triangle PQR$ with $\widehat{P} = 90°$, $\widehat{Q} = 30°$ and $\widehat{R} = 60°$

    c $\triangle LMN$ with LM = 8 cm, MN = 6 cm and $\widehat{M} = 40°$

    d $\triangle DEF$ with DE = 8 cm, DF = 5 cm and $\widehat{E} = 30°$

    e $\triangle XYZ$ with $\widehat{X} = 40°$, $\widehat{Y} = 60°$ and XY = 10 cm

    f $\triangle RST$ with $\widehat{R} = 90°$, RT = 5 cm and ST = 10 cm.

LOCI INVOLVING REGIONS

A locus is the set of points that satisfies a given condition. If this condition involves an inequality, the set of points is a region rather than a line. For example, the locus of a point P that moves in such a way that it is always 10 cm from a fixed point A, is the *circumference* of a circle, centre A, radius 10 cm, whereas the locus of a point P that moves in such a way that it is always less than 10 cm from a fixed point A, is the region *within the circle* centre A, radius 10 cm.

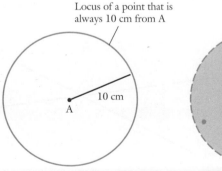

Locus of a point that is always 10 cm from A

10 cm

A

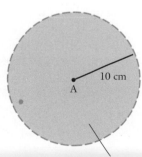

10 cm

A

Locus of a point that is always less than 10 cm from A

A and B are fixed points. Show, by shading in a suitable sketch, the locus of P such that AP < BP.

> Begin by drawing the locus of P such that AP = BP.
> This is the perpendicular bisector of AB.

A •————————————• B

> If P is such that AP < BP. P must be to the left of the line.
> P cannot lie on the line so we show this by using a dotted line.
> If the line were to be included, it would be drawn as a solid line.

1 A is a fixed point. Show, by shading in a suitable sketch, the locus of P such that AP ⩽ 6 cm.

2 A is a fixed point. If 3 cm < AP < 6 cm, illustrate the locus of P.

3 ABC is a triangle. Illustrate the locus of P, within the triangle, such that BP < PC.

4 AB is a fixed line, 10 cm long. Illustrate the locus of P such that P is more than 3 cm from any point on AB but less than 6 cm from any point on AB.

5

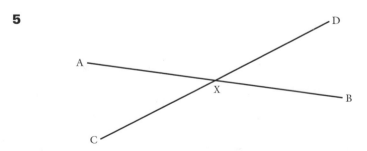

A house P is to be built 100 m from a road AXB and 50 m from a road CXD. Show on a sketch the possible positions for P.

6

X •
Y •

X and Y represent two houses 100 m apart. Sketch the loci that will enable you to shade the area of land that is both nearer to X than to Y and is within 60 m of Y.

7

The diagram shows two mats, A and B, which are strategically placed in an L-shaped school playing field. When the whistle blows a child has to go as quickly as possible to the nearest mat. Trace the diagram and shade the area of the field from which the children are expected to go to mat B. Are there any positions from which a child may be uncertain which mat to go to?

8 ABCD represents a rectangular factory yard measuring 40 m by 30 m. A guard dog is tethered at M, the midpoint of AD, by a chain 20 m long.

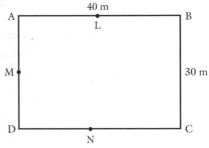

a Make a scale drawing to show the area that the dog can patrol. Shade it ▨.

b A second guard dog is tethered at C, also by a chain of length 20 m. Show the area that this dog can patrol. Shade it ▧.

c Is there any area that can be patrolled by both dogs? If there is, shade it ▦.

d Show a safe route (if there is one) from the factory yard gate N to the workshop entrance L. N is 19 m from D.

e Assuming that the chain always remains taut, find the angle through which the chain turns as it moves from one extreme position to the other for

i the first guard dog　**ii** the second guard dog.

9 The sketch shows Callum's lawn in front of his house. He decides to buy an electric mower and has a choice of a 7 m or a 10 m cable. The only accessible power point P is at ground level, just inside the garage.

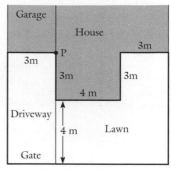

a Make a scale drawing to show the area of lawn he can mow if he buys the shorter cable.

b On a separate diagram show the area that he can mow if he opts for the longer length of cable.

LOCI IN THREE DIMENSIONS

We can extend the work we have done concerning loci in two dimensions to loci in three dimensions.

EXERCISE 18F

1 An electric light is placed at a fixed point A. A point P is such that AP = 4 m. Discuss the locus of P.

2 Two helicopters A and B are hovering, 300 m apart, at the same height. P is a point in space such that PA = PB. Discuss the locus of P.

Our discussions in the previous exercise lead us to two important loci in three dimensions.

1 If a point moves so that it is always at a fixed distance from a fixed point its locus is a sphere.
The fixed point is the centre of the sphere and the fixed distance is its radius.

2 The locus of points that are equidistant from two given points A and B is the plane bisecting AB at right angles.

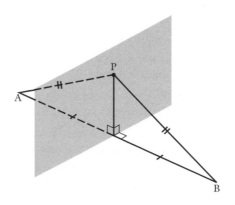

1 a A is a fixed point and a point X moves in space so that the length of AX is 5 cm. Describe the locus of X.

b B is a second fixed point so that AB = 8 cm. A point Y moves in space so that AY = YB. Describe the locus of Y.

c Describe the set of points that form the intersection of the two loci.

2 Describe the locus of a point in space that is always 10 cm from the surface of a spherical ball of radius 5 cm.

3 Describe the locus of a point in space that is always 5 cm from the surface of a hollow sphere of radius 10 cm. What important difference is there between your answer to this question and your answer to question **2**?

4 Assuming that the Moon is a sphere of radius 1750 km find the locus of a point that is always 50 km above the surface of the Moon.

5 ABCD represents a rectangular piece of card with AB = 12 cm and BC = 10 cm.

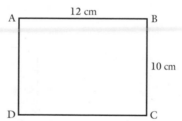

a Describe the locus of A as the card is rotated about the edge DC through 360°.

b Describe the locus of A as the card is rotated about the edge BC through 360°.

c Describe the shape traced out by the diagonal AC as the card is rotated about the edge BC through 360°.

6 A, B and C represent three aeroplanes in space. Describe the locus of points equidistant from A and B and 1000 m from C.

7 Describe the locus of points equidistant from three given points A, B and C.

8 At the beginning of the chapter we discussed some of the factors Jim had to bear in mind when deciding where to position his security light. Eventually he fitted it **3.5** m above the ground on the vertical line through the centre of the garage door. The light would come on when the part of the car nearest to it was 5 m away.

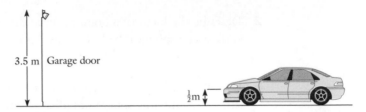

Make an accurate scale drawing to find how far the front of the car is from the garage door at the moment the light comes on. Assume that the point of the car nearest the sensor is $\frac{1}{2}$ m above the ground and is at the front of the car. State any other assumptions that you make.

MIXED EXERCISE

EXERCISE 18H

1 Describe the locus of a point on this page which moves so that it is always 3 cm from the top edge of this page.

2

The rectangle ABCD is rotated through 90° clockwise about A in the plane of the page.
Sketch the locus of **a** D   **b** C

3 Construct a triangle XYZ, in which XY = 12 cm, XZ = 5 cm and YZ = 9 cm. Drop a perpendicular from Z to XY.

4 Two straight lines AXB and CXD intersect at X such that $A\widehat{X}C = 90°$.

a Construct the locus of points that are 4.5 cm from X.

b Construct the locus of points equidistant from AXB and CXD.

c In how many points do these loci intersect? How far is each point from X?

5 ABCD is a rectangle measuring 10 cm by 8 cm.

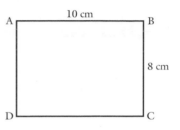

Describe

a the locus of points that are 8 cm from the *plane* of the rectangle.

b the locus of points in space that are equidistant from A and D.

c the intersection of these two loci.

INVESTIGATION

When a rugby player scores a try it may be converted into a goal. The goal kick must be taken from a point P, on the line that runs parallel to the side lines and which goes through the point T, where the try was scored.

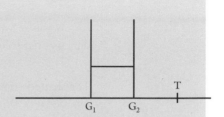

Assuming that the goal kicker has a long enough kick, the best chance of kicking the goal is when the angle G_1PG_2 is greatest.

Investigate the angle G_1PG_2 for different positions of P when T is 15 m from G_2. Find the position of P when G_1PG_2 is greatest.

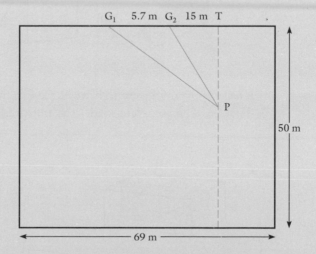

If you would like a bigger challenge find the locus of the best position for P for different positions of T. Assume that P is within the half of the field shown, that G_1G_2 is 5.7 m, the width of the field is 69 m and the distance from the goal line to the half-way line is 50 m.

SOLIDS

PLANES OF SYMMETRY

This solid is called a *pyramid*.
It has an isosceles triangle for a base.

Just as some two dimensional figures have lines of symmetry so a three dimensional solid can have *planes* of symmetry.

The shaded triangle lies in the plane of symmetry of this pyramid.

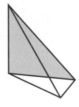

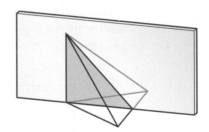

If half the pyramid were to be reflected in a mirror, as in the diagram, we should see the complete solid. The mirror is the plane of symmetry.

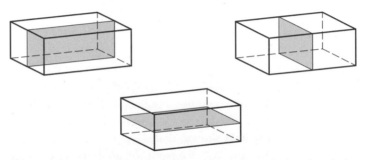

A solid may have more than one plane of symmetry. A cuboid, for instance, has three, while a sphere has an infinite number.

1 How many planes of symmetry has each of the following solids?

a

d

b

e

c

f

2 Some solids have only approximate symmetry. A human head for example, or a leaf, is only roughly symmetrical, with one plane of symmetry. A book is roughly symmetrical with two or three planes of symmetry (ignoring the printing).

How many approximate planes of symmetry do each of the following solids have?

a Aeroplane   **b** Cup   **c** Plate

d Chair   **e** Bicycle

f Name other solids with only one approximate plane of symmetry.

g Name other solids with more than one approximate plane of symmetry, giving the number of planes in each case.

PLANS AND ELEVATIONS

If we look down on a *solid* we can see its *plan*. For instance, if we look down on this metal waste-paper bin we see that its plan consists of two concentric circles; we can see the rim and the inside edge of the base.

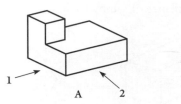

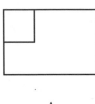

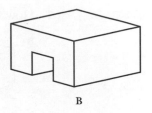

1   A   2   B

The plans of the two solids above are given below.

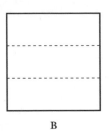

A   B

Edges in sight are drawn with plain lines, but hidden edges are drawn with broken lines.

We could also draw a side view or *elevation* of the wastepaper bin. This is not a picture with perspective, but a diagram. The outline shape we see from a distance is a trapezium so we draw this shape.

A view of solid A in the direction of arrow 1 has a broken line because there is a hidden edge.

A view of solid A from the direction of arrow 2 has all the edges either in view or behind edges which are in view so there are no broken lines.

Plans and elevations are also called *projections*.

EXERCISE 19B

1 Sketch the plans of the following solids.

a

b

c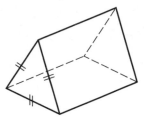

d

e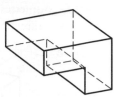

f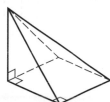

g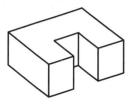

h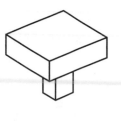

2 Sketch the two elevations of each of the solids, looked at in the directions indicated by the arrows.

a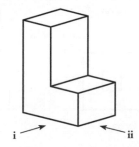

b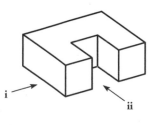

c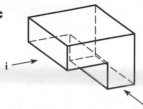

3 The following solids are built up of cubes with edges of length two centimetres. For each solid draw accurately a plan and elevations in the directions indicated by the arrows.

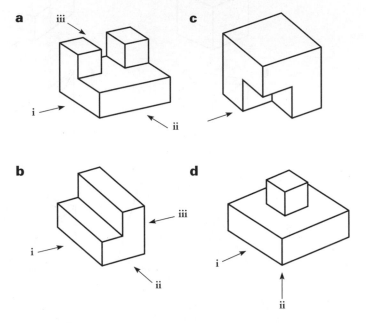

4 The plan and one elevation of a solid (in the direction of the arrow on the plan) are given. Sketch the solid. (There may be more than one.)

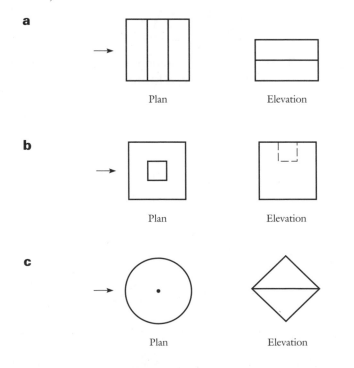

5 This is a classic puzzle. Sketch or describe the one solid which will pass, with no space to spare, through the three holes drawn below.

  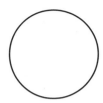

SECTIONS

If we have a solid made of modelling clay we can slice it into two parts.

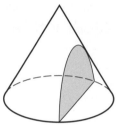

The cut surface so produced is called a *section*. We have already encountered the *cross-section* of a prism.

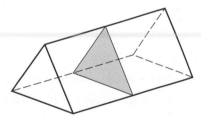

If we cut a prism anywhere with a slice parallel to its ends we produce the same section, in this case a triangle. The prism has a *uniform* (i.e. constant) cross-section.

If we cut a square pyramid with a slice parallel to its base, the resulting section is a square but its size depends on the distance of the cut from the base.

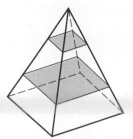

1 The diagram shows a cube of side 4 cm.

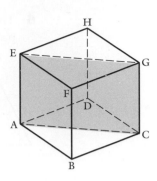

a Sketch the shaded section.

b Use Pythagoras' theorem to find the length of EG.

c Find the area of the shaded section.

d Is the shaded plane a plane of symmetry?

e Sketch the section EBCH. How does it compare with the shaded section?

2 The diagram shows a cube of side 6 cm.

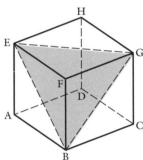

a Sketch the shaded section.

b Find the lengths of the sides of the section.

c Find the area of the section. (If necessary draw the section accurately.)

3 The diagram shows a cube of side 5 cm. ABCD is horizontal. P and Q are midpoints of EF and FG. The cube is cut into two parts by a vertical slice through PQ.

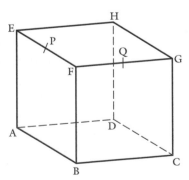

a Sketch the resulting section.

b Find the area of the section.

c Find the volumes of the two solids into which the cube has been cut.

4

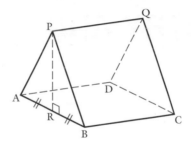

The diagram shows a triangular prism. AB = 6 cm,
AP = PB = 5 cm and PQ = 8 cm. ABCD is horizontal.

a Draw the section given by a vertical cut through PQ. Find the
area of the section.

b Draw the section given by a horizontal cut through the midpoint
of AP and find its area.

c Find the surface area of the prism.

d Find the volume of the prism.

e Find the volume of the smaller of the two solids into which the cut
in part **b** divides the prism.

f Which of the cuts described in parts **a** and **b** is in a plane of
symmetry?

5

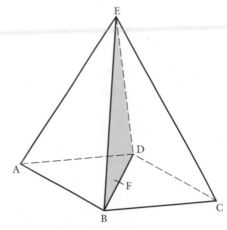

The diagram shows a pyramid on a square base. Its base is horizont.
and E is vertically above F, the centre of the base. AB = 8 cm and
EF = 10 cm.

a Sketch section EBD. What type of triangle is it?
Find BD and the area of the section.

b Sketch the vertical section through EF, parallel to BC.
Find its area.

GEOMETRY AND PROOF

20

Jim has drawn about thirty different triangles of all shapes and sizes. For each one he measured the three angles and found their sum. The results he obtained varied from 178° to 181.5°. It was from demonstrations such as these in Book 7A that it was concluded that the sum of the angles of *any* triangle is 180°.

Could it be that this method is 'jumping to conclusions' and is unsatisfactory for many reasons? After all, it is impossible to draw a line, for a line has no thickness and if it did not have thickness we could not see it! Furthermore it is impossible to measure angles with absolute accuracy. The protractor Jim uses is probably capable of measuring angles at best to the nearest degree. The only conclusion that Jim can draw from his results is 'it seems likely that the angle sum of any triangle is 180°' but he must remain aware that 'proof' by examining particular cases leaves open the possibility that somewhere, as yet unfound, there lurks an exception to the rule.

Jim needs to know if the result can be *proved* to be true for every triangle. If it can, several other results follow. For example, if the angle sum of any triangle is 180°, the angle sum of the four angles in every quadrilateral must be 360° since one diagonal always divides a quadrilateral into two triangles.

This chapter shows how certain geometric properties can be proved and how other properties follow from them.

DEDUCTIVE PROOF

Learning geometrical properties from demonstrations gives the impression that each property is isolated. However geometry can be given a logical structure where one property can be deduced from other properties. This forms the basis of deductive proof; we quote known and accepted facts and then make logical deductions from them.

For example, if we accept that

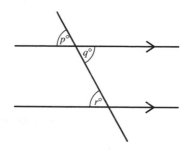

- vertically opposite angles are equal
- corresponding angles are equal,

then, using just these two facts, we can prove that alternate angles are equal.

In the diagram $p° = q°$ (vertically opposite angles)

$p° = r°$ (corresponding angles)

$\Rightarrow\ q° = r°$

Therefore the alternate angles are equal.

The symbol $\Rightarrow$ means 'implies that' and indicates the logical deduction made from the two stated facts.

This proof does not involve angles of a particular size; $p°$, $q°$ and $r°$ can be any size. Hence this proves that alternate angles are *always* equal whatever their size.

As a further example of deductive proof we will prove that in *any* triangle, the sum of the interior angles *is* $180°$. Note that angles on a straight line by definition, add up to $180°$.

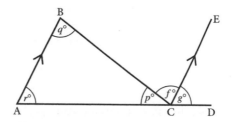

If $\triangle ABC$ is any triangle and if AC is extended to D and CE is parallel to AB then

$$p° + f° + g° = 180° \qquad \text{(angles on a straight line)} \qquad (1)$$
$$f° = q° \qquad \text{(alternate angles)} \qquad (2)$$
$$g° = r° \qquad \text{(corresponding angles)} \qquad (3)$$
$$\Rightarrow\ p° + q° + r° = 180°$$

i.e. the sum of the interior angles of *any* triangle is $180°$.

The statements on page 411 also lead to another useful fact about angles in triangles:

$$(2) \text{ and } (3) \quad \Rightarrow \quad f° + g° = q° + r°$$

i.e. an exterior angle of a triangle is equal to the sum of the two interior opposite angles.

Because this proof does not involve measuring angles in a particular triangle it applies to all possible triangles thus closing the loophole that there may exist a triangle whose angles do not add up to 180°.

Notice how this proof uses the property proved in the first example, that is, this proof follows from the previous proof. The angle sum property of triangles can now be used to prove further properties.

Euclid was the first person to give a formal structure to Geometry. He started by making certain assumptions, such as 'there is only one straight line between two points'. Using only these assumptions (called axioms), he then proved some facts and used those facts to prove further facts and so on. Thus the proof of any one fact could be traced back to the axioms.

However when *you* are asked to give a geometric proof you do not have to worry about which property depends on which; you can use *any* facts that you know. One aspect of proof is that it is an argument used to convince other people of the truth of any statement, so whatever facts you use must be clearly stated.

It is a good idea to marshal your ideas before starting to write out a proof. This is most easily done by marking right angles, equal angles and equal sides etc. on the diagram.

The exercises in this chapter give practice in writing out a proof.

For the next exercise the following facts are needed;

- vertically opposite angles are equal,
- corresponding angles are equal,
- alternate angles are equal,
- interior angles add up to 180°,
- angle sum of a triangle is 180°,
- an exterior angle of a triangle is equal to the sum of the interior opposite angles,
- an isosceles triangle has two sides of the same length and the angles at the base of those sides are equal,
- an equilateral triangle has three sides of the same length and each interior angle is 60°.

EXERCISE 20A

In a triangle ABC the bisectors of angles B and C intersect at I.
Prove that $\widehat{BIC} = 90° - \frac{1}{2}\widehat{A}$

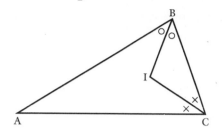

In triangle BIC

$\widehat{BIC} + \widehat{IBC} + \widehat{ICB} = 180°$ (angles in a $\triangle$)

i.e.

$\widehat{BIC} + \frac{1}{2}\widehat{B} + \frac{1}{2}\widehat{C} = 180°$ (BI bisects $\widehat{B}$ and CI bisects $\widehat{C}$) (1)

But $\widehat{A} + \widehat{B} + \widehat{C} = 180°$ (angles in a $\triangle$)

i.e. $\widehat{B} + \widehat{C} = 180° - \widehat{A}$

so $\frac{1}{2}\widehat{B} + \frac{1}{2}\widehat{C} = 90° - \frac{1}{2}\widehat{A}$

Substituting in (1)

$\widehat{BIC} + 90° - \frac{1}{2}\widehat{A} = 180°$

i.e. $\widehat{BIC} = 90° + \frac{1}{2}\widehat{A}$

1

Prove that $\widehat{ACD} = \widehat{ABC} + \widehat{DEC}$.

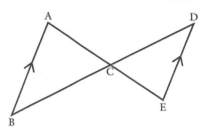

2

Prove that $\widehat{ACB} = 2\widehat{CDB}$.

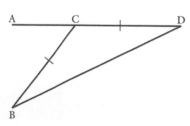

3

Prove that AD bisects $\widehat{BAC}$.

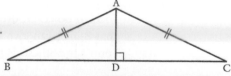

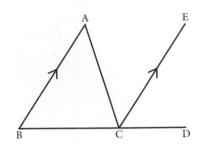

4 CE bisects $\widehat{ACD}$ and CE is parallel to BA. Prove that $\triangle ABC$ is isosceles.

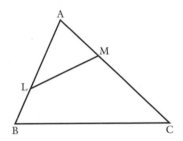

5 $\widehat{AML} = \widehat{ABC}$. Prove that $\widehat{ALM} = \widehat{ACB}$.

SHOWING THAT A HYPOTHESIS IS FALSE

We saw in the last section that drawing a few triangles and measuring the angles led us to say that 'it looks as though' the angles of any triangle add up to 180°. At that stage we had a *hypothesis*, which we then *proved* to be true for any triangle.

It is also important to be able to show that certain hypotheses are in fact false.

Suppose that students were asked to investigate the relationship between the number of lines drawn across a circle and the number of regions that the circle is divided into by those lines.

1 line
2 parts

2 lines
4 parts

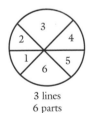

3 lines
6 parts

These three drawings led John to the hypothesis that *n* lines drawn across a circle give *2n* regions.

Jocelyn however, drew the lines this way, showing that 3 lines can give 7 regions, and therefore that John's hypothesis is false.

Jocelyn used a *counter example* to disprove the hypothesis.

Now consider the hypothesis 'the square root of an even number is itself even'. This can be shown to be untrue using this counter example: $\sqrt{6} = 2.449\ldots$ which is not an even number.

You may like to see if you can find a counter example to disprove that all prime numbers are odd.

Not every hypothesis can be either proved or shown to be false. In mathematics there are several in this category that are well known, one being Goldbach's conjecture. That is that every even number greater than or equal to 6, can be written as the sum of two odd prime numbers. At the time of writing no one has yet proved this to be true; on the other hand no one has found a counter example.

EXERCISE 20B

In questions **1** to **4** see if you can find a counter example to disprove each hypothesis.

1 The square root of a positive number is always smaller than the number.

2 If the side of a square is x cm long, the number of units of area of the square is always different from the number of units of length in the perimeter.

3 The diagonals of a parallelogram never cut at right angles.

4 The sum of any two angles in a triangle is always greater than the third angle.

Questions **5** and **6** give 'proofs' that are obviously invalid since they lead to untruths. Find, in each case, the flaw in the argument.

5 It is a fact that $$4 - 10 = 9 - 15$$

Adding $\frac{25}{4}$ to each side gives $$4 - 10 + \frac{25}{4} = 9 - 15 + \frac{25}{4}$$

Factorising $$\left(2 - \tfrac{5}{2}\right)\left(2 - \tfrac{5}{2}\right) = \left(3 - \tfrac{5}{2}\right)\left(3 - \tfrac{5}{2}\right)$$

i.e. $$\left(2 - \tfrac{5}{2}\right)^2 = \left(3 - \tfrac{5}{2}\right)^2$$

Take the square root of each side $$2 - \tfrac{5}{2} = 3 - \tfrac{5}{2}$$

Add $\frac{5}{2}$ to each side $$2 = 3 \quad \text{which is nonsense.}$$

6 Let $\qquad\qquad\qquad\qquad x = y$

and obviously $\qquad\qquad\quad x^2 - xy = x^2 - xy$

Now $\qquad\qquad\qquad\qquad x = y$ so $xy = y^2$

i.e. line 2 can be rewritten $\quad x^2 - xy = x^2 - y^2$

Factorise $\qquad\qquad\qquad x(x - y) = (x - y)(x + y)$

Divide both sides by $(x - y)$ $\qquad x = x + y$

but $x = y$, so $\qquad\qquad\qquad x = x + x$

i.e. $\qquad\qquad\qquad\qquad\qquad x = 2x$

i.e. $\qquad\qquad\qquad\qquad\qquad 1 = 2$ which is nonsense.

CIRCLE FACTS

In Book 8A we defined parts of a circle such as radius, diameter, circumference and arc. In this chapter we summarise the facts we already know and have used in previous chapters, extend a few of them and then investigate some of the basic geometric properties of a circle.

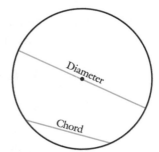

Every point on a circle is the same distance from its centre. This distance is called the *radius* of the circle.

Sometimes we use the word 'circle' to include the space inside the curve. When we do this we call the curve itself the *circumference* of the circle.

A straight line joining any two points on the circumference is called a *chord*.

Any chord passing through the centre of a circle is called a *diameter*.

Any part of the circumference is called an arc. If the arc is less than half the circumference it is called a *minor arc*; if it is greater than half the circumference it is called a *major arc*.

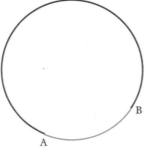

A minor arc AB

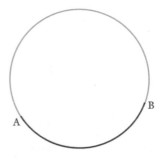

A major arc AB

The shaded area is enclosed by two radii and an arc. It is called a *sector*.

A chord divides a circle into two regions called segments. The larger region is called a *major segment* and the smaller region is called a *minor segment*.

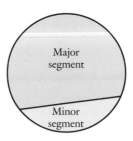

TANGENTS

The line PQ cuts the circle at A and B.

AB is a *chord*.

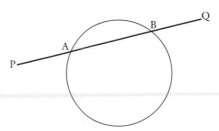

Imagine that this line is pivoted at P. As PQ rotates about P, we get successive positions of the points A and B. As PQ moves towards the edge of the circle, the points A and B move closer together, until eventually they coincide.

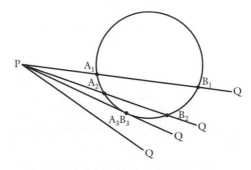

When PQ is in this position it is called a *tangent* to the circle and we say that PQ touches the circle. (When PQ is rotated beyond this position it loses contact with the circle and is no longer a tangent.)

We therefore define a tangent to a circle as a straight line which touches the circle.

The point at which the tangent touches the circle is called the point of contact.

PT is a tangent to the circle.
T is the point of contact.

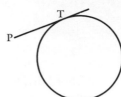

The *length of a tangent* from a point P outside the circle is the distance between P and the point of contact. In the diagram the length of the tangent from P to the circle is the length PT.

1

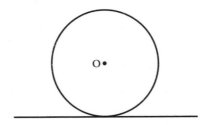

The diagram shows a disc, of radius 20 cm, rolling along horizontal ground. Describe the path along which O moves as the disc rolls.

At any one instant,

a how many points on the disc are in contact with the ground

b how far is O from the ground

c how would you describe the line joining O to the ground and what angle does it make with the ground?

2 Copy the diagram and draw any line(s) of symmetry.

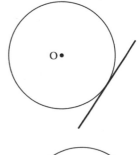

3 Copy the diagram and draw any line(s) of symmetry.

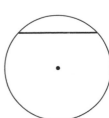

4

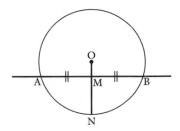

a Show that the chord AB is perpendicular to the radius ON which bisects AB. (Join OA and OB.)

b Now imagine that the chord AB slides down the radius ON. When the points A and B coincide with N, what has the line through A and B become? What angle does this line make with ON?

The investigational work in the last exercise suggests that

> a tangent to a circle is perpendicular to the radius drawn from the point of contact.

The general proof of this property is an interesting exercise in logic. We start by assuming that the property is *not* true and end up by contradicting ourselves. (This is called 'proof by contradiction'.)

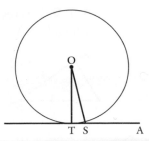

TA is a tangent to the circle and OT is the radius from the point of contact.

If we *assume* that $\widehat{OTA}$ is *not* $90°$ then it is possible to draw OS so that OS *is* perpendicular to the tangent, i.e. $\widehat{OST} = 90°$.

Therefore $\triangle OST$ has a right angle at S.

Hence   OT is the hypotenuse of $\triangle OST$

i.e.   OT > OS

$\therefore$ S is inside the circle, as OT is a radius.

$\therefore$ the line through T and S must cut the circle again.

But this is impossible, as the line through T and S is a tangent.

Hence the assumption that $\widehat{OTA} \neq 90°$ is wrong, i.e. $\widehat{OTA}$ *is* $90°$

EXERCISE 20D   Some of the questions in this exercise require the use of trigonometry.

The tangent from a point P to a circle of radius **4.**2 cm is 7 cm long. Find the distance of P from the centre of the circle.

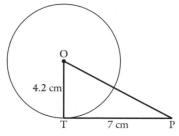

$\widehat{OTP} = 90°$ (tangent perpendicular to radius)

$OP^2 = OT^2 + TP^2$ (Pythagoras' theorem)

$ = (4.2)^2 + 7^2$

$ = 17.64 + 49 = 66.64$

$\therefore$ $OP = 8.163\ldots$

The distance of P from O is 8.16 cm, correct to 3 s.f.

In questions **1** to **5**, O is the centre of the circle and AB is a tangent to the circle, touching it at A.

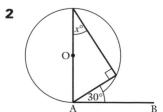

1

Find OB and CB.

2

Find the marked angle.

3

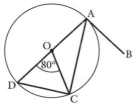

Find **a** DÂC **b** BÂC

6

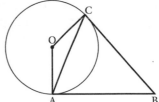

AB and BC are tangents to the circle centre O touching it at A and C.
Show that △ABC is isosceles.

4

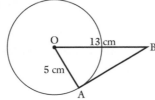

Find AB and OB̂A.

7

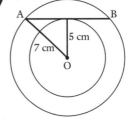

AB is a chord of the larger circle and a tangent to the smaller circle. If O is the centre of both circles, find the length of AB.

5

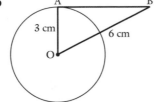

Find AB̂O.

8

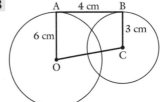

AB is a tangent to the circles with centres O and C, touching them at A and B respectively.
Find OC.

CONSTRUCTIONS

EXERCISE 20E

(Remember to draw a rough sketch before doing the construction.)

1 Draw a circle of radius 5 cm. Label the centre O and mark a point T on the circumference. Construct the tangent to the circle at T.
(Use the fact that the radius OT is perpendicular to the tangent.)

2 **a** The diagram shows a circle, centre O, inscribed in a square
(i.e. the sides of the square are tangents to the circle).
The radius of the circle is 2 cm.
Find the length of a side of the square.

b Draw a square of side 8 cm. Construct the inscribed circle of the
square.

For questions **3** and **4**, use a ruler and compasses only to construct the
figure. Measure the marked length.

3
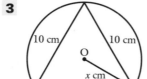

4
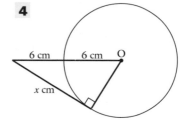

SUMMARY 5

TRIGONOMETRY

In a right-angled triangle

the *tangent of an angle*

$$= \frac{\text{side opposite the angle}}{\text{side adjacent to the angle}}$$

the *sine of an angle*

$$= \frac{\text{side opposite the angle}}{\text{hypotenuse}}$$

the *cosine of an angle*

$$= \frac{\text{side adjacent to the angle}}{\text{hypotenuse}}$$

or more briefly,

$$\tan \widehat{A} = \frac{\text{opp}}{\text{adj}} = \frac{BC}{AB}, \quad \sin \widehat{A} = \frac{\text{opp}}{\text{hyp}} = \frac{BC}{AC}, \quad \cos \widehat{A} = \frac{\text{adj}}{\text{hyp}} = \frac{AB}{AC}$$

LOCI

A locus is the shape given by all the possible positions of the points satisfying a given rule.

The locus of a point that moves so that it is at a fixed distance from a given point is the circumference of a circle.

The locus of a point that moves at a constant distance from a fixed straight line is a pair of parallel lines. In the case of a line of fixed length, semicircular ends join the lines.

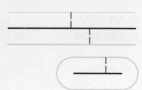

The locus of a point that moves so that it is equidistant from two fixed points, A and B, is the perpendicular bisector of the line joining AB.

The locus of a point that moves so that it is equidistant from two intersecting straight lines is the pair of bisectors of the angles between the lines.

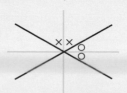

CONSTRUCTIONS The diagrams show 'ruler and compass only' constructions.

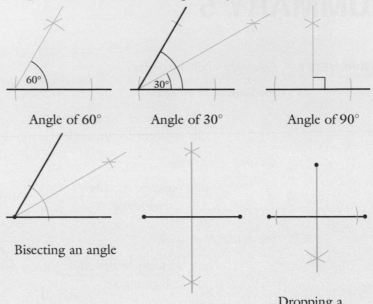

Angle of 60°   Angle of 30°   Angle of 90°

Bisecting an angle   Perpendicular bisector of a line   Dropping a perpendicular from a point to a line

PLANE OF SYMMETRY

A plane of symmetry divides a solid into two identical shapes.
A plane of symmetry of this cylinder is shaded.

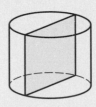

PLAN AND ELEVATION

A plan is the view of an object from directly above it.
An elevation is a vertical view of an object from a given direction.

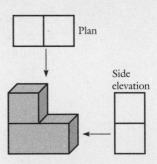

SECTIONS

A section through a solid is the shape of the face that would be revealed if the solid was sliced in a given direction.
The diagram shows the section of the cylinder when it is cut along the lines shown.

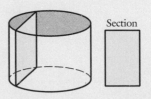

CIRCLES AND
TANGENTS

A line joining two points on the circumference of a circle is a chord.
A line that touches a circle is a tangent.
The point where a tangent touches a circle is called the point of contact.
A segment of a circle is the part of the circle cut off by a chord.

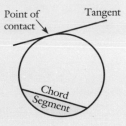

TANGENT
PROPERTY

A tangent to a circle is perpendicular to the radius drawn through the point of contact.

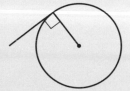

REVISION
EXERCISE 5.1
(Chapters 16
and 17)

1 Find, correct to 3 significant figures, the tangent of

a 23°   **b** 47°   **c** 36.2°   **d** 72.9°

2 Find, correct to one decimal place, the angle whose tangent is

a 0.62   **b** 3.4   **c** 1.333   **d** $\frac{1}{5}$

3 In △ABC, $\widehat{B} = 90°$, $\widehat{A} = 59°$ and BC = 18 cm. Find AB.

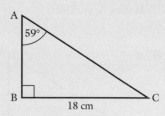

4 Find the angle A in each diagram.

a

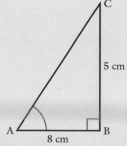

b

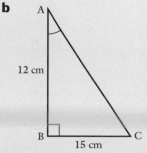

c

Triangle with B at top, right angle at B, 3.4 cm from B to C, 5.7 cm from A to B, C to the right, A at bottom.

d

Triangle with A at top left, angle at A, 7.4 cm from A to B, right angle at B, 13 cm from B to C, C at top right.

5 The bearing of a town Q from a town P is 041.7°. R is 14 km due north of P and R is due west of Q.

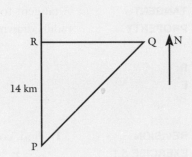

a How far is Q east of R?

b Use Pythagoras' result to find the distance of P from Q.

6 Find, correct to 3 significant figures

  a sin 67°   **b** cos 24°   **c** sin 42.7°   **d** cos 62.8°

7 Find, correct to 1 decimal place, the angle A if

  a $\sin \widehat{A} = 0.474$ **c** $\sin \widehat{A} = 0.8682$

  b $\cos \widehat{A} = 0.926$ **d** $\cos \widehat{A} = 0.5432$

8 a

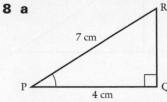

Find $\widehat{P}$.

b

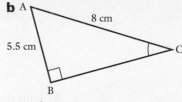

Find $\widehat{C}$.

9 a

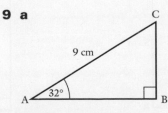

Find BC and AB.

b

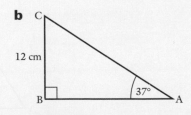

Find AC.

10 From a point A on the ground, 3 m from the foot of a vertical pole, the angle of elevation of the top is 52°.

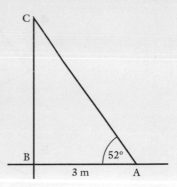

a How high is the pole?

b A supporting wire runs from the top of the pole to A. How long is this wire?

REVISION EXERCISE 5.2 (Chapters 18 to 20)

1 AB and BC represent two roads connecting three villages A, B and C. Describe the locus of points on the ground that are

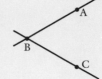

a equidistant from A and B

b equidistant from AB and BC.

2 Construct a triangle ABC in which AB = 10.5 cm, BC = 6.4 cm and AC = 8.6 cm. Construct the perpendicular bisector of AB and the bisector of AB̂C. Mark the point X where they cross. Measure AX.

3

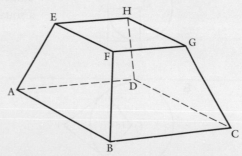

The diagram shows the *frustum* of a pyramid (i.e. a pyramid with its top sliced off). ABCD and EFGH are both horizontal squares. EA = FB = GC = HD = 6 cm, AB = 10 cm and EF = 6 cm.

a How many planes of symmetry are there?

b Sketch the face FBCG. What shape is it?

c Sketch the section FHDB and draw it accurately.

d Sketch the vertical section through FE.

4 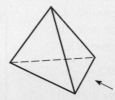 The four faces of this solid are all equilateral triangles.
Sketch **a** the plan
b the elevation in the direction of the arrow.

5 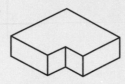 **a** How many planes of symmetry does this solid have?

b Sketch the plan of this solid.

6 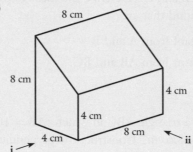 Draw the two elevations of this solid looked at in the direction of the arrows.

8 cm

8 cm

4 cm

4 cm

4 cm 8 cm **ii**

i

7 O is the centre of the circle. BT and AT are tangents to the circle. Prove that OATB is a square.

8 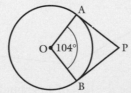 O is the centre of the circle.
AP and BP are tangents to the circle. Find $A\hat{P}B$.

9 PT and PS are the two tangents from P to a circle centre O.
PT = 8 cm and OP = 10 cm.
Find the length of
a OT **b** RP

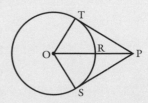

10 CAB is the tangent at A to a circle
centre O.
Find the marked angles.

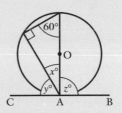

**REVISION
EXERCISE 5.3
(Chapters 16
to 20)**

1 a Find, correct to 3 significant figures **i** tan 56° **ii** tan 37.7°

b Find, correct to 1 decimal place, the angle whose tangent is

i 0.492   **ii** $\frac{5}{8}$ **iii** $2\frac{1}{7}$

c In △ABC, $\widehat{B} = 90°$, $\widehat{A} = 36°$ and AB = 6 cm. Find the
length of BC.

2 The diagram shows the cross-section of a
ridge tent. The tent is 1.4 m wide at
ground level and $B\widehat{A}C = A\widehat{B}C = 69°$.
How high is the ridge of the tent above the
ground?

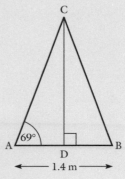

3 a In △ABC, $\widehat{B} = 90°$, AB = 3.8 cm and AC = 4.5 cm.
Find $\widehat{A}$ and $\widehat{C}$.

b In △PQR, $\widehat{Q} = 90°$, RQ = 5.6 cm and $\widehat{P} = 29°$ cm.
Find PR.

4 a Find AC.

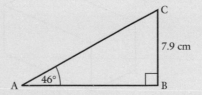

b ABCD is a rectangle in which DC = 4 cm and the diagonal
BD = 15 cm. Find, correct to the nearest whole number
i $D\widehat{B}C$ **ii** BC.

5 a Construct a rectangle ABCD such that AB = 12 cm and
BC = 8 cm.

b Draw the locus of points equidistant from AB and BC.

c Draw the locus of points equidistant from A and B.

d Mark the point P that lies on both the loci referred to in parts **b**
and **c**. Measure PC.

6 a Draw a line AB that is 10 cm long. Construct the locus of a point P so that the area of △ABP is 30 cm².

b ABCD is a rectangle measuring 12 cm by 10 cm.

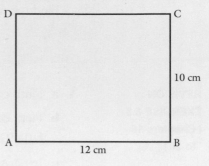

i Describe the locus of a point in space that is 8 cm from A.

ii Describe the locus of a point that is 6 cm above the plane of the rectangle.

iii Describe briefly the shape of the intersection of these two loci.

7

Plan   Elevation

The plan and one elevation of a solid are given above. Which of these could this solid be?

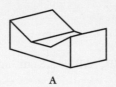 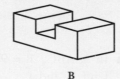 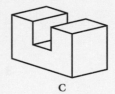

A   B   C

8

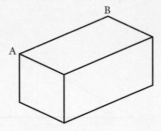

This cuboid is made by sticking two cubes together.

a How many planes of symmetry does it have?

b Sketch the section through AB that is also a plane of symmetry.

9 PT is a tangent at T to a circle centre O.

Find, giving reasons   **a** QR̂T   **B** OQ̂T   **c** RQ̂T

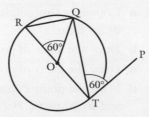

10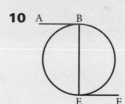
A   B

E   F

AB and EF are tangents to the circle.

BE is a diameter.

Prove that AB is parallel to EF.

REVISION EXERCISE 5.4 (Chapters 1 to 20)

1 **a** Use dot notation to write as a decimal **i** $\frac{11}{30}$ **ii** $\frac{7}{11}$

  b Express as a fraction **i** $0.\dot{7}$ **ii** $0.0\dot{7}$

  c Find $\left(6\frac{5}{8} - 4\frac{3}{10}\right) \times 2\frac{2}{3}$

2 **a** A greengrocer buys a box of 150 Seville oranges for £13.50 and sells them at 14 p each. Find his percentage profit.

  b Find the compound interest on £750 invested for 2 years at 4.55%.

3 Expand **a** $4a(b - 3c)$ **c** $(4x - 1)(5 + x)$

        b $(x - 6)(x + 2)$ **d** $(5p + 2q)(5p - 2q)$

4 **a** Factorise **i** $a^2 + a^3$ **iii** $x^2 + 10x + 24$

         ii $x^2 - 4x - 21$ **iv** $x^2 - y^2$

  b Find the value of $7.54^2 - 2.46^2$ without using a calculator.

5 **a** If $T = x(y - 3z)$ find T when $x = 3$, $y = 4$ and $z = \frac{3}{4}$.

  b Make the letter in brackets the subject of the formula.

    i $a = b - c \ (b)$ **ii** $4z = 3x - 2y \ (x)$ **iii** $A = b + \frac{1}{5}c \ (c)$

  c Find, in terms of n, an expression for the nth term of the sequence 5, 8, 13, 20, . . .

  d If $A = 2\pi r(r + h)$ and $h = 2r$ find

    i A in terms of r **ii** A in terms of h.

6 **a** Solve the equations.

    i $x^2 + 5x - 36 = 0$ **ii** $6x^2 - 18x = 0$ **iii** $63 = 16x - x^2$

  b I think of a positive number x. If I square it and add this to the number I first thought of, the total is 56. Find the number I first thought of.

7 The diagram shows the cross-section of a 3 m length of angle iron. Find

  a the area of the cross-section in **i** cm^2 **ii** m^2

  b its volume in m^3.

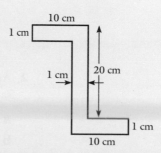

8 a Show that triangles ABC and ADE
   are similar.

 b Given that AD = 12 cm,
   BC = 6 cm and DE = 8 cm find
   BD.

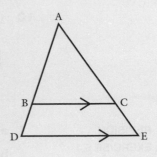

9 a In $\triangle$ABC, $\widehat{B} = 90°$, $\widehat{C} = 32°$ and AC = 12.5 cm.
   Find   **i** AB   **ii** BC.

 b In $\triangle$PQR, $\widehat{Q} = 90°$, $\widehat{P} = 63°$ and PQ = 4.3 cm. Find
   i PR   **ii** QR   **iii** the perpendicular distance from Q to PR.

10 Prove $A\widehat{C}D = 2B\widehat{A}C$

**REVISION
EXERCISE 5.5
(Chapters 1 to 20)**

1 Viv is asked to choose a date at random from the year 1998.
Calculate the probability that the day is

 a a day in April   **c** a day in November

 b a Sunday in February   **d** a day in April or November

 e either a day in January or a Sunday in February.

Would any answers have been different if the chosen year was 1996
instead of 1998? Explain your answer.

2 At a steady speed a car uses 8 litres of fuel to travel 96 km. At the
same speed

 a what distance could be travelled if 7 litres of fuel is used

 b how many litres would be needed to travel 240 km?

3

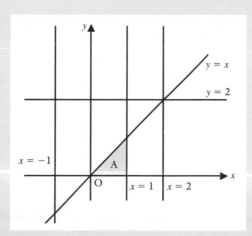

a Use inequalities to describe the region marked **A** in the diagram.

b Copy the diagram onto squared paper. One region on the diagram is defined by the inequalities $x \geqslant 1$, $x \leqslant 2$, $y \geqslant x$, $y \leqslant 2$. Mark this region **B**.

4 The heights of two varieties of broad bean plant were measured five weeks after planting and the following data was obtained.

Variety A

Height of plant, h cm	$0 \leqslant h < 5$	$5 \leqslant h < 10$	$10 \leqslant h < 15$	$15 \leqslant h < 20$
Frequency	3	5	25	13

Variety B

Height of plant, h cm	$0 \leqslant h < 5$	$5 \leqslant h < 10$	$10 \leqslant h < 15$	$15 \leqslant h < 20$
Frequency	4	14	17	6

a Find the median, the range and the interquartile range for the heights of each new variety after five weeks.

b Use your results from part **a** to compare the heights of the two varieties after five weeks.

5 Solve the simultaneous equations

a $2x + y = 16$
$3x + 2y = 26$

b $3x + 5y = 8$
$5x - 3y = 36$

6 a The graph representing $y = 2x - x^2 - x^3$ could be

A **B** **C** **D**

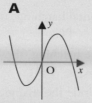

b Three cars were observed approaching a set of traffic lights. The graphs show how the speeds of each car changed. Describe the way in which the speed of each car changed.

i   **ii**   **iii**

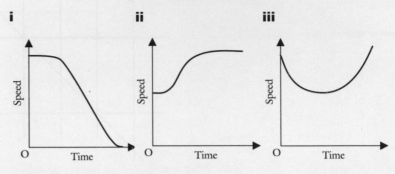

7 a On x- and y-axes scaled from -3 to 16 plot the points $A(6, 6)$, $B(5, 4)$, $C(7, 3)$ and $A_1(6, 10)$, $B_1(3, 4)$, $C_1(9, 1)$.

b Find the centre and scale factor of the enlargement that transforms $\triangle ABC$ onto $\triangle A_1B_1C_1$.

c A translation of $\triangle A_1B_1C_1$ is defined by the vector $\begin{pmatrix} -3 \\ 2 \end{pmatrix}$ and the resulting figure is reflected in the line $x = 6$ to give $\triangle A_2B_2C_2$. Write down the coordinates of the vertices of $\triangle A_2B_2C_2$.

8 a Find PQ.

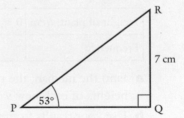

b In $\triangle ABC$, $\hat{B} = 90°$, $AB = 6.4\,cm$ and $BC = 4.5\,cm$. Find $\hat{A}$.

9 a Construct $\triangle ABC$ such that $AB = AC = 9\,cm$ and $BC = 7.5\,cm$.

b Illustrate, using horizontal hatching, the locus of P within the triangle, such that $AP < PC$.

c Illustrate, using oblique hatching, the locus of Q within the triangle, such that $C\hat{B}Q < A\hat{B}Q$.

d Are there any points inside the triangle that satisfy both loci?

10

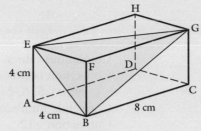

a Sketch the section through BEG.

b Find the length
of **i** EB **ii** BG.

INDEX